LiDAR Technologies and Systems

LiDAR Technologies and Systems

Paul McManamon

SPIE PRESS
Bellingham, Washington USA

Library of Congress Cataloging-in-Publication Data

Names: McManamon, Paul F., 1946– author.
Title: LiDAR technologies and systems / Paul McManamon.
Description: Bellingham, Washington, USA : SPIE Press, [2019] | Includes
  bibliographical references and index.
Identifiers: LCCN 2018053788 | ISBN 9781510625396 (hard cover ; alk. paper) |
  ISBN 1510625399 (hard cover ; alk. paper) | ISBN 9781510625402 (PDF) |
  ISBN 1510625402 (PDF) | ISBN 9781510625419 (ePub) | ISBN 1510625410
  (ePub) | ISBN 9781510625426 (Kindle/Mobi) | ISBN 1510625429
  (Kindle/Mobi)
Subjects: LCSH: Optical radar.
Classification: LCC TK6592.O6 M36 2019 | DDC 621.3848–dc23
LC record available at https://lccn.loc.gov/2018053788

Published by

SPIE
P.O. Box 10
Bellingham, Washington 98227-0010 USA
Phone: +1 360.676.3290
Fax: +1 360.647.1445
Email: books@spie.org
Web: http://spie.org

The content of this book reflects the work and thought of the author. Every effort
has been made to publish reliable and accurate information herein, but the
publisher is not responsible for the validity of the information or for any
outcomes resulting from reliance thereon.

Cover imagery collected by Lincoln ladar sensor and provided by the Active Optical
Systems Group, MIT Lincoln Laboratory.

Printed in the United States of America.
First Printing.
For updates to this book, visit http://spie.org and type "PM300" in the search field.

Contents

Preface

About six years ago, I co-taught a semester-long course in LiDAR technology with Dr. Ed Watson at the LiDAR and Optical Communications Institute (LOCI) of the University of Dayton. At the time, there was a book that covered part of what I wanted to teach, but it did not cover all of the areas I thought should go into the course. There were a couple of other books that had interesting LiDAR-related material, but no book that covered all of the topics that I thought were needed. Since then, I have done a number of week-long, or almost-week-long, courses. One of those I co-taught with Gary Kamerman, and a number of them with Ed Watson. Shortly after teaching that 2012 semester-long course, I started writing the *Field Guide to LiDAR*, which was published by SPIE Press in 2015. I thought a shorter book would be easier to write than a longer one. I was wrong. The Field Guide came out great, but its format with one topic per page made it a challenging type of book to write. Also, when I finished writing the Field Guide, I still did not have a really good *text* book on LiDAR technology and systems. Thus, the decision to write this book grew out of the need for a good teaching reference for a longer course on LiDAR. The Field Guide is great as a quick reference, with of all the equations in one place and each topic concisely presented, but it does not provide enough background or detail to be a text book. This book presents an in-depth coverage of LiDAR technology and systems, and the Field Guide serves as a reminder of the essential points and equations once you already understand the technology.

I learned a lot writing the *Field Guide to LiDAR*, and then writing this book. When I considered all of the topics that should be covered in this book, there were some I knew really well, and some I knew less well. The neat thing I found about writing a book like this is that, before I could effectively explain a particular concept, I needed to clearly understand the concept. To this end, the comparison paper I recently wrote[1] with Ed Watson, Andrew Huntington, Dale Fried, Paul Banks, and Jeffrey D. Beck taught me a lot about receivers. I have included sections from that paper in this book. The paper on the history of laser radar in the U.S.[2] that I wrote with Milt Huffacker and Gary Kammerman, and the more recent paper[3] "Laser radar: historical prospective—from the East to the West," which I wrote with Vasly Molebny, Ove Steinvall, T. Kobayashi, and W. Chen, both provide a good summary of the history of LiDAR. Chairing the 2014 United States National

Academy of Sciences study on laser radar[4] helped me learn more. Of course, decades of experience monitoring LiDAR development for the Air Force taught me a lot as well.

Once I started working on this book, I had two students take a self-study course with me in LiDAR. Both students read early versions of the manuscript and developed possible problems to include at the end of each chapters. Dr. Abtin Ataei was the first student to do this, and Andrew Reinhardt was the second. A special thanks goes to Abtin Ataei for doing a final check on the Problems and Solutions. For the last chapter on LiDAR applications, I felt I did not know the 3D mapping area as well as I should. Dr. Mohan Vaidynathan, my former post-doctorate, works for Harris Corporation (now merged with L3 Technologies) on one of the commercial 3D mappers and volunteered to make an input. Admittedly, he is an advocate of the Geiger-mode version of 3D mapping, as he should be, given where he works, but I knew that. He and his colleagues provided significant input. To balance things out, I did request information from Teledyne Optech and RIEGL as well.

MIT/Lincoln Lab has a nice library of 3D LiDAR images. I would like to thank them for providing one of those images for the book cover.

Finally, I thank Dara Burrows, Senior Editor at SPIE Press, whose tireless work editing this book has made it happen.

This has been an educational experience, and I am pleased with the way the book has turned out. I hope you enjoy it, and I hope many people can use it to learn more about LiDAR technology and systems. I enjoyed writing it, and as I mentioned, learned a lot in certain areas. Perhaps, once in a while it happens that an author learns almost as much by writing a book as a reader learns by reading that book!

Paul McManamon
May 2019

1. P. F. McManamon, P. S. Banks, J. D. Beck, D. G. Fried, A. S. Huntington, and E. A. Watson, "Comparison of flash lidar detector options," *Opt. Eng.* **56**(3), 031223 (2017) [doi: 10.1117/1.OE.56.3.031223].
2. P. F. McManamon, G. Kamerman, and M. Huffaker, "A history of laser radar in the United States," *Proc. SPIE* **7684**, 76840T (2010) [doi: 10.1117/12.862562].
3. V. Molebny, P. F. McManamon, O. Steinvall, T. Kobayashi, and W. Chen, "Laser radar: historical prospective—from the East to the West," *Opt. Eng.* **56**(3), 031220 (2016) [doi: 10.1117/1.OE.56.3.031220].
4. National Academy of Sciences, *Laser Radar: Progress and Opportunities in Active Electro-Optical Sensing*, P. F. McManamon, Chair (Committee on Review of Advancements in Active Electro-Optical Systems to Avoid Technological Surprise Adverse to U.S. National Security), Study under Contract HHM402-10-D-0036-DO#10, National Academies Press, Washington, D.C. (2014).

Chapter 1
Introduction to LiDAR

1.1 Context of LiDAR

LiDAR uses electromagnetic (EM) waves in the optical and infrared wavelengths. It is an active sensor, meaning that it sends out an EM wave and receives the reflected signal back. It is similar to microwave radar, except at a much shorter wavelength. This means that it will have much better angular resolution than radar but will not see through fog or clouds. It is similar to passive electro-optical (EO) sensors in wavelengths, except that it provides its own radiation rather than using existing radiation, and has many more sensing modes due to control over the scene illumination. LiDAR brings its own flashlight and can therefore see at night using near-infrared wavelengths, whereas passive EO sensors have limited capability in the near infrared at night because of insufficient available near-infrared radiation. This means that LiDAR can have increased angular resolution associated with the shorter wavelengths and still operate 24 hours per day.

Figure 1.1 is a diagram showing the EM spectrum that puts LiDAR (and EO devices, in general) into the broader EM–wave context. We see that the visible and infrared spectra have shorter wavelengths than radiowaves and microwaves, but longer wavelengths than x rays and gamma rays. This is a log scale, so the change in wavelength is large. A typical tracking microwave radar might have a frequency of 10 GHz, which corresponds to a wavelength of 3 cm. This is called X-band radar. A typical search radar may have a frequency of 1 GHz and a wavelength of 30 cm. This is called L-band radar. Since a typical eye-safe LiDAR will have a frequency of 200 THz and a wavelength around 1.5 μm, a typical LiDAR will have a wavelength about 20,000 times smaller than the X-band tracking radar, and 200,000 times smaller than the L-band search radar, with corresponding increases in carrier frequency. X rays and gamma rays will be orders of magnitude shorter in wavelength and higher in frequency than visible or infrared EM radiation.

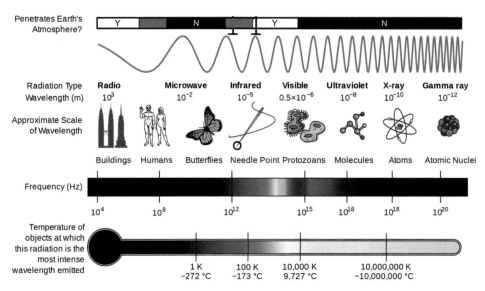

Figure 1.1 The electromagnetic spectrum and the size of common objects compared to the size of a wavelength (adapted from Wikipedia[1] for a NAS study.[2])

In general, microwave radar engineers talk in frequency, and optical engineers talk in wavelength. The relationship between wavelength and frequency is

$$c = \lambda \nu, \tag{1.1}$$

where c is the speed of light, λ is the wavelength, and ν is the frequency.

Fog particles are from 1 to 100 μm in diameter, with most particles in the 10- to 15-μm region. Rain drops range from 0.5 to 5 mm in diameter. Truck spray on the highway is likely to have diameter sizes between those of fog particles and rain drops, which will be relevant for autonomous vehicle navigation LiDAR. If the wavelength of an EM wave is significantly longer than the particle size, the wave just flows around particles with little or no attenuation. A 30-cm (or even a 3-cm) microwave signal is significantly larger than rain droplets so will not be significantly attenuated by fog or rain. A 1.5-μm wavelength LiDAR signal will, however, be highly scattered by clouds, fog, or rain. Millimeter-wave radar has a frequency of 95 GHz and a wavelength of 3.16 mm. You can see that millimeter waves at 95 GHz have a wavelength larger than fog but smaller than many rain droplets, so they see through fog well, but not as well through some rain. Fog is worse for LiDAR than rain because fog is made of more particles. A cloud and fog are essentially the same except that fog is sitting on the ground. LiDAR does not penetrate very far through either one. It can penetrate thin clouds with so-called ballistic photons, which are not scattered in the cloud; however, this is exponential decay over a very short range, so LiDAR will not penetrate any but the thinnest clouds. Even a 10-μm LiDAR would encounter many particles that are as large as or larger than the wavelength so would be significantly attenuated in fog and clouds.

Another comparison for LiDAR is against passive EO sensors. To consider passive EO sensors, we need to think about which EM wavelengths have available radiation that a passive sensor can use. Blackbody radiation is a key aspect of this. Blackbody radiation provides the main signal for passive EO sensors and is the main background noise source for LiDARs. We should therefore be aware of blackbody radiation, which will be covered in the chapter on receivers.

LiDAR is often used as an imaging sensor. It can be for two-dimensional (2D) imaging, similar to the eye or a traditional passive EO sensor, or 3D imaging, where range is measured in each pixel. 3D pixels are usually referred to as voxels. Traditional grayscale, showing intensity variation of the reflected light and therefore a variation in object reflectivity, can also be measured. The color of an object can be measured if more than one wavelength of light is used in the LiDAR. Velocity can be measured either directly using the Doppler shift in frequency due to motion, or by multiple measurements of position. Micro-motion can be measured using the Doppler shift, which is usually measured with a coherent LiDAR. A coherent LiDAR beats the return signal against a local oscillator (LO), which is a sample of the outgoing laser signal. This allows us to measure the phase of the return signal. Optical frequencies are much higher than what any detector can measure, but detectors can measure the beat frequency between two optical waves—the return signal and the LO. This will be further discussed later in this chapter. Coherent LiDAR has an issue with speckle because speckle is a form of interference. When you see laser light reflected from a wall, you will usually see light and dark areas in the reflected spot. This is caused by constructive and destructive interference of narrowband light. A speckled image can be hard to interpret, but if you add many different speckle images with a diversity of speckle, the light and dark areas average out and the image is easier to interpret. Light from the sun or a light bulb has many optical frequencies, so we do not see speckle. The speckle is averaged out. While speckle is usually considered to be an obstacle to laser-based imaging, and usually is averaged out using multiple samples, speckle can also be a feature used in coherent LiDAR to provide additional information about an object.

LiDAR has broad application throughout the military and civilian sectors. Every object has a finite number of observable features that can be exploited by a remote EO sensing system. These features are broadly categorized into the following five types: geometry, surface character, plant noise, effluents, and gross motion. Geometric sensing characterizes the shape of the object in one, two, or three dimensions, including the fully geometrically resolved intensity distributions of the object. Surface character includes roughness, the spectral and directional distribution of scattered energy, and polarization properties. Plant noises include a variety of vibrations and cyclical motions attributed to the operation of the target. These could be, e.g., signatures associated with piston or turbine engines, transmissions, or other moving components. Effluents include exhaust air, gases, and waste heat. The gross motions are movements of the system, including system translation, rotation, or articulation. All of these types of object discriminants can be sensed by various forms of LiDAR.

LiDAR is a great sensor for identifying objects. One of the reasons for this is that LiDAR can operate at wavelengths similar to what the eye is used to seeing. Due to the reflectance characteristics of microwave radar, it is difficult for the average human to recognize objects using radar images. Scattering in these wavelengths is more specular, resulting in images that are mostly a series of bright spots. People can readily identify objects using visible optical radiation. That is what our eyes use, so we are very familiar with how objects look in the visible wavelength region. One difficulty of passive sensors is they are limited by available radiation. This means that we cannot image objects at extremely long range, as is possible with active sensors such as radar and LiDAR, if we use powerful illumination beams. Also, an active sensor can see at night by illuminating the scene at shorter wavelengths than the available night time radiation, providing a better diffraction-limited resolution at night for a given aperture size.

Another main benefit of LiDAR compared to passive EO sensors is its extremely rich phenomenology due to control of the illumination. Passive sensors provide a 2D image with grayscale or color (or both) in each pixel element. Passive sensors can provide many colors, which can aid in detection or in material identification. If passive sensors frame at a high rate, temporal fluctuations can be observed. A LiDAR can do everything a passive imager can do, although sometimes at a greater cost in photons, sensor size, sensor power, or sensor weight. A LiDAR can provide a 2D angle/angle image but can also provide a 3D image with angle/angle/range information. A LiDAR can directly measure range in each pixel because it controls when light is emitted so can measure range based on time of flight to and from the object in a given pixel. A 3D image can also have grayscale and color if we have enough signal. Coherent LiDAR can measure velocity very accurately. There is a LiDAR mode called laser vibrometry, which can see vibration modes of an object. Reflective surface properties can change the LiDAR return because of speckle. Synthetic-aperture radar (SAR) provides a 2D image in range and cross-range. A synthetic-aperture LiDAR synthetically constructs a larger receive aperture by measuring the field (intensity and phase) of the return in the receive aperture at a given instant. It then moves both the transmit and receive apertures and takes a series of measurements. Because the field is captured in each location, a larger pupil plane image can be assembled and Fourier transformed to obtain a higher-resolution image. Large effective-pupil-plane images can also be synthesized using multiple receive and/or transmit apertures, a process sometimes referred to as multiple-input, multiple-output (MIMO). LiDAR's rich phenomenology make it an extremely capable sensor.

1.2 Conceptual Discussion of LiDAR

Next, we discuss a conceptual diagram of a LiDAR, shown in Fig. 1.2. A waveform generator creates the laser waveform needed to obtain range or velocity information. A laser would usually provide the illumination, although it is possible to have a LiDAR without a laser. Prior to the invention of the laser, a few LiDAR systems were built. In theory, we could use any light source, but since its invention, the laser has been the light source of choice for LiDAR. This can be a single laser, a seeded laser, or a master oscillator with one or more following laser amplifiers. It could also be an array of lasers. We will talk about those possibilities in Chapter 5. There needs to be a transmit optical aperture that emits the light. This can be the same aperture as the receive aperture, or a different aperture. If it is the same aperture, we call it a monostatic LiDAR, where mono means one aperture. If the transmit and receive apertures are different, we call it a bistatic LiDAR, meaning two apertures. We have to aim the laser at the target we want to view, so we need a pointing method and a way to figure out where to point.

Light has to traverse a medium, usually atmosphere, to arrive at a target. LiDAR can be transmitted through the vacuum of space. LiDAR can also be used through water instead of air or space. Water has significant absorption and sometimes significant scattering, so LiDAR only works well over short ranges in water, possibly on the order of a meter or tens of meters. We do not image though kilometers of water. People usually use blue/green wavelengths to image through water because that is the best spectral window for transmission through

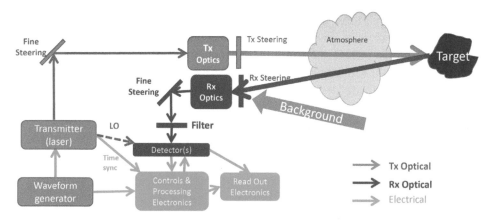

Figure 1.2 LiDAR Conceptual diagram (adapted from Ref. 3).

water. The medical community has also used very short-range LiDAR through human tissue. Usually medical LiDAR used in humans is in a tomographic imaging mode, where the laser illuminates from one angle, but multiple angles of the scattered light are then captured to develop a picture of the human tissue. There is significant scattering in the human body, and just like in water, some wavelengths transmit better than others. Red, the color of blood, transmits relatively well through the body. If you take a red laser pointer and a green laser pointer into a dark room and put each against the bottom of your index finger in turn, you will see much more red light coming through your finger than green light. LiDAR light bounces off the target and traverses the medium again until it is captured by the receiver optical aperture in the pupil plane, which also needs to point at the target. If we use a lens to focus the light captured by the receive aperture, then we convert to the image plane.

Alternatively, if we can capture or estimate the field at the aperture (the pupil plane), we can Fourier transform the field at the pupil plane to form an image. One of the nice things about capturing the full field (phase and intensity) in the pupil plane is that a larger pupil-plane aperture can be synthesized and then Fourier transformed to make a higher-resolution image. To measure the full field in the pupil plane rather than just the intensity requires that we beat the return signal against another signal that (as mentioned) we call the local oscillator (LO). Using a detector array, we can make an image with a single pulse (flash imaging), but we also can scan a small optical beam and develop an angle/angle image over time while using one or a small number of detectors.

Optical wavelengths are at very high frequencies, so we cannot directly measure phase. No detector is fast enough to detect 200 THz (1.5-μm light) or other similar very high frequencies. To detect temporal phase, we can use an LO that interferes with the return signal on the detector. We refer to the LO beating against the return signal. We can then measure the beat frequency, which allows us to measure the phase of the return light as well as the intensity (we call this coherent LiDAR). If the LO is perfectly stable, the phase change in the beat frequency is the same as the phase of the carrier frequency, which is the returned signal, so in coherent LiDAR, we can capture phase. We need a timing signal from the transmitter to determine range by measuring time of flight of the laser pulse to and from the target. We know the speed of light in vacuum and in air, so we can calculate the distance to an object. The signal generated by the detector is digitized and then processed to make an image or to generate information such as velocity or vibration frequency based on the return Doppler shift.

1.3 Terms for Active EO Sensing

In this book we use the term LiDAR because it is most popular, but other terms that have been used for active EO sensing are shown in Fig. 1.3. Historically,

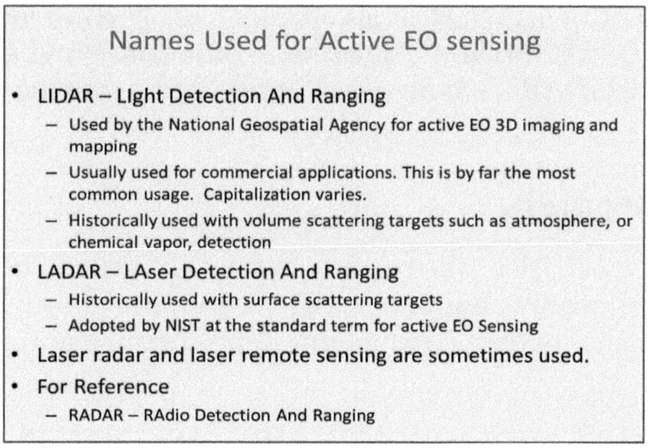

Names Used for Active EO sensing

- **LIDAR – LIght Detection And Ranging**
  - Used by the National Geospatial Agency for active EO 3D imaging and mapping
  - Usually used for commercial applications. This is by far the most common usage. Capitalization varies.
  - Historically used with volume scattering targets such as atmosphere, or chemical vapor, detection
- **LADAR – LAser Detection And Ranging**
  - Historically used with surface scattering targets
  - Adopted by NIST at the standard term for active EO Sensing
- **Laser radar and laser remote sensing are sometimes used.**
- **For Reference**
  - RADAR – RAdio Detection And Ranging

Figure 1.3   Terms for active EO sensing (reprinted from Ref. 2).

the term LiDAR has been used by the active EO sensing community in conjunction with measuring volume-based targets such as aerosols, and ladar has been used in conjunction with surface-based reflections. Sometimes people would refer to soft targets for volume scattering and hard targets for surface scattering. LiDAR has been the term used more often in commercial applications and therefore has more widespread usage. There have also been variations in the letters that are capitalized in the acronym. The most popular version is lidar, then LIDAR, followed by LiDAR, and LADAR, but LiDAR has recently grown in popularity due to its use in the commercial sector, such as auto LiDAR. Often the various terms are used almost interchangeably, depending on who is using them. This can cause confusion.

Figure 1.4 is a Google Ngram that charts the usage of LiDAR versus ladar.[4]

You can see that the term LiDAR, grouping all the various capitalizations, was much more widely used than the term ladar during the period

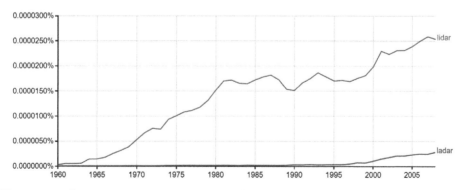

Figure 1.4   Google Ngram comparing use of the term LiDAR to use of the term ladar, ignoring capitalization. The graph stops at 2008.[4]

covered by the Ngram. As a result, people are more likely to understand what we mean with respect to active EO sensing if we use the term LiDAR. Finally, this book uses LiDAR as both a technology and a specific instrument or system.

1.4 Types of LiDARs

Many types of LiDARs will be discussed in Chapter 4, but this section provides a short preview. Range in a LiDAR is measured by timing how long it takes light to hit an object and return. Range resolution is given by

$$\Delta R = \frac{c}{2B}, \tag{1.2}$$

where ΔR is the range resolution, c is the speed of light in the intervening medium, and B is the system bandwidth, which could be limited by the transmit signal bandwidth or the receiver bandwidth (which can be limited by the detector or its electronics). Angle information is detected by one or a number of detectors moving in angle, or by an array of detectors sampling the field of view (FOV). Angular resolutions can either be limited by the detector angular subtense (DAS) or by diffraction. Figure 1.5 shows various sampling possibilities; the middle curve represents the case where the DAS is about the same as the point spread function (PSF) of the optical aperture. We could also oversample the PSF as shown on the right, or under sample it as shown on the left.

A LiDAR that only measures one dimension (range) is called a range-only LiDAR or a 1D LiDAR. Another LiDAR is like a passive imager in that it measures both azimuth and elevation, but does not measure range in each pixel. A 2D LiDAR can be range gated to a certain range region of interest. This can reduce noise and distractions from ranges that are not in the region of interest. For example, in fog, backscatter from closer ranges could obscure the return from the range of interest. A 2D LiDAR provides its own

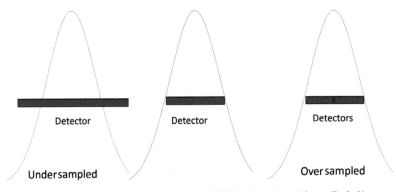

Figure 1.5 Various sampling possibilities (reprinted from Ref. 3).

illumination, as do all LiDARs, which is an advantage compared to passive imagers. A common type of LiDAR today is a 3D LiDAR, which measures angle/angle/range. A 3D image can be just three dimensions with no grayscale, or it can measure grayscale as well, meaning that the relative intensity of the returns vary across the image. Some areas of the image reflect more strongly than other areas. The number of grayscale levels is given by

$$N_{gs} = 2^N, \qquad (1.3)$$

where N is the number of bits of grayscale, and N_{gs} is the number of levels of grayscale. Table 1.1 shows that we will have 64 levels of grayscale with 6 bits of grayscale and 8 levels with 3 bits. In direct detection systems, two types of information are typically recovered. One is the range from the sensor to the target on a pixel-by-pixel basis, which is often called a point cloud. Here, the range information is gathered as a function of position (often through some form of timing circuitry) on the receiver focal plane. Hence, the contrast information that is provided from pixel to pixel is a variation in range information. The other type of information that can be gathered is reflectance information. The contrast from pixel to pixel in this case is the number of returned photons for each pixel (e.g., irradiance). This is a measurement of grayscale.

Polarization and color (wavelength) are other dimensions that can be measured. A coherent LiDAR, which measures the phase of the return signal as well as amplitude, can measure the velocity of an object by measuring the Doppler shift, a change in return frequency that results from an object's velocity. Historically, people would use a train approaching and leaving a station to explain the frequency change due to velocity to or away from an object. This worked because at the time people were familiar with the change in pitch as the train approached and passed the station. Angle measurement can be made using an aperture that is synthesized by sampling the pupil plane field

Table 1.1 Number of grayscale levels N_{gs} versus number of bits of grayscale N.

N	N_{gs}
1	2
2	4
3	8
4	16
5	32
6	64
7	128
8	256
9	512
10	1025

in many locations and then constructing a larger pupil plane field sample, which can be Fourier transformed to obtain the image. Synthetic-aperture LiDAR (SAL) is one form of a synthesized LiDAR aperture. Other forms can include multiple physical apertures on receive only or on both receive and transmit. As mentioned earlier, LiDARs that use multiple receive and transmit apertures are called MIMO, for multiple input, multiple output. To date, the term MIMO has been used much more in microwave radar then in LiDAR.

1.4.1 Some LiDARs for surface-scattering (hard) targets

LiDAR can be used for many different applications because it is such a rich sensing approach. 3D LiDAR is used for 3D mapping. 2D LiDAR images look very much like conventional EO images. As mentioned, we can have range-only, or 1D, LiDAR. We can also have coherent Doppler LiDAR that measures velocity and vibration. A well-known application of LiDAR is the driverless car. All of the DARPA Grand Challenge finishers used at least one LiDAR. Almost all of the other driverless car developers are using LiDAR. Admittedly, the Tesla driverless car in which a passenger was killed did not have LiDAR, but it was not designed for true driverless function. Drivers are supposed to maintain alertness in case they are needed. Police speed detection is another common LiDAR use. LiDAR can pick out one vehicle in a cluttered traffic situation using a system that can be mounted in a small camera.

Small UAVs such as the Amazon UAV use LiDAR. The Microsoft® Kinect game system is a relatively new, very widespread use of a simple LiDAR. Forestry LiDAR is unique in its ability to measure the vertical structure of forest canopies, map the ground beneath the forest, and estimate canopy bulk density. LiDAR can map flood plains and can map the ground in coastal areas in and out of the water.

Transportation corridors can be 3D mapped to support high-scale engineering mapping accuracy. The military can use 3D map data for route planning to avoid being seen from a certain location, and for precise object identification. LiDAR's high accuracy means that a quick survey will give precise volumetric measurements for oil and gas exploration and for quarrying.

LiDAR can provide the data for cellular network planning to determine line of sight and viewshed for proposed cellular antenna. LiDAR allows any physical object to be re-created in a computer environment, or with a 3D printer, so the advent of 3D printing should increase demand for LiDAR. LiDARs can make recording the scene of accidents and crimes quick, easy, and precise. LiDAR has been used in archeology mapping. LiDAR is a useful tool when designing and constructing new buildings and in determining the location of specific items within the interior of a room for licensing regulations and to provide proof of compliance. LiDAR can enable surveys to be taken of places that may be considered too dangerous for humans to enter. A robotic vehicle can be sent down sewers and can take detailed surveys of the interior of the system. This is only a small subset of the surface-scattering LiDAR applications, which are limited only by people's imagination and inventiveness.

1.4.2 Some LiDARs for volume-scattering (soft) targets

Wind speed can be measured either by the Doppler shift along the path from the LiDAR or from multiple accurate range measurements. Doppler LiDAR is used to measure wind speed along the beam by measuring the frequency shift of the backscattered light. This can be for military applications such as air drop of cargo or to map the wind fields around an airport or around a set of wind turbines, or near sailboats in a sailboat race. Differential absorption LiDAR (DIAL) can be used to detect range-resolved trace amounts of gases such as ozone, carbon dioxide, or water vapor. The LiDAR transmits two wavelengths: an online wavelength that is absorbed by the gas of interest and an offline wavelength that is not absorbed. For example, ARPA-E has a recent program called MONITOR (Methane Observation Networks with Innovative Technology to Obtain Reductions) to detect methane gases. The differential absorption between the two wavelengths is a measure of the concentration of the gas. Raman LiDAR is also used for measuring the concentration of atmospheric gases. Raman LiDAR exploits inelastic scattering to single out the gas of interest. A small portion of the energy of the transmitted light is deposited in the gas during the scattering process, which shifts the scattered light to a longer wavelength by an amount that is unique to the species of interest.

Laser-induced fluorescence (LIF) LiDARs are similar to Raman LiDARs in that they only require that the entity to be detected be present in the atmosphere or on the surface of a target of interest. Unlike Raman LiDARs, however, the LIF systems seek to create real transitions in the entity through excitation to higher electronic levels. Laser-induced breakdown spectroscopy (LIBS) is part of the volume-scattering discussion because it involves first vaporizing a small amount of material and then using spectroscopy to identify the vaporized material. The Curiosity spacecraft on Mars uses a LIBS sensor to remotely determine rock type.

1.5 LiDAR Detection Modes

An optical detector does not respond to the 200- to 600-THz carrier frequency of 1.5- to 0.5-μm light. When light hits a detector, it generates a voltage equal to the square of the intensity of the impinging light. In ideal direct detection, only the LiDAR return hits the detector, causing a response equal to the square of the intensity of the impinging light. In coherent LiDAR, the return signal beats against a sample of the emitted signal, which we call the local oscillator (LO). In this case, the detector can respond to the real portion of the beat frequency, or difference frequency, between the return signal and the LO:

$$I = 2\,E_{\text{sig}}\,E_{\text{LO}}\,\cos[-j(\omega_{\text{sig}} - \omega_{\text{LO}})], \qquad (1.4)$$

where it can be seen that, in temporal heterodyne detection, the LO and return signal are spatially aligned to interfere on the detector. For the temporal heterodyne mode, the LO and the return signal must be spatially matched, or else high–spatial-frequency fringes will lower heterodyne mixing efficiency. This means that both the direction of propagation as well as the phase front of each beam must be matched. The frequency of the LO is offset so that we can tell the direction of target velocity and to reduce $1/f$ noise. If there is no frequency offset, temporal heterodyne LiDAR is called homodyne LiDAR. This beating is shown in Fig. 1.6.

In spatial heterodyne detection, also called digital holography, the LO and the return signal are slightly misaligned, as shown in Fig. 1.7. For simplicity, this figure does not show the slightly offset beams after they are combined and before they hit the detector. The tilt between the LO and the return signal creates fringes that can be sampled by a detector array. If the angular tilt is too large, the fringes will be too rapid and will average out across a detector. In phase shift interferometry, the reflected wave is mixed with a LO of the same frequency. The interferogram is measured on a 2D detector array. Multiple interferograms are recorded in which a known piston phase has been applied to the LO. Three or more interferograms are processed to extract the field amplitude and phase.

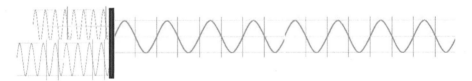

Figure 1.6 Beat frequency for temporal heterodyne detection (reprinted from Ref. 3).

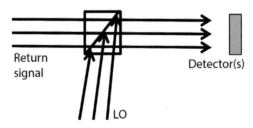

Figure 1.7 Angularly offset LO for spatial heterodyne detection, or digital holography (reprinted from Ref. 3).

1.6 Flash LiDAR versus Scanning LiDAR

Spatial coherence uses the angular width of the laser beam. For LiDARs that have a single detector, the laser beam must illuminate the area covered by the single detector. If the LiDAR uses a detector array, which we refer to as flash LiDAR, then the laser illuminates an area as large in angle as the detector array, so the laser beam is wider in angle. The diffraction limit provides the smallest possible laser beam divergence. The full beam width, half maximum diffraction limit is

$$\theta \approx \frac{1.03\,\lambda}{D}, \tag{1.5}$$

where D is the diameter of the transmit aperture. If the transmit aperture is the same aperture, or the same size, as the receive aperture, then for flash imaging we do not need high spatial coherence. Our transmit beam does not need to be single mode because the illuminated area can be much larger than the diffraction limit. Flash imaging is shown in Fig. 1.8.

Many LiDAR illumination beams are Gaussian, but a flat topped beam is often desirable. A super-Gaussian beam has a flatter top than a Gaussian beam, allowing for a more uniform illumination pattern. For a Gaussian beam where $N = 2$, and for a super-Gaussian beam, where $N > 2$,

$$f(x) = ae^{-\frac{(x-b)^N}{2c^2}}, \tag{1.6}$$

where a is the magnitude of the Gaussian beam, b is the offset from zero, and c is the width of the Gaussian beam. Figure 1.9 shows a Gaussian beam and a super-Gaussian shape. With a finite-sized transmit aperture, you can clip less energy from the laser beam with any flatter-topped beam shape, including variously powered super-Gaussian shapes. A Gaussian beam concentrates its energy in the middle of the point spread function (PSF) more than a super-Gaussian, or flat top, PSF, but it also spreads some energy into a wider angle, where it might be clipped by an aperture. A flat-top beam spreads energy across the aperture, but then the energy decreases quickly, so not as much energy is clipped by an aperture.

Figure 1.8   Flash imaging (reprinted from Ref. 3).

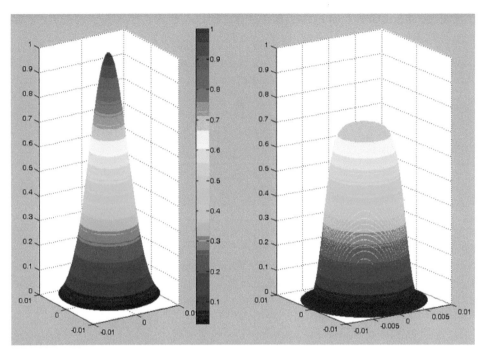

Figure 1.9 Gaussian (left) versus super-Gaussian (right) beam shapes (reprinted from Ref. 3).

1.7 Eye Safety Considerations

Laser radiation can damage the eye by burning the retina after magnification, or by burning the surface of the eye (see Fig. 1.10). The eye has significant magnification when it focuses light on the retina, so the damage threshold for light entering the eye and focusing on the retina is much lower than the damage threshold for burning the surface of the eye. Lasers beyond $\sim$1.5 μm,

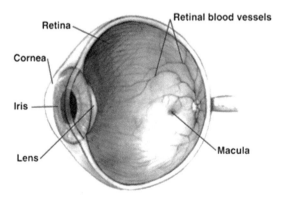

Figure 1.10 A schematic image of the eye (from Wikipedia: Laser safety. Created by Han-Kwang Nienhuys and released under license CC-BY-2.5).

or below about 0.4 μm, are safer because water in the eye absorbs wavelengths in these regions, preventing light from focusing on the retina. You can still burn the surface of the eye, but without the large increase in irradiance resulting from focusing the light. It is rare for LiDARs to operate below 0.4 μm for better eye safety, but it is common for them to operate at 1.5 μm or longer. A slight decrease in the maximum allowed flux levels is seen as you move to operation wavelengths longer than 1.5 μm. This is because the eye absorbs the light that is closer to the surface at longer wavelengths, so the light is absorbed in a smaller volume of eye tissue. A smaller volume will heat up sooner. At 1.5 μm, the light is all absorbed before it hits the retina, but it is absorbed in a volume of most of the eye. At 10 μm, water is more absorptive, so it is absorbed near the surface of the eye.

Figure 1.11 shows the maximum permissible exposure (MPE) for various wavelengths. MPE is given in units of energy per area (J/cm^2). Often this level of exposure in ten seconds is used as a threshold. A scanning laser hits the eye only for a brief time period, whereas the laser in a flash-illuminated LiDAR illuminates the eye for a much longer time period. Because a scanning laser does not illuminate the eye for as high a percentage of time, the laser can be more powerful without exceeding the eye safety threshold. From Fig. 1.11, we can see the effect of operating in a higher–eye-safety wavelength region. The maximum permissible laser flux is around 1.5 μm. Depending on the laser pulse width, the reduction in allowed laser flux can be anywhere from 3 to 8 orders of magnitude at 1.5 μm compared to that of the visible light that will hit the retina. The time numbers in Fig. 1.11 are pulse widths. As mentioned, people generally assume a 10-s exposure of the eye and calculate the total energy absorbed by the eye, and then compare this to Fig. 1.11 for a given pulse width. If the laser scans, it can

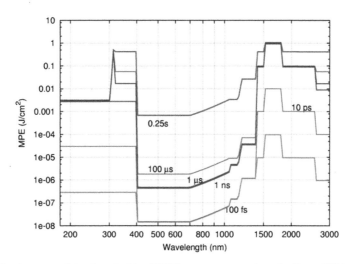

Figure 1.11 Laser safety thresholds (MPE) versus wavelength (from Wikipedia: Laser safety. Created by Dahn and released under license CC-BY-2.5).

reduce the total energy absorbed by the eye because the laser will only be illuminating the eye during a portion of the time.

1.8 Laser Safety Categories

Since the early 1970s, lasers have been grouped into four classes and a few subclasses according to wavelength and maximum output power. The classifications categorize lasers according to their ability to produce damage to people. There are two classification systems: the "old" system, which was used before 2002, and the "revised" system, which started being phased in after 2002. The latter reflects the greater knowledge of lasers that has been accumulated since the original classification system was devised and permits certain types of lasers to be recognized as having a lower hazard than was implied by their placement in the original classification system. Figure 1.12 shows the laser safety classes:

- Class 1 lasers are safe unless focused.
- Class 2 lasers only cover the spectral range of 400–700 nm. They are safe because of the blink reflex, which only occurs at those spectral regions because you won't blink if you don't see it. Class 2 lasers are 1 mW continuous wave (cw) or less.
- Class 3R lasers are less than or equal to 5 mW cw. They are safe if handled properly.
- Class 3B laser operators may need goggles. These lasers can be up to 30 mW.
- Class 4 lasers can burn the skin as well as damage the eye.

Note that some laser pointers are more than 5 mW. Many of those you can order on the Internet are higher than 5 mW. Be careful!! For LiDAR, the

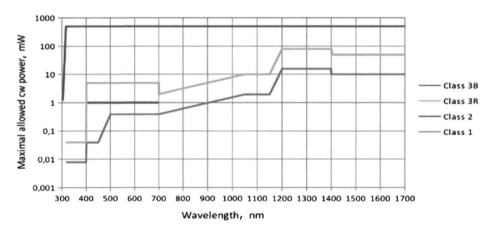

Figure 1.12 Laser safety classes (from Wikipedia: Laser safety. Created by Dahn and released under license CC-BY-2.5).

MPE values discussed previously are more important than the laser classifications because they are given per area of aperture.

1.9 Monostatic versus Bistatic LiDAR

A way to isolate the transmit and receive functions is to use a different telescope for each function such that the backscatter in the optical system is eliminated. The only backscatter you then have to worry about comes from scattering off of the aerosols close to the LiDAR. If the transmit and receive apertures are two separate apertures, we call that a bistatic LiDAR, as shown in Fig. 1.13. The transmit, or illumination, aperture does not have to be the same size as the receive aperture, nor in the same place. A laser designator is a bistatic LiDAR illuminator with the seeker as the receiver, even though people do not think of laser designator systems as being a bistatic LiDAR.

One reason for using monostatic LiDARs is to save weight and space by not having a second aperture for illumination; however, with bistatic illumination often you can have an illumination aperture that is much smaller than the receive aperture, reducing the size, weight, and power impact of having two apertures. This is the case for flash LiDAR, where the area illuminated is larger than the area viewed by a single DAS. Having a small transmit aperture for illumination is especially important for coherent LiDAR so that the phase across the illuminated region is constant. Also, as will be discussed in Chapter 4, multiple-input, multiple-output (MIMO) LiDARs provide the benefits of having more than one transmitter and more than one receiver. Any MIMO LiDAR is inherently bistatic. Often in a MIMO LiDAR, the transmit apertures will be smaller than the receive apertures.

A monostatic LiDAR uses the same aperture for transmit and receive. As a result, we need to provide isolation between the transmitter and receiver, or backscatter from the transmitted pulse will blind the receiver. Various methods of providing isolation are discussed next.

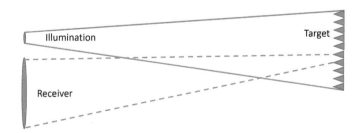

Figure 1.13   Bistatic LiDAR illumination diagram (reprinted from Ref. 2).

1.10 Transmit/Receive Isolation

High-power backscattered laser radiation can blind or damage the receiver, so we need to isolate the high-power transmitted laser waveform from the receiver. One way to isolate the receiver from the transmitter is to have a bistatic LiDAR with separate apertures. If the LiDAR has a pulsed-cycle, or low-duty cycle, transmitter, then another way to isolate the receiver from the transmitter is to keep the receiver off while the laser is transmitted. With the receiver off, we do not have to worry about the amount of isolation between the transmitter and the receiver. We will see, however, that there are reasons we might want a high–duty-cycle or continuous-wave (cw) waveform. In this case, we need a way to prevent emitted laser power from being backscattered into the receiver. The most common way to do this is to use polarization (see Fig. 1.14). If we transmit a linearly polarized laser beam, we can isolate it using a polarizing beamsplitter. A quarter-wave plate will convert linear polarization to circular polarization. When that circular polarization bounces off of an object, most of the light will reverse its handedness. On return, the opposite-handedness circular-polarization beam will be converted to the opposite linear polarization of the laser and will be transmitted through the polarizing beamsplitter to the receiver. This polarization method of transmit/receive isolation can achieve up to 40–45 dB of isolation with careful design. Isolation of up to 40–45 dB is a rejection factor of between 10,000 and 30,000. The backscatter that hits the camera will come from the quarter-wave plate and the transmit telescope. Quality engineering will reduce this scatter.

1.11 Major Devices in a LiDAR

There are three main components in a LiDAR: the laser source, the receiver, and the optical systems for pointing the LiDAR. Each of these components are briefly discussed next and will be discussed in detail in Chapters 5, 6, and 7.

1.11.1 Laser sources

As stated earlier, while it is possible for a LiDAR to be built with a light source other than a laser, this is highly unlikely. At the current time, virtually

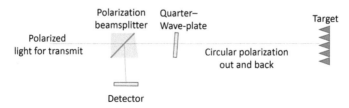

Figure 1.14   Transmit/receiver isolation approach using polarization (reprinted from Ref. 3).

all LiDARs use either a diode laser or a diode-pumped solid state laser. Diode-pumped solid state lasers can be divided into bulk solid state lasers and fiber lasers. Diode lasers can be very efficient and can be inexpensive. However, they do not store energy so cannot be Q switched, they tend to have a broad laser line width, and they tend to have a broad beam. Because of these characteristics, it is often useful to use diode lasers to pump a solid state medium that can be Q switched to obtain higher peak power. The solid state laser may also have narrower linewidth and beam divergence closer to the diffraction limit. Fiber lasers tend to be limited in peak power because of limited gain area in the fiber. This means that fiber lasers are normally used with higher–duty-cycle LiDARs.

1.11.2 Receivers

LiDAR receivers can be a single detector or an array of detectors. In the early days of LiDAR, we often used single detectors, but more recently, high-bandwidth arrays have become available that can measure range in each pixel based on the time of return of a reflected laser pulse. Also, some LiDAR implementations do not require high-bandwidth detectors. To increase the signal-to-noise ratio (SNR) in the receiver, two main approaches have been used. For coherent LiDAR, increasing the intensity of the LO increases the SNR in the receiver. For any LiDAR, but primarily for those used for direct detection, we can make use of gain on the receiver to increase the receiver SNR. While the gain can come prior to detection, it is usually done in a detector or a detector array after detection by generating multiple electrons per received photon. Linear-mode avalanche photodiodes (LMAPDs) have a linear relationship between the number of received photons and the number of generated electrons, while Geiger-mode APDs always generate a maximum number of electrons when one or more photons is received.

1.11.3 Apertures

We need optics in a LiDAR, and we need a method of pointing both the transmitter and the receiver. As discussed above, a single aperture can be used for both the transmitter and the receiver, or these apertures can be separate. When we point an optical system for transmit or for receive, we use either mechanical or nonmechanical pointing approaches. A simple pointing scheme can be a mirror that can be tilted. There are many mechanical approaches to pointing an optical system. The effect of these mechanical pointing systems is to change the tilt of the optical wavefront. If we can change the tilt of the optical wavefront without moving a mechanical device, we call this nonmechanical beam pointing. Over the past few decades, many nonmechanical beam-pointing approaches have been explored.[5,6] Some of these approaches use optical phased arrays, which are similar to the phased arrays in microwave systems that can change the tilt of the phase front of an electromagnetic wave. In Chapter 7 both mechanical and nonmechanical beam-pointing approaches will be discussed in detail.

1.12 Organization of the Book

This book has 12 chapters. The introduction you have just read is the first chapter. Chapter 2 looks back at LiDAR history. This chapter summarizes some of the information in Ref. 7 and includes some information from Ref. 8 that is not included in Ref. 4. The author of this book coauthored Ref. 6 and was a very active coauthor in Ref. 5.

Chapter 3 develops the LiDAR range equation and link budget. It shows how to calculate the number of returned photons when the LiDAR aperture, laser power, atmospheric conditions, and other parameters are known.

Chapter 4 discusses the types of LiDAR. Chapter 5 discusses lasers for LiDAR application and LiDAR waveforms. Chapter 6 discusses LiDAR receiver technology; to calculate the range of a given LiDAR, you will need information from both Chapters 3 and 6. Chapter 7 discusses apertures, pointing technology, and LiDAR optics.

Chapter 8 addresses LiDAR processing, and Chapter 9 presents LiDAR testing. Chapter 10 discusses metrics for evaluating LiDAR, and Chapter 11 discusses some LiDAR applications.

Problems and Solutions

1-1. Calculate how many 1550-nm photons will be contained in 1 nJ of energy.

1-1 Solution:

$$c = 3 \times 10^8 \, \text{m/s},$$
$$h = 6.63 \times 10^{-34} \, \text{J s},$$
$$\lambda = 1550 \, \text{nm},$$
$$E = 1 \, \text{nJ}.$$

First, consider how much energy is in a single photon:

$$E_\text{p} = \frac{hc}{\lambda}.$$

Second, consider how much energy we are given in this problem: $E = 1$ nJ. Therefore,

$$N = \frac{E}{E_p} = \frac{E\lambda}{hc} = \frac{(1 \times 10^{-9}\,\text{J})(1.55 \times 10^{-6}\,\text{m})}{(6.626 \times 10^{-34}\,\text{J s})(3 \times 10^8\,\text{m/s})},$$

$$N = 7.798 \times 10^9 \text{ photons.}$$

1-2. Assuming that the sun is a blackbody, calculate the ratio of flux from the sun at a wavelength of 900 nm to flux from the sun at 1550 nm.

1-2 Solution:

$$T = 6050\,\text{K},$$

$$\lambda_1 = 900\,\text{nm},$$

$$\lambda_2 = 1550\,\text{nm},$$

$$k = 1.38 \times 10^{-23}\,\text{J/K},$$

$$c = 3 \times 10^8\,\text{m/s},$$

$$h = 6.63 \times 10^{-34}\,\text{J s.}$$

Blackbody flux is given as

$$L_1 = \frac{2\,hc^2}{\lambda_1^5}\left[\exp\left(\frac{hc}{\lambda_1\,kT}\right) - 1\right]^{-1},$$

$$L_2 = \frac{2\,hc^2}{\lambda_2^5}\left[\exp\left(\frac{hc}{\lambda_2\,kT}\right) - 1\right]^{-1}.$$

Next, consider the ratio of the two fluxes:

$$\frac{L_2}{L_1} = \frac{\left(\frac{2hc^2}{\lambda_2^5}\left[\exp\left(\frac{hc}{\lambda_2 kT}\right) - 1\right]^{-1}\right)}{\left(\frac{2hc^2}{\lambda_1^5}\left[\exp\left(\frac{hc}{\lambda_1 kT}\right) - 1\right]^{-1}\right)} = \frac{\lambda_1^5\left[\exp\left(\frac{hc}{\lambda_1 kT}\right) - 1\right]}{\lambda_2^5\left[\exp\left(\frac{hc}{\lambda_2 kT}\right) - 1\right]},$$

$$\frac{L_2}{L_1} = \frac{(9 \times 10^{-7}\text{m})^5\left[\exp\left(\frac{(6.63\times10^{-34}\text{J s})(3\times10^8\text{m/s})}{(9\times10^{-7}\text{m})(1.38\times10^{-23}\text{J/K})(6050\,\text{K})}\right) - 1\right]}{(1.55 \times 10^{-6}\text{m})^5\left[\exp\left(\frac{(6.63\times10^{-34}\text{J s})(3\times10^8\text{m/s})}{(1.55\times10^{-6}\text{m})(1.38\times10^{-23}\text{J/K})(6050\,\text{K})}\right) - 1\right]}$$

$$= 0.2371.$$

So, the ratio of blackbody flux from the sun at 1550 nm to that at 900 nm is 23.71%.

1-3. For a 300-deg blackbody, calculate the ratio of flux at a wavelength of 10 μm to the flux available at a wavelength of 4.8 μm. Then compare the flux at 4.8 μm to the flux at 3.8 μm.

1-3 Solution:

$$T = 300\,\text{K},$$
$$\lambda_1 = 10\,\mu\text{m},$$
$$\lambda_2 = 4.8\,\mu\text{m},$$
$$\lambda_3 = 3.8\,\mu\text{m},$$
$$k = 1.38 \times 10^{-23}\,\text{J/K},$$
$$c = 3 \times 10^8\,\text{m/s},$$
$$h = 6.63 \times 10^{-34}\,\text{J s}.$$

Blackbody flux is given as

$$L_1 = \frac{2hc^2}{\lambda_1^5}\left[\exp\left(\frac{hc}{\lambda_1 kT}\right) - 1\right]^{-1},$$

$$L_2 = \frac{2hc^2}{\lambda_2^5}\left[\exp\left(\frac{hc}{\lambda_2 kT}\right) - 1\right]^{-1},$$

$$L_3 = \frac{2hc^2}{\lambda_3^5}\left[\exp\left(\frac{hc}{\lambda_3 kT}\right) - 1\right]^{-1}.$$

Next, consider the ratio of the two fluxes:

$$\frac{L_2}{L_1} = \frac{\left(\frac{2hc^2}{\lambda_2^5}\left[\exp\left(\frac{hc}{\lambda_2 kT}\right) - 1\right]^{-1}\right)}{\left(\frac{2hc^2}{\lambda_1^5}\left[\exp\left(\frac{hc}{\lambda_1 kT}\right) - 1\right]^{-1}\right)} = \frac{\lambda_1^5\left[\exp\left(\frac{hc}{\lambda_1 kT}\right) - 1\right]}{\lambda_2^5\left[\exp\left(\frac{hc}{\lambda_2 kT}\right) - 1\right]},$$

$$\frac{L_2}{L_1} = \frac{(1 \times 10^{-5}\text{m})^5\left[\exp\left(\frac{(6.63\times10^{-34}\text{J s})(3\times10^8\text{m/s})}{(1\times10^{-5}\text{m})(1.38\times10^{-23}\text{J/K})(300\,\text{K})}\right) - 1\right]}{(4.8 \times 10^{-6}\text{m})^5\left[\exp\left(\frac{(6.63\times10^{-34}\text{J s})(3\times10^8\text{m/s})}{(4.8\times10^{-6}\text{m})(1.38\times10^{-23}\text{J/K})(300\,\text{K})}\right) - 1\right]}$$

$$= 0.2371.$$

So, the ratio of blackbody flux from an object of 300 K temperature at 10 μm to that at 4.8 μm is 21.37%.

Similarly,

$$\frac{L_3}{L_2} = \frac{\lambda_2^5\left[\exp\left(\frac{hc}{\lambda_2 kT}\right) - 1\right]}{\lambda_3^5\left[\exp\left(\frac{hc}{\lambda_3 kT}\right) - 1\right]},$$

$$\frac{L_3}{L_2} = \frac{(4.8 \times 10^{-6}\text{m})^5\left[\exp\left(\frac{(6.63\times10^{-34}\text{J s})(3\times10^8\text{m/s})}{(4.8\times10^{-6}\text{m})(1.38\times10^{-23}\text{J/K})(300\,\text{K})}\right) - 1\right]}{(3.8 \times 10^{-6}\text{m})^5\left[\exp\left(\frac{(6.63\times10^{-34}\text{J s})(3\times10^8\text{m/s})}{(3.8\times10^{-6}\text{m})(1.38\times10^{-23}\text{J/K})(300\,\text{K})}\right) - 1\right]}$$

$$= 0.2309.$$

So, the ratio of blackbody flux from an object of 300 K temperature at 4.8 μm to that at 3.8 μm is 23.09%.

1-4. For a 10-nm–wide spectral band at a wavelength of 1.5 μm, what is the flux per nanometer in one square meter $B_{\text{sqm-nm}}$ on Earth?

1-4 Solution: The flux per unit area on Earth is given by

$$B_{\text{sqm-nm}} = \frac{LR_{\text{s}}^2}{R_{\text{e}}^2},$$

where R_{s} is the radius of the sun, R_{e} is the radius of the earth, and L is the blackbody radiation of the sun, which is given as

$$L = \frac{2hc^2}{\lambda^5} \left[\exp\left(\frac{hc}{\lambda kT}\right) - 1 \right]^{-1},$$

$$L = \frac{2(6.63 \times 10^{-34} \text{ J s})(3 \times 10^8 \text{ m/s})^2}{(1.5 \times 10^{-6} \text{ m})^5}$$
$$\left\{ \exp\left[\frac{(6.63 \times 10^{-34} \text{ J s})(3 \times 10^8 \text{ m/s})}{(1.5 \times 10^{-6} \text{ m})(1.38 \times 10^{-23} \frac{\text{J}}{\text{K}})(6050 \text{ K})} \right] - 1 \right\}^{-1},$$

$$L = 4.035 \times 10^{12} \text{ W/m}^3.$$

$$B_{\text{sqm-nm}} = d\frac{LR_{\text{s}}^2}{R_{\text{es}}^2} = 10 \text{ nm} \frac{(4.035 \times 10^{12} \text{ W/m}^3)(6.96 \times 10^8 \text{ m})^2}{(1.47 \times 10^{11} \text{ m})^2}$$
$$= 0.00905 \text{ W/m}^2$$

1-5. Calculate the MPE needed for a 1550-nm wavelength laser to be eye safe in the following configurations:

(a) At the exit aperture. The aperture is 1 square cm in area.
(b) At a spot 10 m from the aperture. The beam will spread to cover 20 deg in elevation and 3 deg in azimuth. The beam will scan back and forth in azimuth 10 times per second from the edge of the beam at 30-deg azimuth to the opposite edge of the beam at –30-deg azimuth.
(c) In the same geometry as (b), but with a laser wavelength of 900 nm.

For cases (a) through (c), assume 3-ns pulse widths.

1-5 Solution:
Our solution uses Fig. 1.11, which is repeated here. The plot shows that, for 1-ns pulses, the energy threshold is 1 J for 1550 nm and about 1×10^{-6} J for 900 nm. As stated in Section 1.7, Fig. 1.11 charts energy threshold per square cm for a 10-s exposure. For case (a), the exit aperture is 1 cm^2; therefore, we have an average power of 100 mW over 10 s to reach the 1-J/cm^2 threshold. For case (b), the beam is 20 deg $\times$ 3 deg and scans back and forth. Any given location is only illuminated a portion of the time. We actually can say that the

energy is spread over 20×60 deg. At 10 m away, this is an area of 3.49 m $\times$ 10.47 m, or 365,540 square cm. This is a lot of area. Our laser eye damage threshold is therefore very high: 3.65×10^5 W. If we switch to 900 nm for the same case, the laser damage threshold is 0.365 W or 365 mW. This shows the significant impact of wavelength on MPE.

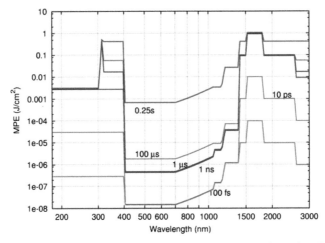

1-6. A monostatic direct-detection LiDAR uses a wavelength of 1550 nm.

(a) Calculate the diameter of the transmit aperture for a divergence of less than 0.002 deg.

(b) Plot the radial intensity profile of the optical beam in orders of $N = 2$, 4, 6, 8 and 10.

1-6 Solution:

(a) Using Eq. (1.5), the aperture diameter is about 4.6 cm:

$$\theta = \frac{1.03\lambda}{D} \rightarrow D = \frac{1.03\lambda}{\theta} = \frac{1.03(1.55 \times 10^{-6}\text{m})}{0.002\,\text{deg}(\pi/180\,\text{deg})} = 0.0457\,\text{m} = 4.57\,\text{cm}.$$

(b) The plot of the radial intensity profile for the optical beam in orders of $N = 2$, 4, 6, 8, and 10 is shown below:

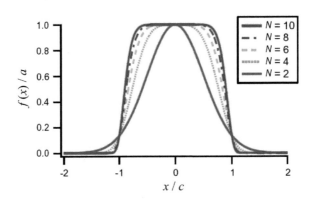

1-7. What is the ratio of the separation in returns of the reflectivity between 14-bit and 8-bit digital returns for a uniform distribution of reflectivity between 9% and 12%?

1-7 Solution:

$$r = \frac{(0.12 - 0.09)}{2^8} = \frac{(0.12 - 0.09)}{256} = 0.0001719.$$

The LiDAR is required to discern a variation of 0.0001719 for 8 bits of grayscale.

$$r = \frac{(0.12 - 0.09)}{2^{14}} = \frac{(0.12 - 0.09)}{16384} = 1.8311 \times 10^{-6}.$$

The LiDAR is required to discern a variation of 1.8311×10^{-6} for 14 bits of grayscale.

Thus, the ratio of variation in reflectivities between a 14-bit and an 8-bit digital receiver is $0.0001719/1.8311 \times 10^{-6} = 0.01065$. The variation in reflectivity for a 14-bit receiver with a uniform distribution of reflectivity is 1.065% of that of an 8-bit receiver with the same distribution.

1-8. How many photons irradiate the entire surface of the earth from the sun at 540 nm? How much power arrives over a year?

1-8 Solution:

$$\lambda = 540\,\text{nm},$$
$$R_e = 6.96 \times 10^8\,\text{m},$$
$$R_{es} = 1.47 \times 10^{11}\,\text{m},$$
$$R_e = 6.37 \times 10^6\,\text{m},$$
$$c = 3 \times 10^8\,\text{m/s},$$
$$h = 6.63 \times 10^{-34}\,\text{J s}.$$

The number of photons per square meter on Earth's surface is given by

$$N = \frac{L R_s^2 \lambda}{R_{es}^2 \, hc}.$$

However, the total number of photons over Earth's surface is given by

$$n = NA = 4\pi R_e^2 N = 4\pi R_e^2 \frac{L R_s^2 \lambda}{R_{es}^2 \, hc}.$$

The blackbody radiation of the sun is given by

$$L = \frac{2hc^2}{\lambda^5}\left[\exp\left(\frac{hc}{\lambda kT}\right) - 1\right]^{-1}$$

such that

$$n = 4\pi R_e^2 \frac{LR_s^2\lambda}{R_{es}^2 hc} = 4\pi R_e^2 \frac{R_s^2\lambda}{R_{es}^2 hc}\frac{2hc^2}{\lambda^5}\left[\exp\left(\frac{hc}{\lambda kT}\right) - 1\right]^{-1}$$

$$= 8\pi R_e^2 \frac{R_s^2}{R_{es}^2}\frac{c}{\lambda^4}\left[\exp\left(\frac{hc}{\lambda kT}\right) - 1\right]^{-1},$$

$$n = 8\pi R_e^2 \frac{R_s^2}{R_{es}^2}\frac{c}{\lambda^4}\left[\exp\left(\frac{hc}{\lambda kT}\right) - 1\right]^{-1},$$

$$n = 8\pi(6.37 \times 10^6\,\text{m})^2 \frac{(6.96 \times 10^8\,\text{m})^2}{(1.47 \times 10^{11}\,\text{m})^2}\frac{(3 \times 10^8\,\text{m/s})}{(5.40 \times 10^{-7}\,\text{m})^4}$$

$$\left[\exp\left(\frac{(6.63 \times 10^{-34}\,\text{Js})(3 \times 10^8\,\text{m/s})}{(5.40 \times 10^{-7}\,\text{m})(1.38 \times 10^{-23}\,\text{J/K})(6050\,\text{K})}\right) - 1\right]^{-1},$$

$$n = 6.5664 \times 10^{45}\text{ photons m}^{-1}\text{s}^{-1},$$

$$P = nE_p d = nd\frac{hc}{\lambda} = (6.5664 \times 10^{45})(1 \times 10^{-9}\,\text{m})$$

$$\left[\frac{(6.63 \times 10^{-34}\,\text{Js})(3 \times 10^8\,\text{m/s})}{(5.40 \times 10^{-7}\,\text{m})}\right] = 2.419 \times 10^{18}\,\text{W}.$$

So, 2.419 million TW arrive each year from solar irradiation on the Earth's surface. Note that the actual value would be less by one-half (day versus night): 1.209 million TW annually over Earth's surface.

1-9. What distance would a 1-km^2 target need to be in order to be fully filled by laser illumination from an aperture of diameter 5 cm and wavelength of 1550 nm, assuming a diffraction-limited beam? Check to make sure you are in the far field.

1-9 Solution: The divergence is

$$\theta = \frac{1.03(1550\,\text{nm})}{5\,\text{cm}} = 31.9\,\mu\text{rad}.$$

Thus, if we assume a circular target and a circular aperture/beam,

$$r_{\text{target}} = \frac{\sqrt{1\,\text{km}^2}}{\pi} = 318.3\,\text{m}.$$

The distance z traveled by the light path must equal the radius of the target divided by the tangent of the divergence angle:

$$z = \frac{r_{target}}{\tan \theta} = 9.969 \times 10^6 \, \text{m} = 9969 \, \text{km}.$$

Note that this is roughly between low Earth orbit and geostationary orbit as measured from a ground station. This also highlights an inherent problem in satellite characterization using active sources of illumination (LiDAR or radar) beyond low Earth orbit, i.e., at distances of a few hundreds of kilometers.

The far field is at

$$R = \frac{2D^2}{\lambda},$$

which means that the far field is 3.2 km away, so the target is in the far field.

1-10. Assume two lasers, both having a pulse width of 1 ns. One laser is 10-W average power, made up of a 10-kHz repetition rate with 1 mJ/pulse. The second laser has the same repetition rate but has only 10 μJ/pulse, so is a 100-mW average power laser. Both are at the 1550-nm wavelength. The first laser illuminates an area of 60 deg by 30 deg continuously in a flash LiDAR system. The second laser has a beam width of 1 deg but scans in azimuth over 60 deg. It also has an elevation coverage of 30 deg. The beam of the first laser is 60 deg wide by 30 deg high. The beam of the second laser is 1 deg wide by 30 deg high and has a scan rate of 1 Hz. For both lasers, calculate the flux (in W/cm^2) illuminating the eye at a range of 100 m, assuming that we are in the far field. Then compare each flux value against the eye-safety threshold.

1-10 Solution: For laser 1, we have 10-W average power, so the laser will emit 10 J in 10 s. The area covered is $A = 2 \times 100 \times \tan(30 \, \text{deg}) \times 1 \times \tan(15 \, \text{deg})$. This is 115 m by 58 m, or an area of 6188 m^2. For laser 2, we have an area of 201 m^2. Therefore, the flux is 1/6188 for laser 1, and 0.1/201 for laser 2 in W/m^2. The flux in W/cm^2 is 1.6 for laser 1, and 5 for laser 2, but laser 2 is only illuminating the area for 1/60th of the time. Therefore, the total energy/cm^2 in case 1 over 10 s is 16 J/cm^2. For laser 2, it is 0.83 J/cm^2. Laser 1 is not eye safe, even though it puts out more power. Laser 2 is eye safe. It is below the threshold of 1 J/cm^2 for 1550 nm.

References

1. Wikimedia Commons page on EM Spectrum. Adapted from File:EM Spectrum3-new.jpg by NASA. The butterfly icon is from the P icon set, P biology.svg. The humans are from the Pioneer plaque, Human.svg. The

buildings are the Petronas towers and the Empire State Buildings, both from Skyscrapercompare.svg. Released under license CC BY-SA 3.0, https://commons.wikimedia.org/w/index.php?curid=2974242.

2. National Academy of Sciences, *Laser Radar: Progress and Opportunities in Active Electro-Optical Sensing*, P. F. McManamon, Chair (Committee on Review of Advancements in Active Electro-Optical Systems to Avoid Technological Surprise Adverse to U.S. National Security), Study under Contract HHM402-10-D-0036-DO#10, National Academies Press, Washington, D.C. (2014).

3. P. F. McManamon, *Field Guide to LiDAR*, SPIE Press, Bellingham, Washington (2015) [doi: 10.1117/3.2186106].

4. Google Books Ngram Viewer: https://books.google.com/ngrams/graph?content=lidar%2Cladar&year_start=1960&year_end=2015&corpus=15&smoothing=3&share=&direct_url=t1%3B%2Clidar%3B%2Cc0%3B.t1%3B%2Cladar%3B%2Cc0.

5. P. F. McManamon, T. A. Dorschner, D. L. Corkum, et al., "Optical phased array technology," *Proc. IEEE* **84**(2), 268–298 (1996).

6. P. F. McManamon, P. J. Bos, M. J. Escuti, et al., "A review of phased array steering for narrow-band electrooptical systems," *Proc. IEEE* **97**(6), 1078–1096 (2009).

7. V. Molebny, P. F. McManamon, O. Steinvall, T. Kobayashi, and W. Chen, "Laser radar: historical prospective—from the East to the West," *Optical Engineering* **56**(3), 031220 (2016) [doi: 10.1117/1.OE.56.3.031220].

8. P. F. McManamon, G. Kamerman, and M. Huffaker, "A history of laser radar in the United States," *Proc. SPIE* **7684**, 76840T (2010) [doi: 10.1117/12.862562].

Chapter 2
History of LiDAR

LiDAR has a rich history of over 50 years. It started most of its development in the early 1960s, shortly after the invention of the laser. There had been some earlier LiDAR development prior to the invention of the laser, but the laser has been a critical enabler for LiDAR development. Wind sensing and laser designation were developed starting in the 1960s. I call laser designation a form of bistatic LiDAR because we have an illuminator, which is called the designator, and receiver, which is on the seeker. The illuminator and receiver are separated, so it is a bistatic system. Admittedly, the term LiDAR stands for light detection and ranging, and in a designator there is no ranging, but a designator does use a laser for illumination and a receiver for detection, so we will include it in the discussion of history. LiDAR has become relatively inexpensive and reliable, and has very rich phenomenology, making it competitive compared to alternative sensor technologies, such as passive EO sensing or microwave radar. LiDAR started operating in the visible region (ruby laser), then appeared in the near infrared (Nd:YAG lasers), and next in the thermal infrared (CO_2 laser). Many LiDARs are now being developed in the eye-safe, short-wave infrared (SWIR) region ($\sim$1.5 μm and beyond).

Numerous publications accompanied the maturing of LiDAR, bringing to life new journals and new professional meetings, symposia, and conferences, such as the Laser Radar Technologies and Applications conference, which is managed by SPIE as part of its Defense and Commercial Sensing symposium, OSA's Laser Sensing and Communications (LSC) conference, and the Coherent Laser Radar Conference (CLRC). Special LiDAR courses were developed.[1] LiDAR became a topic for new fundamental books[2–5] and reports,[6] as well as a subject taught at universities.[7] The range of most LiDAR applications is from micrometers[8] to tens of kilometers, although a laser beam has been bounced off of the Moon from the Earth. This was a very long-range laser rangefinder.

Overviews on the history of LiDAR development concerning Europe,[9] USA,[10] the former Soviet Union (USSR),[11,12] Japan,[13] and China appeared. Recently, a summary was published compiling the history of LiDAR around the world.[14] Much of the content of this chapter is drawn from either the USA article[10] or the recent worldwide article.[14] History does not change, so once it is recorded the only issue is how it is presented.

Initially, LiDARs used various lasers, until CO_2 became the popular choice for coherent LiDAR and Nd:YAG for laser rangefinders and designators. CO_2 reigned as the coherent LiDAR of choice from the early 1970s until the late 1980s or early 90s. Most CO_2 LiDARs were at 10.6 μm, although to avoid atmospheric CO_2 absorption, some CO_2 LiDARs used different isotopes. These could be carbon 13, lasing at 11.18 μm, or oxygen 18, lasing around 9.1 μm. CO_2-based navigation LiDAR was deployed in the 1980s and 1990s but with the advent of GPS became less popular. Even in the 1970s, the 3D nature of LiDAR allowed automatic recognition of certain objects. As an example, the Cruise Missile Advanced Guidance (CMAG) program automatically recognized terrain and correlated terrain features with location.

2.1 Rangefinders, Altimeters, and Designators

2.1.1 First steps of rangefinders

A laser rangefinder is a very simple LiDAR. It uses a single detector to determine the range to a target based on the round-trip time-of-flight of a laser pulse to and from the object. Because we know the speed of light, we can calculate range. The idea of using short pulses of light to measure distance was initiated by A. A. Lebedev in 1933.[15] The use of short pulses is one method that provides excellent range resolution. In 1936, this early LiDAR prototype used a specially developed interference modulator to obtain short pulses of light. A range of up to 3.5 km was measured with an accuracy of 2–3 m. During 1963–64, laser-based rangefinders were developed using ruby and gallium arsenide in the same lab that was used by Lebedev in the 1930s: the Vavilov Optics State Institute (GOI) in Leningrad (now St. Petersburg).

In the late 1960s and early 1970s, rangefinders, proximity fuses, and weapon guidance 'LiDARs' were the first military laser systems to be developed. The early ruby lasers were high cost with poor efficiency and eye safety issues. Later, short-pulse, high-energy, highly collimated, monochromatic beams became available when Q-switched lasers revolutionized LiDAR capabilities. These LiDARs still had the issue of low efficiency and lack of eye safety.

In Sweden, laser research with a ruby laser started at the Swedish Defense Research Establishment (FOI) in 1961. In Swedish industry, the pioneers were ASEA and LM Ericsson. In 1968, Ericsson delivered laser rangefinders to the Swedish Coastal Artillery for operational use. Bofors AB developed laser rangefinders for the infantry cannon vehicle IKV 91 and for the BOFI system in cooperation with Hughes Aircraft. Later, ASEA developed cloud altimeters for the civilian and military markets. During the early 70s, Bofors developed a successful anti-aircraft missile beam-rider system, the RBS 70, which later was modified into the RBS 90 and sold worldwide. The world's first laser beam-rider surface-to-air missile (SAM) to enter service was developed at Bofors and contained laser sensing. During the late 70s, Ericsson developed laser-based proximity fuses for the Sidewinder missile. Ericsson also developed a laser-tracking system, but that laser application was soon overtaken by video trackers.

In Norway, the Norwegian Defense Research Establishment (FFI) transferred their knowledge to Simrad Optronics AS, which became famous for building their laser rangefinder into a handheld binocular. More recently, the company developed a family of rangefinders for target location and fire control.

In the United Kingdom, the Royal Signals and Radar Establishment (RSRE) pioneered military laser development. An excellent review on the early laser rangefinder development was published by Forrester and Hulme.[16] These authors claim that the LF-2 ruby tank laser sight was the first laser system in the world to be in large-quantity production. It was developed primarily for use with the Chieftain main battle tank for the British army, but was widely used on other tanks, such as the Vickers MBT (main battle tank), Centurion, Scorpion, and Chieftain derivatives. The equipment that was originally used was a ruby laser with a spinning prism Q switch, although a later version was used a passive Q-switched YAG (LF-11).

In 1968, Ferranti (now Leonardo S.p.a.) developed the world's first fully stabilized laser system incorporating a Nd:YAG laser rangefinder and marking target seeker (LRMTS). The marking target seeker is a unit on the aircraft that locks onto a laser-designated target. This equipment was a part of the weapon-aiming systems of Jaguar, Harrier, and Tornado aircraft. The 1.06-μm laser transmitter power was generated by an electro-optically Q-switched Nd:YAG laser capable of operating at 10- or 20-Hz repetition rate.

One of the first Soviet laser rangefinders BD-1 (Fig. 2.1) was described in a 2007 SPIE Press book.[17] Another example of an early laser rangefinder is the KTD 2-2. The technology of laser rangefinders and laser designators is represented by numerous operational instruments described in an earlier publication.[18] They are installed on Su and MiG helicopters and airplanes (see Fig. 2.2). In France, Thales Research and Technology (TRT) delivered a Nd:YAG laser rangefinder to the French Official Services in 1967. The laser rangefinder was installed on the AMX 13 tank for field tests.[19] In the 1970s, Thomson and Cilas developed an airborne target illumination system that demonstrated the ability to be used on a single-seat airplane fighter, and that was later developed into the ATLIS targeting pod.

Germany also developed laser rangefinders and target designators based on a Nd:YAG laser by Zeiss and ELTRO GmbH. Military laser research was performed by FGAN–Forschungsinstitit für Optik. Many of their early publications were dedicated to laser propagation in the atmosphere.[20]

Figure 2.1 Examples of early laser rangefinder (LRF) development. First row, left to right: Test equipment for the first Ericsson LRF (1965); LRF for the Swedish Coastal Artillery (1968); Simrad handheld LP-7. Second row, left to right: LRF KTD 2-2 (Polyus, USSR); LRF BD-1 (Institute # 801, USSR); Ferranti CO_2 TEA LRF (reprinted from Ref. 14).

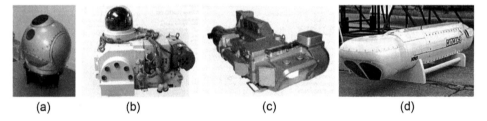

(a)   (b)   (c)   (d)

Figure 2.2 Examples of the Soviet/Russian rangefinder/designator LiDARs: (a) Samsheet-50 rangefinder/designator for Ka-52 helicopter; (b) 31E-MK EO system for Su-30 fighter; (c) Shkval series for Su-25T, Su-25TM/Su-39, and Ka-50; and (d) pod-mounted Sapsan-E for air-to-surface MiG and Su missions (reprinted from Ref. 14).

2.1.2 Long-distance rangefinders

The first laser ranging to the moon was achieved by MIT Lincoln Lab in 1962 using a 50-J/pulse ruby laser. More-precise laser ranging to the moon was achieved in 1969 by NASA Goddard Space Flight Center in Greenbelt, Maryland. NASA used a retroreflector (Fig. 2.3) that was positioned on the Moon by the Apollo 11 astronauts. By beaming laser pulses from Earth at the reflector, scientists were able to determine the distance to that spot on the Moon to an accuracy of about 3 cm using a 60-J/pulse ruby laser. This range measurement was more accurate than the 1962 measurement because the return signal was higher, increasing the SNR. The laser reflector consists of 100 fused-silica half-cubes mounted in a 46-cm square aluminum panel.

Figure 2.3 A corner cube retroreflector placed on the Moon in 1969 imaged using an early, high-power ruby laser (reprinted from Ref. 10).

Additional retroreflector packages were landed on the lunar surface by NASA during the Apollo 14 and Apollo 15 missions. French-built retroreflector packages were soft landed on the lunar surface by Soviet landers.[21]

A retroreflector package was also used in the Japanese Retroreflector in Space (RIS) project, which was installed on the ADEOS satellite [by the National Space Development Agency of Japan (NASDA), currently known as the Japan Aerospace Exploration Agency (JAXA)] and launched in August 1996. The retroreflector had a curved-mirror surface to compensate for the velocity aberration due to the satellite movement. A ground-based transmitter–receiver system used a TEA CO_2 laser and a HgCdTe detector with the 1.5-m–diameter high-precision tracking telescope at the Communication Research Laboratory (CRL) [now called the National Institute of Information and Communications Technology (NICT)] in Kokubunji, Tokyo. The absorption spectra of atmospheric ozone were successfully measured in the 10-μm–wavelength region using the reflection from the RIS.[22]

In the USSR, long-distance measurement with lasers began in 1962 by the Vympel Design Bureau in cooperation with the Lebedev Physical Institute. In 1967, the first experiments allowed laser ranging of a Tu-134 airplane equipped with optical retroreflectors.[11] Ten years later, a giant LiDAR LE-1 was tested at the Sary Shagan range in Kazakhstan. The main goal for this LiDAR was its use in antimissile defense. It tracked the Molniya satellite and measured the distance to it without use of a retroreflector.

The USSR LE-1 had a multichannel transmitter (49×4) and a multichannel receiver with an array of 196 range-gated photomultipliers, each having its own optical system. The switch is made of a block of 4 optical wedges rotating at 80 Hz. The LE-1 multibeam was controlled by means of a fast 2D scanner consisting of two mirrors driven by stepping motors. The mirrors kept the beam stable during the transmit–receive cycles. Each laser consisted of a master oscillator/power amplifier with identical ruby crystals and a KDP EO switch. The output pulse energy was 1 J, the pulse repetition frequency of each laser was 10 Hz, and the pulse duration was 30 ns. The optical train was designed by Geofizika Central Design Bureau in cooperation with the Vavilov Optical Institute. The Cassegrain telescope with 1.2 m of main mirror aperture and a 21-arc-min field of view provides target tracking in the upper hemisphere. It was designed and manufactured by LOMO. The telescope drives operation with angular velocity up to 5 deg/s and angular acceleration up to 1.5 deg/s^2 with 5-arc-min dynamic error. Velocity aberration is compensated by the tilt of a mirror. The LE-1 LiDAR could detect a 1-m^2 target at 400 km.

2.1.3 Laser altimeters

Time-of-flight measurement is a keystone of LiDAR altimeters. The Lunar Orbiter Laser Altimeter (LOLA), built by NASA, was designed to characterize landing sites and to provide a precise global geodetic grid on the Moon.[23] LOLA's primary measurement is surface topography. The instrument provides ancillary measurements of surface slope, roughness, and reflectance. LOLA is a multibeam laser altimeter that operates at a wavelength of 1064.4 nm with a 28-Hz pulse repetition rate. A single laser beam is split by a diffractive optical element into five output beams [Fig. 2.4 (left)], each of which has a 100 mrad divergence and illuminates a 5-m–diameter spot from the mapping orbit, resulting in a total sampling rate of the lunar surface of 140 measurements per second. Backscattered pulses are detected by the receiver, which images the five-spot pattern onto separate optical fibers, each of which relays the received signal to a distinct silicon avalanche photodiode detector. An example of lunar profiles is shown in Fig. 2.4 (top right).

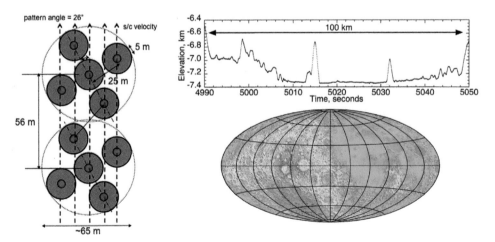

Figure 2.4 Lunar Orbiter Laser Altimeter (LOLA): (left) five output beams; (top right) lunar profiles; and (bottom right) topographic map of the Moon (reprinted from Ref. 14).

The laser spots form a cross-pattern on the lunar surface, with each beam separated by an angle of 500 mrad and rotated 26 deg about the nadir axis with respect to the spacecraft forward velocity vector. The sample pattern permits calculation of surface slopes along a range of azimuths. A topographic map of the Moon is shown in Fig. 2.4 (bottom right).

The Institute of Space and Astronautic Science (ISAS), Japan developed various laser altimeters for space and astronautical science use. Figure 2.5 shows some details of the Hayabusa mission using the Hayabusa LiDAR launched in 2005.[24,25] Figure 2.5(a) is a photograph of the asteroid Itokawa (size: 540 m × 270 m × 210 m) taken by an imager, Fig. 2.5(b) shows a flight model of ISAS. It operates at 1064 nm with 8 mJ, 1 Hz, 12.5 cm aperture, 3.7 kg, 24 cm × 23 cm × 25 cm). Figure 2.5(c) shows relative elevation measured by the Hayabusa LiDAR. It has a range resolution of ±1 m. The precise elevation measurement of the asteroid was the first in the world. The Hayabusa-2 LiDAR was launched in 2015 targeting a near-Earth asteroid named 1999 Ju3.

2.1.4 Laser designators

An Nd:YAG laser at 1.064-μm wavelength was the laser of choice for laser designators and rangefinders. Laser designators were introduced in the late 60s and are still being used today. Whether laser designators do ranging or not, they use almost all of the same components as laser rangefinders. Initially, Nd:YAG lasers were flashlamp pumped. Figure 2.6(a) is a photograph of the first laser designator, and Fig. 2.6(b) shows the first laser-guided bomb. Since its inception in 1968, the Paveway series of laser-guided bombs has revolutionized tactical air-to-ground warfare. These semi-active laser-guided munitions, which home on reflected energy directed on the target, drastically reduce the number of munitions required to destroy a target. The universal nature of the laser guidance kits enabled current general-purpose bombs to be readily converted to precision, tactically effective munitions.

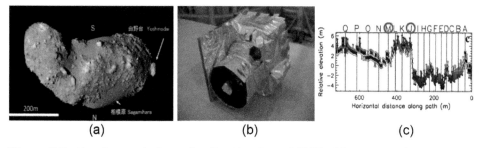

(a)   (b)   (c)

Figure 2.5 Hayabusa mission using the Hayabusa LiDAR: (a) a photo of the asteroid Itokawa taken by an imager; (b) a flight model; and (c) relative elevation and horizontal distance measured by the Hayabusa LiDAR (reprinted from Ref. 14).

(a)   (b)

Figure 2.6 (a) First laser target designator (1969) and (b) the first laser-guided bomb (1968) from the Paveway series (reprinted from Ref. 10).

Figure 2.7 The Thanh Hoa Bridge, also called the Dragon's Jaw, over the Song Ma river in Vietnam (reprinted from Ref. 10).

Figure 2.7 shows the Thanh Hoa Bridge, aka Dragon's Jaw, in Vietnam. This bridge was finally destroyed by a laser designator in 1972 after withstanding many attacks by conventional bombs. The U.S. Air Force had lost 11 planes attacking this bridge and had conducted 871 sorties against it. The bridge was destroyed with four sorties and no losses the first time a laser designator was used.[26]

NATO forces successfully employed Paveway during Operation Desert Fox and the subsequent patrol of the no-fly zones over Iraq because of its pinpoint accuracy and reduced chances of collateral damage. The Paveway III is the third generation, which provides the optimum operational flexibility through the use of an adaptive digital auto pilot, a large field-of-regard, and a highly sensitive seeker. It adapts to conditions of release, flies the appropriate midcourse, and provides trajectory shaping for enhanced warhead effectiveness. When used in conjunction with the BLU-109 or BLU-113 penetrator warheads, Paveway optimizes not only the trajectory and impact angle, but also the angle of attack.

2.1.5 Obstacle avoidance applications

In Europe, interest in imaging LiDAR seeker technology was more limited compared to the U.S. One example of an operational obstacle avoidance system for helicopters is the German Hellas system[27] developed by EADS in the early 2000s. The Hellas system was designed to warn pilots of obstacles in due time and to detect thin wires up to 1 km from the platform. It also has a brown-out recovery system. During the takeoff and landing, helicopters can encounter serious brown-out or white-out problems in dusty, sandy, and snowy areas as the downwash of the rotor blades creates dust clouds around the helicopter. With the help of LiDAR, an augmented enhanced synthetic vision of the landing area with surroundings is provided to the pilot based on range image data as well as on the altimetry and inertial reference information.

Other LiDAR obstacle avoidance systems were developed in Israel and the U.S. 3D LiDAR imaging holds potential for many other applications, including robotics, terrain visualization, augmented vision, reconnaissance, etc. Different types of the 3D laser sensors have been proposed and implemented.[28–30] To minimize the transmitted energy, a pulsed 3D laser sensor with 2D scanning of the transmitting beam and a scanless receiver was proposed by Mitsubishi Electric Corporation.[31] The system configuration is shown in Fig. 2.8. The laser is a 1.5-μm pulsed fiber laser with pulse energy of 2 μJ and duration of 10 ns. The beam diameter of the collimated beam is about 1 mm. The aperture diameter of the receiving optics is 15 mm. The FOV of the receiver is about 6×6 deg. The signal processor generates an intensity image, a range image, and, consequently, a 3D image by combining with angle information for each pixel. To enable the system to obtain correct

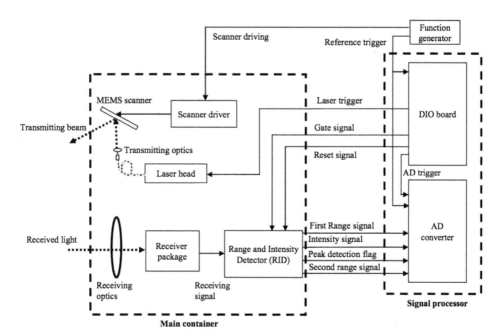

Figure 2.8 Schematical configuration of the obstacle avoidance system by Mitsubishi Electric Corporation (DIO is digital input/output; MEMS is microelectromechanical system; AD is analog-to digital) (reprinted from Ref. 14).

3D images, the processing includes a peak detection flag that indicates whether the range and intensity detector (RID) recognizes the received signal.

2.2 Early Coherent LiDARs

2.2.1 Early work at MIT/Lincoln Lab

MIT/Lincoln Lab did quite a bit of the early work in LiDAR. For strategic LiDAR, a workhorse was the MIT/Lincoln Lab Firepond system, shown in Fig. 2.9. Firepond used a 20-W (11.15 μm) master oscillator, mechanically chopped at 4–10 Hz to 35-μs pulses of 700 μJ (each). This is an example of a CO_2 LiDAR that avoided the standard 10.6-μm line to avoid atmospheric CO_2 absorption. Firepond could create range-Doppler images, as will be discussed later in the book. Range-Doppler images can provide range information and can use Doppler velocity information for cross-range. An example of a range-Doppler image is shown in Fig. 2.10.

2.2.2 Early coherent LiDAR airborne applications

By mixing the received signal with an optical local oscillator, the full field can be measured, including both phase and amplitude information, as compared to direct detection, where we only measure the amplitude (intensity) of the return laser signal. We call this type of LiDAR coherent LiDAR, which will be further discussed in Chapter 4. Coherent LiDARs make use of the spatial and temporal coherence properties of laser radiation. A typical application for

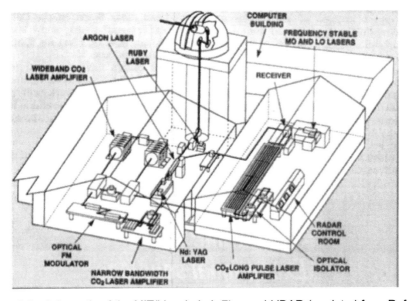

Figure 2.9 Schematic of the MIT/Lincoln Lab Firepond LiDAR (reprinted from Ref. 10).

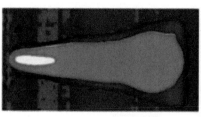
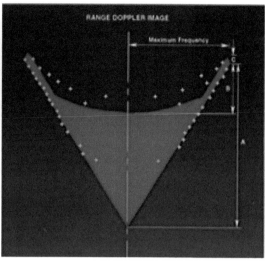

Figure 2.10 Example of a range-Doppler image (reprinted from Ref. 10).

coherent LiDAR is velocity measurement because, by measuring phase, we can directly measure the Doppler frequency shift.

Airborne coherent LiDARs have been used for ground imaging, obstacle warning, terrain following, as well for as wind sensing, including backscatter measurements. CLARA was a French–UK system developed for hard target (cables, ground surface, etc.) measurements.[32] This equipment followed successful trials of the LOCUS (Laser Obstacle and Cable Unmasking System) pulsed CO_2 LiDAR jointly developed by two groups within the former GEC Marconi. In another project, SFENA in France, a frequency-modulated, continuous-wave (FMCW) LiDAR was demonstrated for terrain following and terrain avoidance of combat aircraft.[33] The airborne LATAS (Laser True Airspeed System) LiDAR developed by Royal Signals and Radar Establishment (RSRE) was mainly used for true air speed and atmospheric backscatter measurement at 10.6 μm but also demonstrated intensity imaging of natural terrain and manmade objects.[34]

Figure 2.11 shows a transverse-flow laser velocimeter that was the first commercial laser Doppler velocimeter instrument (circa 1969). The instrument projected two coherent beams into the fluid flow field. The two beams interfered to create a fringe or spatial modulation pattern in the fluid. Particles or inhomogeneities scatter light. The frequency of the modulation of the scattered light is proportional to the particles' velocity. This type of laser instrument does not measure range, so a purist could say that, like laser designators, this is not a LiDAR; however, again it uses many of the same components. Figure 2.12 is early CO_2 imagery of a tank.

Figure 2.13 shows some early CO_2 3D LiDAR imagery from around 1983. Figure 2.13(a) shows ground test images collected by a short-pulsed,

Figure 2.11 Transverse-flow laser velocimeter (reprinted from Ref. 10).

scanned, CO_2, heterodyne, 3D imaging LiDAR. In the first range gate (at a distance of approximately 300 ft to 320 ft), the tree and foliage at the right and the foreground are visible. In the second range gate (from about 340 ft to 360 ft), the power lines are visible to the left of the tree but also behind it. In the second range gate (from about 360 ft to 380 ft), bushes behind closer foliage are discernable. Additional terrain returns are clearly visible in the center and to the left of the image. This is the first known demonstration of foliage penetration by LiDAR. Figure 2.13(b) shows a long-range image of a bridge complex on the Mississippi river collected by a short-pulsed, scanned,

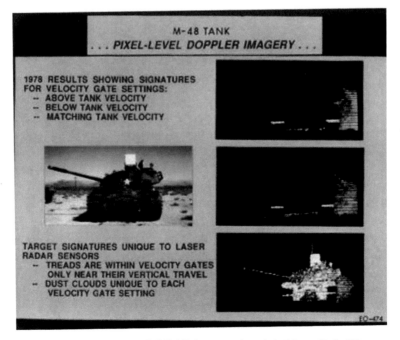

Figure 2.12 Early 3D LiDAR imagery (reprinted from Ref. 10).

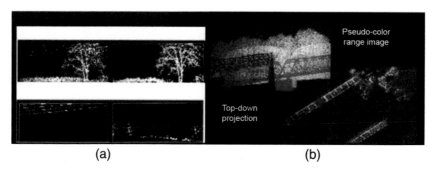

Figure 2.13 Early 3D LiDAR imagery [part (a) reprinted from Ref. 10].

CO_2, heterodyne, 3D imaging LiDAR (circa 1983). The original image is shown in the top left and is presented in spherical, sensor-centric coordinates. Range has been color coded. The image includes a gantry over a river lock in the foreground, a railroad bridge, an automotive bridge, and trees on the far side of the river. The image at the bottom right is a top-down (i.e., north, east, up) transformation of the same data with color-coded elevation. This is the first known example of 3D imagery used for topographic target representation. The clutter at the top right of the north-east–up image is the foliage that was aliased by the transformation algorithm into the wrong location.

2.2.3 Autonomous navigation using coherent LiDAR

From 1979 until 1984 a DARPA program called the Autonomous Terminal Homing (ATH) program was developed, as highlighted in Fig. 2.14. ATH was a predecessor to the Cruise Missile Advanced Guidance (CMAG) program, which resulted from the need for improved guidance and flight control of autonomous cruise missiles. CMAG was conducted by the Air Force Avionics Laboratory (Wright-Patterson Air Force Base), a predecessor organization to the current Air Force Research Laboratory (AFRL), following DARPA's ATH program and the AFATL (Eglin AFB) Backbreaker/Optionbreaker programs. ATH collected active, 3D laser imagery and passive, thermal IR imagery simultaneously from both down-looking and forward-looking perspectives. It also demonstrated laser Doppler velocimetry updating of inertial navigation for improved navigational accuracy. Some pictures and diagrams from CMAG are shown in Fig. 2.15.

Figure 2.16 shows the operational result from the CMAG program, the AGM-129 advanced cruise missile (ACM), hanging in the Air Force Museum. Note the window on the underside of the missile. The ACM uses laser sensor updates to give it high navigational accuracy. Full-scale production began in 1983, and the first production cruise missile was delivered in 1987. Production contracts were awarded to General Dynamics and McDonnell Douglas.

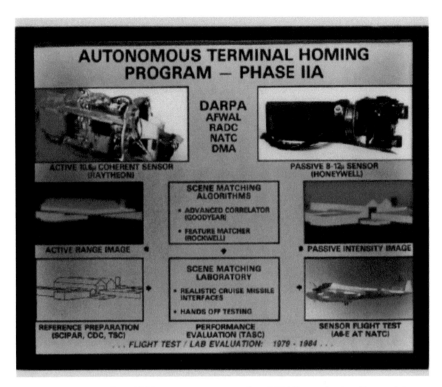

Figure 2.14 The DARPA ATH program, comparing LiDAR against a thermal imager for terminal navigation (reprinted from Ref. 10).

Figure 2.15 Photographs and diagrams relating to CMAG (reprinted from Ref. 10).

Figure 2.16 Photograph of the AGM-129 ACM in the Air Force Museum at Wright-Patterson Air Force Base (reprinted from Ref. 10).

The Quiet Knight program (Fig. 2.17), initiated in 1988, was designed to provide terrain following, terrain avoidance, and obstacle avoidance, and it was supposed to be covert. A LiDAR was an ideal sensor for these purposes. At the time, the U.S. Air Force planned to fly at low levels to avoid surface-to-air threats.

2.2.4 Atmospheric wind sensing

Although Section 2.2.4 is part of the topic of early coherent LiDARs, both early coherent- and noncoherent-LiDAR approaches to wind sensing are discussed here in order to keep the wind-sensing discussion in one place.

2.2.4.1 Early coherent LiDAR wind sensing

Beginning in 1966 and during the early 70s, wind-sensing research using LiDAR became more prominent. Applications included detection of clear air turbulence ahead of an airplane and detection of wind hazards around an airport, such as atmospheric microbursts over the runways causing strong wind shear, or aircraft wake vortices generated by 'heavy' aircraft. Raytheon received a contract from NASA to develop a 20-W CO_2 laser in 1965. The first flight test was conducted on the NASA Ames Research Center's CV-990 aircraft in 1970 and again in 1972.

Figure 2.17 Quiet Knight program (reprinted from Ref. 10).

Wind sensing is a valuable application for coherent LiDARs.[35] Measurements include various ground-based programs, such as local wind field measurement and wake vortex investigation at airfields. Airborne systems are used to measure true airspeed, pressure error, and wind shear warning, and to collect atmospheric backscattering over the North and South Atlantic. In the 90s, the European Space Energy supported a spaceborne wind LiDAR in the Atmospheric Laser Doppler Instrument (ALADIN) program. Pioneering wind LiDAR work in Europe was performed at RSRE (UK), DLR (Germany) and at the Laboratoire de Météorologie Dynamique, Ecole Polytechnique (France). For a review on early coherent LiDAR work in Europe, one can refer to the paper by Vaughan et al.[36] The book edited by Killinger and Mooradian contains several articles dedicated to the first coherent LiDARs in the USA.[37]

Coherent Doppler LiDAR (CDL) appeared to be a useful instrument for atmospheric wind sensing from ground, air, or space. CDLs observe large volumes of atmosphere with high spatial and temporal resolution, making these data important for many applications.

The first CDL was reported in 1970 by Huffaker et al.[38] for wake vortex detection using a 10.6-μm cw CO_2 laser. In the late 1990s, Mitsubishi Electric commercialized eye-safe, compact, fiber-based LiDARs and midrange (8 km) wind LiDARs (see Figs. 2.18 and 2.19) after developing a high-power fiber amplifier of Er:Yb:glass with an output power of 3.3-kW, a width of 580 ns, and a repetition rate of 4 kHz.[39] The ultralong-range wind LiDAR exceeding 30 km was also realized using this amplifier, with a range resolution of 300 m.[40] An all-fiber CDL was also developed based on a new concept of automatic

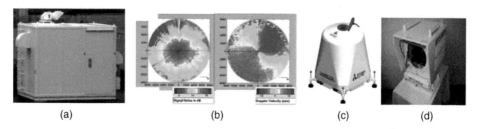

Figure 2.18 (a) Coherent Doppler LiDAR system; (b) displayed diagrams of signal noise and Doppler velocity; (c) all-fiber Doppler LiDAR system and (d) its data display unit (reprinted from Ref. 14).

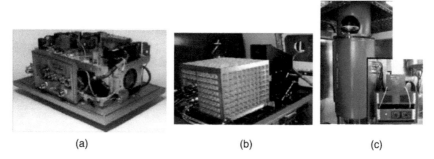

Figure 2.19 Wind sensors: (a) NASA optical air turbulence sensor (2 μm, 1 mJ, 1 kHz, 5-cm aperture, chiller); (b) NASA ACLAIM turbulence warning (2 μm, 8–10 mJ, 100 Hz, 10 cm aperture, chiller); (c) MAG-1A WindTracer (2 μm, 2 mJ, 500 Hz, 10-cm aperture, heat exchanger) (reprinted from Ref. 14).

parameter control that is adaptive to atmospheric conditions. These systems have been used in various industrial applications, such as meteorological monitoring and wind survey for wind power generation and aviation safety.

In the late 1990s and early 2000s Coherent Technologies, Inc. (CTI) led the charge to develop wind sensing, with sponsorship mostly from the Air Force and NASA. CTI (now part of Lockheed Martin) also developed commercial products, like the wind tracer. Multiple WindTracer® systems have been deployed at airports and research facilities for monitoring the wind shear in the airport area and for study of aircraft wake vortices.

Another application for this technology has recently been the monitoring of winds for wind-energy–generating platforms. These units are placed on the tower and monitor the wind and turbulence for better operation of wind power generators. Halo Photonics in the U.K. and Leosphere in France offer these units commercially.

The U.S. Air Force showed interest in wind sensing for airdrops, gunships, and dropping dumb bombs. The flight version of ballistic winds was flown in a near-prototype C-130 Pod System. It was 15 ft^3 and 1000 lbs, and used a solid state, 15-mJ, 2-μm laser.

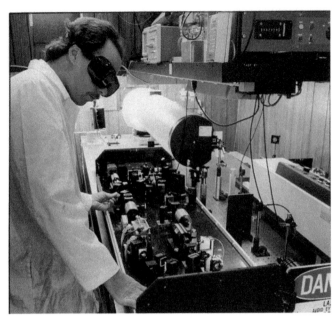

Figure 2.20 The coherent launch-site atmospheric wind sounder (CLAWS) LiDAR (photograph courtesy of Sammy Henderson).

As CO_2 LiDAR use faded in favor of solid state LiDARs, wind sensing also evolved. Figure 2.20 shows an image of the coherent launch-site atmospheric wind sounder (CLAWS) LiDAR, which was a high pulse energy, 1.064-μm wavelength, coherent LiDAR developed in 1991. It was a Nd:YAG master oscillator/power amplifier transmitter. The transmitter output was $\sim$ 1 J/pulse at a pulse repetition frequency (PRF) of 10 Hz, with 0.2- to 5.0-μs actively variable pulse widths. It could measure wind speed and direction up to an altitude of 27 km using a 20-cm–diameter telescope.[41]

CTI had programs with the U.S. Air Force to develop wind-sensing LiDARs for bomb drop, cargo drop, and gunships. These were solid state LiDARs in the 2-μm region. Figure 2.21 is photographs of the ballistic winds LiDARs developed under that program. This picture is from June 1997. CTI also developed a commercial wind-sensing product called the WindTracer.

2.2.4.2 Early noncoherent wind measurement

Low-altitude wind profile measurements with a noncoherent laser rangefinder were demonstrated using a simple balloon-tracking system with small (0.25-m diameter) lightweight balloons.[42] Experiments on balloon trajectories demonstrate that laser range detection ($\pm$0.5 m) combined with azimuth and elevation measurements is a simple, accurate, and inexpensive alternative to other wind-profiling methods. To increase the maximum detection range to

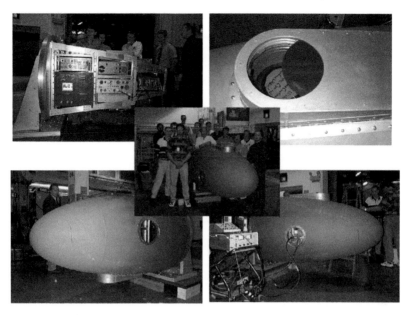

Figure 2.21 Near-prototype C-130 aircraft fuel pod system: 15 cu ft, 1000 lbs, solid state, 15 mJ (reprinted from Ref. 10).

2200 m, a retroreflector tape was attached to the balloons. Nighttime tracking was facilitated by low-power LEDs.

Another example of noncoherent Doppler wind measurement was demonstrated in the paper of Liu et al.[43] Liu is from the Ocean University of Qingdao, China, and some of his collaborators are from NASA Langley Research Center, Hampton, Virginia. A schematic of the LiDAR transmitter, receiver, and frequency control is shown in Fig. 2.22. The master oscillator of the system is a two-wavelength, diode-pumped, cw, single-mode, tunable Nd:YAG seed laser. The output at 1064 nm is used to seed a Continuum® Powerlite 7000 Nd:YAG pulsed laser. The 532-nm output of the seed laser is sent to an iodine filter (Cell 1) to control and lock the seed laser frequency. The frequency precision is maintained to within 0.2 MHz, corresponding to a wind measurement uncertainty of 5.0 cm/s.

A mobile, direct-detection, Rayleigh-scatter Doppler LiDAR (Fig. 2.23) was developed by the University of Science and Technology of China[44,45] to measure wind fields at a range of 15–70 km with height resolution of 0.2 km below 40 km and 1 km above 40 km. This nonscanning system operates at an eye-safe wavelength of 354.7 nm using a frequency-tripled, 50-Hz Nd:YAG laser. A triple channel is used as a frequency discriminator to determine the wind velocity; two of these channels are double-edge channels located in the wings of the thermally broadened molecular backscattered signal spectrum, and the third channel locks the frequency of the outgoing laser at

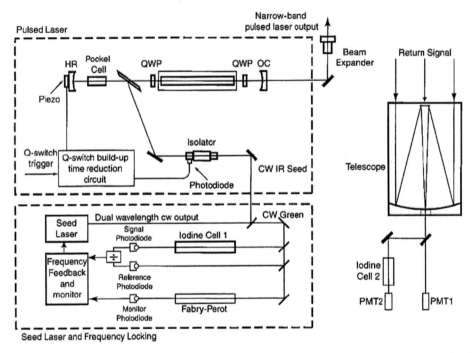

Figure 2.22 Schematic diagram of Continuum's injection-seeded, Q-switched, pulsed Nd:YAG laser transmitter (reprinted from Ref. 14).

the cross-point of the double-edge channels.[46] The scanning system can detect the horizontal wind in four directions (North, South, East, West) by scanning the emitting–receiving system.

2.2.5 Laser vibrometry

One of the attractive applications of coherent LiDAR is vibrometry, which is based on the Doppler effect. Remote noncontact measurements offer potential for civil and military instrumentation. Laser vibrometers typically operate at 1.5 μm, 2 μm, or 10.6 μm. To get spatially resolved vibrational information, scanning and multibeam laser vibrometers are used. Reference 47 focuses on applications in the field of defense and security, such as target classification and identification, including camouflaged or partly concealed targets, and the detection of buried land mines, with some examples of civil medium-range applications. The vibration spectrum of a target was acquired as an important and robust feature necessary for classification and identification purposes. The small target in Fig. 2.24(a) is an ordinary, black, rubber boat with an outboard engine. The boat is approximately 4 × 2 × 2 m. The range is approximately 1 km. The graph in Fig. 2.24(b) displays the frequency spectrum. The vibration frequencies originate from different parts of the target. Figure 5.24(c) is the related

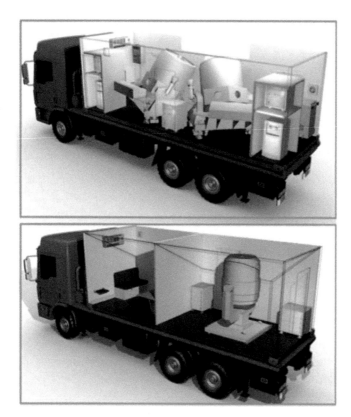

Figure 2.23 Perspective view of two mobile, direct-detection, Rayleigh-scatter Doppler LiDARs (reprinted from Ref. 14).

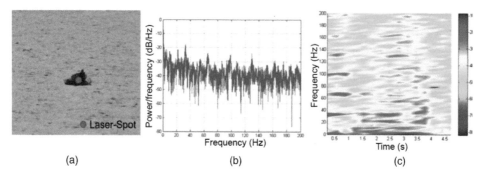

Figure 2.24 Vibration measurements from a rubber boat: (a) visible image with marked laser beam, (b) frequency spectrum of the velocity, and (c) frequency spectrum over time (reprinted from Ref. 14).

spectrogram over a 5-s period. These data were gathered using a 1.54-μm LiDAR with 0.8 mW of power and an aperture diameter of 50 mm. The LiDAR used an InGaAs photodiode.

Besides the CO_2 lasers at 10.6 μm, a number of solid state laser sources based on neodymium at 1.06 μm, semiconductors or erbium fiber at 1.5 μm, and holmium at 2.1 μm have been used as laser vibrometers. There are, however, often environmental conditions associated with poor visibility, turbulence, or high humidity where it might be desirable to operate in the MWIR or LWIR bands to improve system performance. Such conditions are not uncommon at low altitudes and in marine environments. Coherent LiDAR systems in the mid-IR wavelength region can have advantages in low-altitude environments because they are less sensitive to scattering and turbulence than shorter-wavelength LiDARs, and less sensitive to humidity than longer-wavelength LiDARs. A monostatic coherent LiDAR at 3.6 μm based on a single-frequency, optical parametric oscillator was described in Ref. 48. This LiDAR operated outdoors over short ranges using two different stationary trucks with the motors running as targets. The system provided micro-Doppler measurements that were processed to give surface vibration spectra of the stationary, but running, trucks.

2.2.6 Synthetic-aperture LiDAR

Synthetic-aperture radar (SAR) is a mature field that was developed to reconstruct high-resolution microwave images by use of antennas of reasonable size. Synthetic-aperture LiDAR (SAL) started being developed in the later 1990s and early 2000s.[49] The principle of SAL is based on accumulating field information as the radar moves with respect to the target. A large area of pupil-plane, full-field information is then available. For a microwave radar this pupil-plane field information develops a synthetic aperture that is kilometers long in one direction. Because field information contains both phase and intensity, we can Fourier transform this large pupil-plane image to develop a high–angular-resolution image in the dimension with the long synthetic aperture. If the SAR has high bandwidth, it can also provide high range resolution. A SAR therefore can provide high angle resolution in one dimension and high range resolution in a second dimension, providing a high-resolution image. Thus, synthetic-aperture techniques are applicable when there is a relative transverse motion between the radar and the target, with the resolution improvement accomplished by increasing the effective aperture dimension in the direction of motion. SARs operate at microwave and millimeter wavelengths and are extensively used to improve far-field spatial resolution over that provided by conventional diffraction-limited radars. Because microwave radars have limited real beam resolution, there was a large incentive to develop SAR. Microwave wavelengths are relatively large, allowing motion compensation to a fraction of a wavelength.

This technique was extended to the optical/infrared wavelengths with CO_2 laser sources[50,51] and Nd:YAG lasers.[52,53] Because of the difference in wavelength, synthetic apertures can be much smaller at optical wavelengths, but motion compensation must still counteract any phase variations that are more than a fraction of a wavelength. If the wavelength is 3 cm, as in SAR, you will need to compensate for variations in optical path length (OPL) down to <3 mm. If the wavelength is 10 μm, you will need to compensate for OPL changes < 1 μm, and for a 1-μm LiDAR, you will need to compensate for changes < 1 μm. Feasibility studies continued since those initial investigations into SAL without interruption.[54–56] The high-resolution capability of a SAL can be used as an advantage not only for long-range applications (like space-based imaging, for example), but also for short ranges if very high angular resolution is required. Because SAL focusing can be controlled by adjusting the assumed range in the pulse-compression filter, targets with range variations can be imaged with no degradation in along-track resolution. An example of such a short-range application is product quality control when the object has a varying surface relief. Figure 2.25 shows some graphics related to SAL.

2.3 Early Space-based LiDAR

In 1986 McDonell Douglas built the first LiDAR to go into space. It was launched 5 Sep 1986 on a Delta rocket and used a high-power, pulsed Nd:YAG laser, 100-mJ/pulse, with a 100-Hz PRF. This LiDAR was used for on-orbit tracking and terminal guidance for a Kinetic Energy Interceptor for the Strategic Defense Initiative Organization (SDIO).

In 1998 another LiDAR was launched into space: the Video Guidance Sensor (VGS). Figure 2.26 shows a schematic. The Video Guidance Sensor and the Advanced Video Guidance Sensor were developed by NASA as a means to autonomously guide spacecraft through proximity operations, rendezvous, and docking. It consisted of a passive, CCD camera and a laser

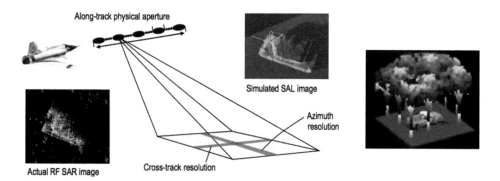

Figure 2.25 (left and center) Principles of synthetic-aperture LiDAR (SAL). (right) Processed SAL data observing a truck through foliage. Information from four separate views is combined in this image (courtesy of Northrop Grumman).

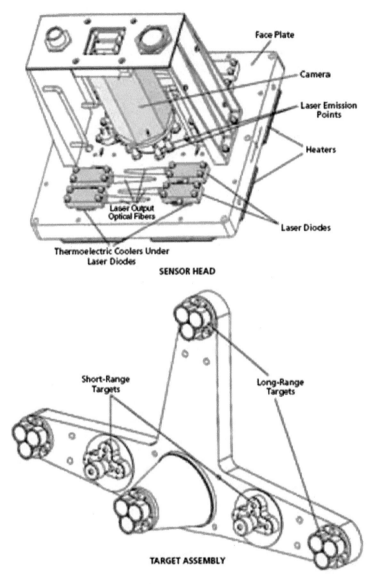

Figure 2.26 Video Guidance Sensor (VGS) / Advanced Video Guidance Sensor (AVGS) (reprinted from Ref. 10).

illuminator (top) on the maneuvering spacecraft and a passive, retroreflector array (bottom) on the target spacecraft.

In the late 1990s, the Jet Propulsion Lab took on the challenge of the autonomous landing of a deep-space probe on the comet Tempel 1. The project required the high-resolution 3D mapping of the comet's surface. Figure 2.27 (top) shows the breadboard of the landing LiDAR, the grayscale image on the bottom shows the intensity, and the right bottom shows the 3D LiDAR images

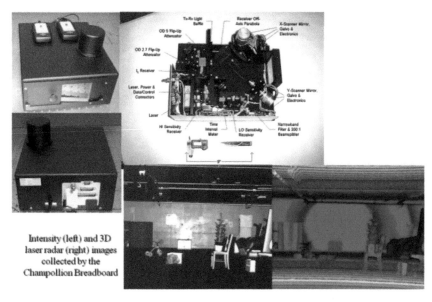

Intensity (left) and 3D
laser radar (right) images
collected by the
Champollion Breadboard

Figure 2.27   Champollion/DS4 Breadboard (reprinted from Ref. 10).

collected by the Champollion Breadboard. The 3D image uses a unique presentation format to display the gross geometric features and the fine structure simultaneously. The gross features are presented in a wrapping, rainbow pseudo-color scale (near to far is red, orange, yellow, green, blue, violet, red, orange,...). The fine structure is presented by taking the lowest two significant bits of the range and using them to modulate the intensity of the color. The effect resembles a fringe pattern superimposed on the pseudo-color image.

Following the Space Shuttle Challenger explosion, NASA initiated standard procedures to inspect the thermal tiles that protect the shuttle from the heat of re-entry. The inspection sensors consist of two LiDARs and a HD TV camera. Damage is routinely detected by the LiDARs, and then the HD pictures are shown to the media. Figure 2.28 shows the sensors and other Shuttle-related images.

In 1988 the Strategic Defense Initiative (SDI) started the LOWKATER program, which was supposed to place a CO_2 LiDAR in space to discriminate between real and decoy re-entry vehicles. An infographic on LOWKATER is shown in Fig. 2.29. Unfortunately, LOWKATER was canceled before it could be flown.

To get a spatial distribution of vibrations on the surface of the investigated object, a scanning vibrometer can be used. Scanning laser vibrometers allow analysis of the structure with a very fine spatial resolution without modifying its dynamic behavior, decreasing the testing time if a large number of measurement points is requested. Based on a spectrogram approach, vibration signatures were obtained from the LACE satellite

Figure 2.28   Shuttle damage detection sensors (reprinted from Ref. 10).

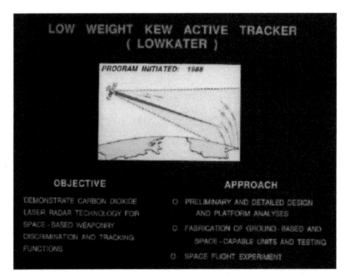

Figure 2.29   Infographic on the SDI-sponsored LOWKATER program (reprinted from Ref. 10).

[Fig. 2.30(a)] in the course of ground-based LiDAR measurements by using a coherent CO_2 laser. A spectrogram-processing approach developed a frequency spectrum of the target. The satellite was equipped with IR germanium retroreflectors on deployable/retractable booms to enhance ground-based IR LiDAR measurements of on-orbit boom vibrations. The data were acquired during, and subsequent to, one of the maneuvers (boom

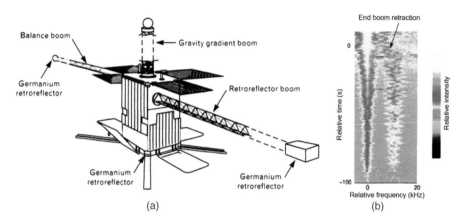

Figure 2.30 (a) Diagram of the LACE satellite. Relative vibration was measured between the germanium retroreflector located on the satellite body and the retroreflector located on the retroreflector boom tip. (b) Doppler time–intensity representation of data aligned to the peak return (reprinted from Ref. 14).

retraction) and indicated the presence of a complex time-varying mode structure. The power spectra of vibrations are shown in a Doppler time–intensity format in Fig. 2.30(b), where they are aligned and displayed along the vertical time axis. The horizontal axis corresponds to Doppler frequency (i.e., velocity).

2.4 Flight-based Laser Vibrometers

The problem of inflight measurements was investigated using a scanning laser Doppler vibrometer to measure vibrations inside the cabin's mockup of the Agusta A109 MKII.[57] The whole area viewed was 430 × 315 mm. In the scanning tests, a 30 × 20 mm grid was used. The comparison was also made with the vibrograms measured when the vibrometer was placed outside the mockup. Figure 2.31 is a summary of the tests at different resonances. Each test refers to the instantaneous amplitude of the velocity component at a certain frequency orthogonal to the surface with the bandwidth ±10 Hz for frequencies up to 1000 Hz and ±30 Hz for frequencies up to 5000 Hz. Vibration sensing was also examined for buried mine detection.[58]

Other applications using coherent LiDAR have included precision navigation to a designated landing site on Earth or on extraterrestrial objects, and rendezvous and docking with orbiting spacecraft, which require accurate information on the vehicle's relative velocity and altitude. A Doppler LiDAR was developed by NASA under the Autonomous Landing Hazard Avoidance Technology (ALHAT) project.[59] The LiDAR precision vector velocity data enabled the navigation system to continuously update the vehicle trajectory toward the landing site.

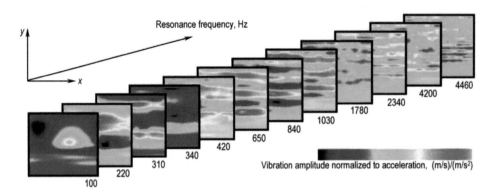

Figure 2.31 Resonance vibration frequencies registered in the cabin of the helicopter mockup (reprinted from Ref. 14).

2.5 Environmental LiDARs

2.5.1 Early steps

In parallel to military laser sensing, the LiDAR research community started to search for applications in atmospheric and ocean sensing. For example, LiDAR observations of the mesosphere were made using a ruby laser as early as 1963 by Fiocco and Smullin.[60] Spatial distribution of aerosols in the troposphere was reported by Collis et al. in 1964.[61] In the U.S., vertical water vapor distribution was studied[62] using a temperature-tuned ruby laser in the first experiment using differential absorption LiDAR. In Japan, a Mie scattering LiDAR was first developed by utilizing a homemade Q-switched ruby laser. Basic relations were analyzed in 1968 on the scattering, extinction and visibility by Inaba et al. at Tohoku University.[63] In the USSR, the atmosphere temperature was studied by Arshinov et al.[64] Laser sensing of the humidity profile of the atmosphere was studied by Zuev et al.[65]

Examples of atmospheric LiDAR systems[66,67] are shown in Fig. 2.32. As reported by H. Hu and J. Qiu,[68] the first Chinese atmospheric LiDAR was completed in 1965. Since then, experimental studies were carried out to investigate stratospheric aerosol and volcanic cloud, stratospheric ozone profile, tropospheric aerosol (including smoke plume density, aerosol extinction coefficient, atmospheric turbulence), sodium layer and Rayleigh scattering in middle atmosphere, seawater temperature, and oil slicks on sea surface, etc. Numerous theoretical studies were also made. Svanberg[69] gives many examples of early LiDAR monitoring of pollutants. Atmospheric and ocean LiDARs involve all kinds of lasers and detectors depending on the goal of the LiDAR. Possible goals might be aerosol or gas sensing, or ocean sensing, such as bottom profiling or water sensing (turbidity, plankton, etc.).

The limited space in this chapter does not allow us to go deeper into atmospheric and ocean LiDAR development. Instead, one can refer to the textbooks and reviews for further reading.[70,71] Reference can be made to two

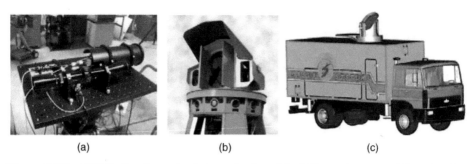

(a)   (b)   (c)

Figure 2.32 LiDARs for atmospheric investigation: (a) all-fiber coherent multifunctional cw LiDAR for range, speed, vibration, and wind measurements at 1.55 μm (FOI, Sweden, 2000); (b) LiDAR automatic system for remote air pollution monitoring in large industrial areas (Institute of Precision Instrumentation, Moscow); and (c) multifunctional LiDAR system (Astrofizika Corporation, Moscow).

other books discussing the propagation of laser radiation in water medium.[72,73] Studies of the environmental LiDARs in the USSR have been published in seminal monographs on propagation of laser radiation in the atmosphere by Tatarskiy[74] and by Zuev.[75] These books were followed by other monographs from the Tomsk scientific school.[76,77] Practical problems associated with the applications in meteorology are discussed in a book for meteorologists.[78] Achievements of the cw FM technologies are described in the book by Agishev.[79]

2.5.2 Multiwavelength LiDARs

Penn State University[80,81] made significant progress in Raman LiDAR by enabling measurements of the optical and meteorological properties of the atmosphere based on vibrational and rotational energy states of molecular species, such as water vapor and ozone, temperature, optical extinction, optical backscatter, multiwavelength extinction, extinction/backscatter ratio, aerosol layers, and cloud formation/dissipation (Fig. 2.33). An angular scattering technique enhances the information by measuring the scattering phase function for aerosols, including the polarization ratio of the scattering phase function, number density versus size, size distribution, identification of multicomponent aerosols, index of refraction, etc. Multistatic aerosol LiDAR and multiwavelength, multistatic LiDAR are good candidates for prospective studies.

2.5.3 LiDAR sensing in China

Studies of yellow sand storm events were one of the practical applications of Mie-scattering LiDAR in China that monitored the transport of dust, aerosol extinction coefficient profiles, regional air pollutant transport, and the urban mixed layer.[82–84] In addition, Anhui Institute of Optics and Fine Mechanics,

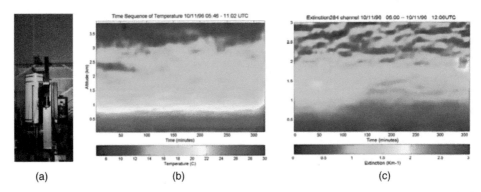

Figure 2.33 (a) LiDAR developed at the Pennsylvania State University and dynamics of (b) temperature and (c) extinction (reprinted from Ref. 14).

AIOFM joined national and international projects on LiDAR monitoring of the atmosphere.[85,86] These projects include the AD-Net (Asian Dust LiDAR observation NETwork)[87] and ADEC (Aeolian Dust Experiment on Climate) impact. A double-wavelength polarization Mie LiDAR was implemented,[88] and 12-year observation of aerosol in Hefei was reported.[89]

The first Mie-scattering LiDAR was implemented to measure stratospheric aerosol in the 1980s at the Institute of Atmospheric Physics (IAP). The stratospheric aerosol enhancement by the volcanic clouds of El Chichon (1982) and Pinatubo (1991) were monitored in Beijing. In the early 1990s, Anhui Institute of Optics and Fine Mechanics (AIOFM) started to develop Mie-scattering LiDAR based on a Nd:YAG laser. A large number of profile data of the Pinatubo volcanic cloud were obtained in Hefei and Beijing.[90]

In 1992, IAP developed a multiwavelength LiDAR to observe stratosphere ozone and aerosol, and high-altitude clouds. The LiDAR used a XeCl excimer laser and a Nd:YAG laser, a 1-m–diameter telescope, and a multichannel detector. The Nd:YAG laser had outputs of 1 J, 300 mJ, and 150 mJ at 1064 nm, 532 nm, and 355-nm wavelengths, respectively. The PRF was 10 Hz. The excimer laser had an output of 140 mJ at 308 nm, and its PRF was 100 Hz.[91] In 1994, AIOFM also developed a LiDAR using the second and third harmonics of Nd:YAG and an excimer laser for stratospheric ozone profile monitoring.[92] A differential absorption LiDAR (DIAL) for observation of pollutant gasses such as SO_2, O_3, and NO_2 was developed at AIOFM.[93,94] A CH_4 and D_2 gas cell was pumped by the fourth harmonic of Nd:YAG.

Vibrational–rotational Raman LiDAR was developed at Xi'an University of Technology,[95] a schematic diagram of which is presented in Fig. 2.34(a). The system employs a pulsed Nd:YAG laser as a light source operating at a frequency-tripled wavelength of 354.7 nm with a 20-Hz repetition rate and an energy output of 250 mJ with a 9-ns pulse duration. Returned signals are collected with a 600-mm Newtonian telescope, and then are coupled into a

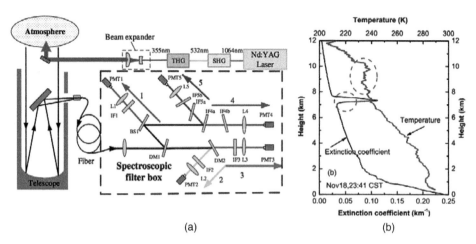

Figure 2.34 (a) Vibrational–rotational Raman LiDAR system developed at Xi'an University of Technology and (b) the resulting extinction and temperature profiles (reprinted from Ref. 14).

multimode optical fiber and guided into the spectroscopic box. This LiDAR system measured profiles of the water vapor mixing ratio by Raman scattering as well as the aerosol extinction coefficient by Mie scattering.[96,97] High-accuracy temperature profiles can be achieved up to a height of 25 km.[98] The effective measurement for atmospheric water vapor can be achieved up to a height of 16 km.[95] Examples of extinction and temperature profiles are given in Fig. 2.34(b).

The first sodium LiDAR for the middle atmosphere in China was developed in 1996 at Wuhan Institute of Physics and Mathematics.[99] Rayleigh and sodium LiDAR was developed by Wuhan University.[100] The LiDAR transmits in both 532-nm and 589-nm wavelengths, and detects Rayleigh backscattering, Raman signal of N_2 and H_2O, and sodium fluorescence. Similar sodium LiDARs were also developed at the University of Science and Technology of China[101] and Wuhan University.[102] Wuhan University developed an iron Boltzmann LiDAR to measure the mesopause temperature.[103] National Space Science Center developed a sodium fluorescence Doppler LiDAR to observe both temperature and wind.[104] At present, five LiDARs (in Wuhan, Hefei, Qingdao, Beijing, and Hainan) are operating to observe the middle atmosphere.

2.5.4 LiDAR sensing in Japan

In Japan, a Raman-scattering LiDAR for sensing air pollution over Japanese industrial areas was proposed in 1969 by H. Inaba and T. Kobayashi at Tohoku University.[105] A nitrogen molecular laser at the UV wavelength of 337.1 nm was developed for the Raman LiDAR. Molecular vibrational Raman spectra of CO_2, O_2, N_2 and H_2O were observed in the clear air, and it was shown for the first time that major air molecules could be separately detected. Various molecules such as O_3, CO, CH_4, and liquid and vapor H_2O were also identified from automobile exhaust gas in air. After this experiment, the mobile scanning Raman LiDAR system was developed by S. Nakahara et al.[106] at Mitsubishi Electric Co. Ltd (MELCO) using a Nd:YAG-laser second-harmonic beam at a wavelength of 532 nm. It was shown that SO_2 molecules in a stack of effluent plume were detectable with 1000-ppm concentration sensitivity at the slant range of 220 m. The Raman LiDAR was also used for humidity sensing by water vapor Raman spectroscopic detection and was gradually extended into sensing of extinction coefficients of atmospheric aerosols by simultaneous measurement of Mie and nitrogen Raman spectra. Figure 2.35 shows two early Raman-scattering LiDARs.

After the demonstration of the high sensitivity of resonant scattering LiDAR in 1969,[107] several sodium (Na) resonant scattering LiDARs were developed and reported by Aruga et al. at Tohoku University[108] and by C. Nagasawa et al. at Kyushu University.[109]

Stratospheric ozone was measured by O. Uchino et al. at the Kyusyu University in 1978 by employing a discharge-pumped XeCl laser at 308-nm wavelength based on the differential absorption technique.[110] In 1988, the Meteorological Research Institute (MRI) developed a mobile LiDAR for simultaneous measurements of ozone, temperature, and aerosols in the stratosphere.[111] These researchers used three Stokes lines (276, 287 and 299 nm) of the stimulated Raman spectrum (SRS) from a carbon dioxide gas cell pumped by a Nd:YAG laser (266 nm).[112]

To clarify the ozone loss mechanisms in the polar stratosphere known as the ozone hole, a XeF excimer laser with 200-mJ pulse energy and 80-Hz repetition rate at 351 nm and 353 nm wavelengths was developed. T. Shibata et al. at the Kyushu University[113] used this laser in 1986 for observation of

(a)   (b)

Figure 2.35 Early Raman-scattering LiDAR system (a) at the observatory of Tohoku University and (b) a mobile, scanning Raman-scattering LiDAR (reprinted from Ref. 14).

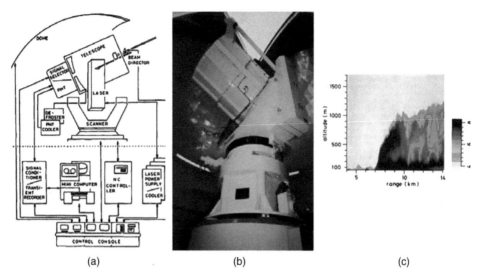

Figure 2.36 Large scanning LiDAR system (NIES): (a) schematic diagram; (b) photo of the LiDAR; and (c) structure of sea breeze front (reprinted from Ref. 14).

molecular density and temperature in the altitude range of the middle atmosphere.

To examine the potential of Mie scattering, a large scanning LiDAR was built in 1979 at the National Institute for Environmental Studies (NIES) in Japan. This system had a pulse energy of 400 mJ at 532 nm and a repetition rate of 25 Hz (Fig. 2.36). The LiDAR was used in various studies on Mie scattering for measuring the structure of the atmospheric boundary layer[114,115] aerosol distribution and optical characteristics.[116]

2.6 Imaging LiDARs

2.6.1 Early LiDAR imaging

Imaging LiDARs usually generate a 3D point cloud of an area by measuring range and time of flight of laser waveforms at a large number of azimuth and elevation positions. To measure range at a large number of angular positions, scanners were initially used. More recently, detector arrays have become available, allowing flash imaging. In flash imaging, detector arrays are used to simultaneously obtain range data at multiple angular locations. Imaging LiDARs can obtain reflectivity, spectral parameters, polarization, Doppler shift, and 3D data. This breadth of data is why we refer to the rich LiDAR phenomenology. Military and security applications include target recognition, target location, aim point selection, tracking, and weapon guidance. Military LiDAR for imaging became of interest during the 1970s and 1980s. Important contributions were made by MIT Lincoln Laboratory (MIT/LL).[117] The Infrared Airborne Radar (IRAR), MIT/LL testbed was primarily an experimental target recognition system[118] capable of detecting and recognizing armored tactical vehicles in a registered range and intensity images provided by a pulsed, infrared, CO_2 forward-looking LiDAR that was carried either on a truck or aboard an aircraft.

Examples of images from MIT/LL are shown in Fig. 2.37. Figure 2.37(a) is an IRAR CO_2 (10.59-μm) LiDAR image of a bridge in which the range to each picture element is coded in color. The data collected in the original oblique view are transformed into an overhead view, as shown in the inset image. This rotation capability is one of the advantages of 3D imaging. This view may be useful for missile seekers that use terrain features for targeting. The Doppler velocity image in Fig. 2.37(b) was collected by a truck-transportable CO_2 LiDAR. The Doppler shifts of each of the approximately 16,000 pixels in the image were extracted by a surface acoustic-wave processor at a frame rate of 1 Hz. Velocity is mapped into color. The ability to sense moving parts on a vehicle provides a powerful means to discriminate targets from clutter. The image in Fig. 2.37(c) made with a GaAs (0.85-μm) LiDAR is an angle-angle-range image of a tank concealed by a camouflage net. The LiDAR utilizes a high-accuracy, sinusoidal, amplitude-modulated waveform while observing the tank in a down-looking scenario. The camouflage net was readily gated out of the image to leave the tank image below the canopy. Figure 2.37(d) is a range-Doppler image of one of the LAGEOS satellites collected by the wideband CO_2 LiDAR. This image was made with a bandwidth of 1 GHz. Doppler velocity resolution is approximately 30 cm/s. Color in the image represents relative signal amplitude.

Early laser-imaging radars were built and tested at Raytheon.[119,120] One example was the tri-service LiDAR (TSLR) containing a CO_2 15-W waveguide laser a galvanometer scanner, a TV, and an InSb IR camera. This system was used in numerous campaigns for evaluation of imaging LiDAR technology. In parallel with the coherent LiDARs, scanning direct-detection LiDARs became more interesting due to somewhat better

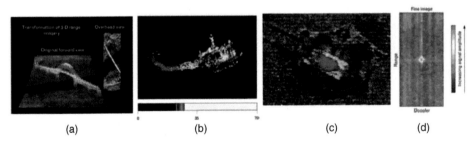

(a)   (b)   (c)   (d)

Figure 2.37 Images from MIT/LL LiDAR systems: (a) CO_2 laser-radar images of a bridge; (b) Doppler velocity image of a UH-1 utility military helicopter; (c) laser-radar image of a tank concealed by a camouflage net; and (d) range-Doppler image of one of the LAGEOS satellites collected by the wideband CO_2 LiDAR at Firepond (reprinted from Ref. 14).

performance in intensity recording (less speckle issues) and better range accuracy and resolution. U.S. examples of scanning, direct-detecting LiDARs in the early 1990s was the Hercules Defense Electronic Systems LiDAR,[121] which was operated at both the 1.047-μm and 1.319-μm wavelengths of Nd:YLF. The 1.319-μm transition was reported to yield about 6–8 dB less peak power than the 1.047-μm transition but offers the advantage of significantly enhanced eye safety characteristics. Another example is the Fibertek Helicopter LiDAR (HLR) system,[122] which was designed primarily for terrain following and wire obstacle avoidance. The HLR system generates eye-safe 1.54-μm radiation using a Nd:YLF transmitter and a KTP OPO. The Nd:YLF laser generates 5-ns pulses at up to 15-kHz PRF. Average power at 1.54 μm was approximately 1.0 W.

Early imaging LiDAR work in Europe included coherent CO_2 systems in France[123] and Sweden.[124–126] Figure 2.38 shows an example of a multifunctional coherent CO_2 LiDAR system developed in the 1980s at FOI. In this system, a pulsed LiDAR with a programmable transmitter was built to study the combined capability of Doppler and range imaging. A new technique was investigated[127] for terrain segmentation based on range data. The basis of the approach was to model the range values obtained from horizontal and vertical scans as a piecewise constant (or linear) signal in random noise.

2.6.2 Imaging LiDARs for manufacturing

In the late 1980s Perceptron, Inc. built a LiDAR for the Ford Motor Company for use in manufacturing. Figure 2.39(b) is a 3D image of a pile of exhaust manifolds and Fig. 2.39(a) is a matching panchromatic photograph. In the 3D image, bright shading corresponds to short range. Robotic manipulators are used to pick up the manifold currently on top of the pile and place it on the assembly line. Note that the most prominent objects in an ordinary photograph are seldom those on top.

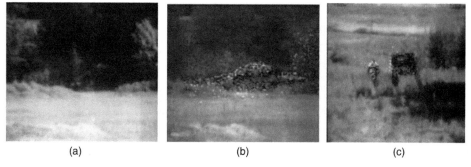

(a) (b) (c)

Figure 2.38 Imaging range and velocity with pulsed CO_2 LiDAR (FOI). (a) TV image of a hidden tank; (b) TV image with a range image overlay from a coherent CO_2 LiDAR; and (c) Doppler pixels overlaid on a TV image (reprinted from Ref. 14).

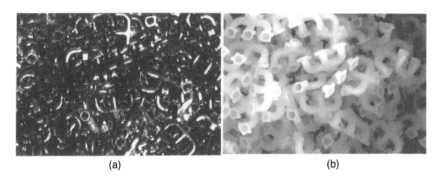

(a)   (b)

Figure 2.39 A LiDAR used for manufacturing: (a) a panchromatic photograph of manufactured parts and (b) a 3D image of the same parts (reprinted from Ref. 10).

2.6.3 Range-gated imaging programs

Two-dimensional, range-gated imaging programs started in the 1990s. The U.S. Air Force had a program called ERASER,[10] whose underlying concept was to use the designator laser already on an aircraft to enhance recognition range. A designator has a shorter wavelength than a thermal imager and can therefore provide better diffraction-limited angle resolution and longer identification ranges compared to a thermal camera. A laser-based system also provides its own illumination so is not as subject to time-of-day variations, such as crossover in thermal imaging. Range-gated imaging also gives a very good target-to-background contrast, especially if the target is seen in silhouette mode.

The U.S. Army studied and picked up on the U.S. Air Force–initiated gated 2D imaging work, and ran modeling and testing of the prototype equipment.[128] The performance of range-gated systems is limited by the sensor parameters as well as the target and atmospherically induced speckles, beam wander, and image dancing.[129,130] Because 2D imagers normally use direct detection, speckle can be mitigated by using a wider laser bandwidth, which speckle averages the signal across the various wavelengths. Close to the range limit, the shot noise restricts the image quality. Frame-to-frame integration is often used for reducing the scintillation and target speckle effects, in which case the image dancing and atmospheric coherence time become of importance. Range-resolved 3D images can be reconstructed from a series of sliding gated imagery.[131]

Gated active 2D imaging was studied by the major defense research laboratories. An overview of the range-gated imaging at FOI including performance modeling was recently published by SPIE.[132] The work included diode-laser–gated systems, 532-nm systems, and, lately, an eye-safe 1.5-μm system using Intevac's tube technology. In later publications,[133] FOI and Fraunhofer IOSB published the results of their studies both of mono- and bistatic configurations with 1.5-μm range-gated systems. Examples of range-gated applications for marine environments include identification of small-surface vessels[134–136] and divers. Gated viewing for underwater imaging has been under test for a long time[137–139] and recently achieved subcentimeter accuracy using range gating.[140] Scanning systems were studied for the detection of underwater objects in front of ships.[141]

Until the First Gulf War, the U.S. Air Force flew aircrafts at low altitudes doing terrain following. With the advent of this war, instead of staying low to avoid anti-aircraft systems, the Air Force flew planes above 20,000 feet to stay out of the range of man-portable missiles. As a result, the Air Force needed to be able to identify targets from a long range. Even at 1000-ft altitude, a 5-km identification range was fine. With the U.S. Air Force aircraft flying at an altitude of 20,000 feet, I defined what I called the '20-km challenge,' which was to identify a tank size target from 20 km. Figure 2.40 shows the dramatic resolution enhancement—with side-by-side images of the same scene—between a 1.06-μm range-gated imager and an LWIR thermal imager. Both the LANTIRN (Low Altitude Navigation and Targeting Infrared for Night) pod and Pave Tack pod had LWIR thermal imagers for target identification.

2.6.4 3D LiDARs

Development of 3D flash LiDARs with intensity and range information in each pixel obtained from one illuminating pulse encouraged research on focal plane array (FPA) detectors. Avalanche photodiode (APD) arrays based on HdCdTe are of high interest and are under development by several actors, such as Raytheon,[142] Leonardo DRS, Sofradir/Leti,[143] and Selex.[144] In addition to linear APD arrays, a lot of attention is being paid to photon-counting detectors of Geiger-mode avalanche photodiodes (GMAPDs). MIT/LL has pioneered development of GMAPDs for such LiDAR programs

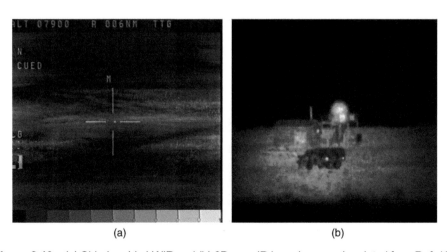

(a)　　　　　　　　　　(b)

Figure 2.40　(a) Side-by-side LWIR and (b) 2D near-IR laser imagery (reprinted from Ref. 10).

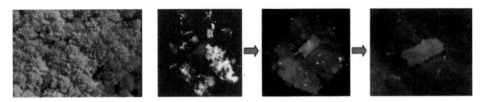

Figure 2.41 Example of data from the Jigsaw program detecting a tank target under trees (reprinted from Ref. 14).

as Jigsaw.[145,146] Jigsaw 3D LiDARs have a poke-through capability to look through the gaps in camouflage and between the leaves in vegetation. MIT/LL developed a LiDAR that produced high-resolution 3D images using short laser pulses and a FPA with 32 × 32 pixel GMAPDs, with independent digital time-of-flight–counting circuits at each pixel.[147,148] Figure 2.41 shows an example of the Jigsaw results. A target under trees can be seen after range cropping of the generated 3D point cloud from the LiDAR.

Foliage poke through has also been studied in other labs. Grönwall et al.[149] propose a sequential approach for detection and recognition of manmade objects in natural forest environments using data from laser-based 3D sensors. Armbruster[150] discusses the exploitation of range imagery for examples drawn from the fields of helicopter obstacle avoidance, object detection in surveillance applications, object recognition at high range, multi-object tracking, and object re-identification in range image sequences. He argues that automatic target recognition (ATR) performance of objects detected by 3D LiDARs can exceed not only 2D ATR performance, but also human vision.

Another U.S. program was the DARPA SPI-3D (Standoff Precision Identification in Three Dimensions) program[151] to "develop and demonstrate the ability to provide precision geolocation of ground targets combined with high-resolution 3D imagery at useful standoff ranges." These dual capabilities were provided using a sensor package composed of commercially available components. It was capable of providing "optical quality precision at radar standoff ranges" and had the ability to overcome limited weapons effects obscuration, and penetrate moderate foliage.

2.6.5 Imaging for weapon guidance

Imaging LiDARs can extend laser weapon guidance capability in a semi-active, or beam-riding, mode. The drawback is increased complexity due to the laser being part of the seeker. Scanning LiDARs have been intensely studied in the USA for missile applications. Target detection and homing were also studied for air-to-air seekers and in space for the SDI program. The largest investments in LiDAR seeker technology have been made for air-to-ground seekers. The Low Cost Autonomous Attack System (LOCAAS) was a demonstrator program[152] driven by the AFRL (in Eglin, Florida). However, the LOCAAS program terminated in mid-2006 without going to production. The seeker was based on scanning LiDAR and generated 3D imagery of targets, with ATR and aim point selection built into the system. The Loitering Attack Missile (LAM)[153] was a project within the U.S. Forward Combat System (FCS) that relied on the LiDAR technology developed for LOCAAS. The paper from Andressen et al.[154] gives some insight into this technology. Figure 2.42 shows the first air-to-air seeker, which had an external diameter of 5 in. and operated at 1.064 μm. This seeker had a field of regard of ±45 deg and could track at angular rates up to 45 deg/s. It was tested at China Lake.

2.6.6 Flash-imaging LiDARs

Much of the imaging LiDAR research today is centered on flash imaging. The technology so far is based mainly on 2D range-gated imaging. The application is mostly targeting, i.e., the recognition of targets at longer ranges than what a cooperating IR camera can provide, thus enabling a longer standoff range for weapon release. Other applications for flash imaging are threat and surveillance/mapping systems combining 3D focal plane technology and scanning (e.g., the Jigsaw system mentioned in Section 2.6.4).

Figure 2.43 shows an airborne flash-imaging LiDAR developed by Northrop Grumman Aerospace Systems.[155] The flash system has a 128×128 pixel array with a readout integrated circuit (ROIC) capable of storing an image every 0.5 ns. The ROIC captures 20 time-lapsed images for each pulse. This can profile a profile in range at each angular position. These time-lapsed images are then used to calculate the range estimates of the target area. This LiDAR uses a flashlamp-pumped Nd:YAG optical parametric oscillator (OPO) at 1.57 μm, with 50 mJ per pulse, a 6.7-ns pulse width, and a 30-Hz

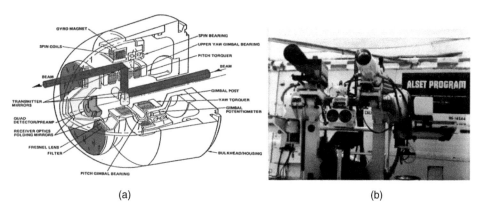

(a)   (b)

Figure 2.42 First air-to-air seeker: (a) schematic and (b) photograph of the system built for lab use (reprinted from Ref. 10).

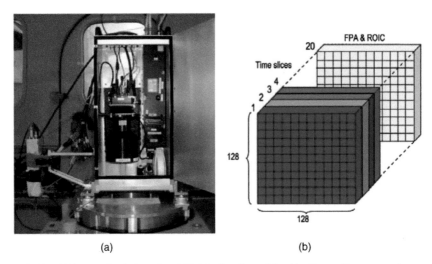

(a)   (b)

Figure 2.43 Airborne flash-imaging LiDAR developed by Northrop Grumman Aerospace Systems: (a) photograph of the system and (b) schematic explanation of the underlying principle (reprinted from Ref. 14).

frame rate. Twenty range-position measurements separated by 2.2 ns in range are taken at each 2D pixel location.

Processing for laser-gated viewing was developed by TNO Defence, Security and Safety (Netherlands)[156] based on Intevac's LIVAR® 4000 laser-gated viewer (Fig. 2.44). A 1.5-μm eye-safe laser and an electron-bombarded CMOS camera for gated viewing (with a minimum gate shift of 1.0 m and a minimum gate width of 20 m) were used.

2.6.7 Mapping LiDARs

2.6.7.1 Terrain mapping

Scanning LiDAR systems are well established for terrain mapping and depth sounding. The civilian market has been leading this field for the last 20 years. Both space and airborne LiDAR systems, as well as terrestrial LiDARs, have been developed by several different vendors. The large number of publications

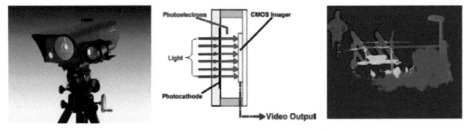

Figure 2.44 Laser-gated viewing using Intevac's LIVAR® 4000 laser-gated viewer (reprinted from Ref. 14).

and applications, as well as the rapid development of hardware, show the large interest in this technology. A book that provides a good overview of topographic laser scanning techniques was edited by Shan and Toth.[157]

Many military applications coincide with civilian applications. The major difference is in the way that data are processed and used. Military systems often have a higher need for area coverage, rate, and/or spatial resolution, and they may have to work at higher altitudes and during the nighttime, and to be more covert. Military mapping systems may also have a need for multifunction operation. Tactical mapping can be combined with surveillance and targeting capabilities. There is also a growing trend involving military LiDAR systems operating from unmanned platforms [unmanned ground vehicles (UGVs)], e.g., an autonomous car; an unmanned aerial vehicle (UAV), commonly known as a drone; or an autonomous underwater vehicle (AUV) for underwater operation, each of which stress the demand on low size, weight, and power consumption.

Wide-area airborne mapping systems using conventional scanning and linear APDs have been developed by several companies (Teledyne Optech, Riegl, and Leica). These LiDARs can fly up to 5–6 km in altitude and have a PRF of up to 1 MHz. Recent trends have made the systems small and compact, and combine laser scanning with digital and hyperspectral cameras. An example of a compact system is the DragonEye developed by Airborne Hydrography AB (AHAB)[158] in Sweden (now Leica GeoSystems AG). The DragonEye has a laser PRF of up to 1 MHz and operates up to 15,600 m in altitude. The sensor head weighs 37 kg, and the control unit weighs 53 kg.

The GMAPD photon-counting technology stems from military development by the U.S. Air Force and DARPA, and enables even faster data collection than the Optec, RIEGL, and other scanning 3D mapping LiDARs. The sensor offered by Harris Corporation[159] claims to allow for larger area mapping than has been possible until now, and to collect data up to 10 times faster and with 10 times higher resolution than with linear LiDAR sensors. The Harris sensor is a flash LiDAR that uses a 32×129 pixel GMAPD FPA. This sensor will be further discussed in Chapter 11.

To raise the sensitivity of LiDAR for automotive applications, Inoue et al. from Toyota used fiber amplifiers for both transmit and receive, after designing an instrument with a sensor head 2 cm^2 in size.[160] The transmission optical system consists of a pulsed fiber laser. The peak output power is 10 kW, and the pulse width is 4 ns. The diameter of the scanning mirror is 10 mm. The optical fiber amplifier has a midway isolator and a bandpass filter to decrease spontaneous emission and improve the conversion efficiency and noise figure.[161] The resonant frequency of the scanning mirror is 100 Hz, and the scanning angle is 40 deg.

Laser scanners are used on unmanned ground vehicles for navigation and obstacle avoidance. The most widespread demonstration of the success of laser scanners was for the UGV112, where laser scanners were the key sensors. Systems from Teledyne Optech[162] and Velodyne LiDAR[163] are examples of commercial road-mapping LiDARs with many potential military applications (for example, generation of synthetic environments, detection of improvised explosive devices, etc.). The accuracy is said to be on the order of centimeters at a maximum range up to 100–250 m. Measurement rates are found between 1 and 2 million points per second. This allows a 6-cm point spacing at a 10-m range for a vehicle velocity of 43 km/h. Figure 2.45 shows a 3D mapping image taken with a Harris GMAPD 3D LiDAR.

Vehicle-borne LiDAR is becoming a very important sensor for self-driving cars, which may be the first widespread commercial application of LiDAR technology. This application will force the system to be small and compact at a cost of a few hundred dollars per LiDAR.

2.6.8 LiDARs for underwater: Laser-based bathymetry

Airborne laser bathymetry (ALB) or hydrography is a technique for measuring the depths of relatively shallow coastal waters.[164] Typical applications include bathymetric surveys of federal navigation channels, large offshore areas, and ports and harbors; and shore protection projects, such as for jetties and breakwaters, coral reefs, beaches, shorelines, and dredge disposal sites.[165,166] Topographic surveys above the water surface can be conducted simultaneously. A LiDAR can acquire data locally for storm surge

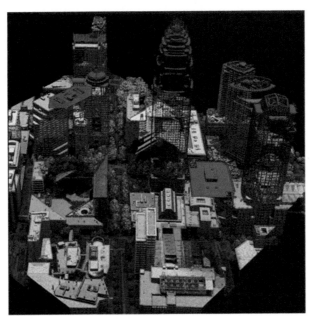

Figure 2.45 Example of 3D urban data collection using a Geiger-mode LiDAR (reprinted from the Harris Corp. web page with permission).

Figure 2.46   Example of a LiDAR bathymetry sensor (HawkEye III from Leica Geosystems). The green-to-blue color is indicative of the water depth (reprinted from Ref. 14).

modeling and sand monitoring. In a 2000 review paper,[167] hardware and software design philosophy is discussed in line with critical design considerations and a history of airborne LiDAR bathymetry development. Figure 2.46 shows a bathymetry image and LiDAR system.

Examples of countries that have developed and used LiDAR bathymetry include USA, Australia, Sweden, Canada, and Russia. In the 1990s, systems became operational in Australia [the Laser Airborne Depth Sounder (LADS[168])], in the U.S. [the Scanning Hydographic Operational Airborne Lidar Survey (SHOALS[169])], and in Sweden (Leica's HawkEye III[170]). The practical depth range for bottom charting is about 3 Secchi depths, i.e., is 5–40 m in coastal waters.

2.6.9 Laser micro-radar

2.6.9.1 Optical coherence tomography

The technique that became known as optical coherence tomography (OCT)[171] is a combination of time-of-flight measurement and interferometry. OCT was implemented using a Michelson interferometer by replacing one of the mirrors with an object under test (Fig. 2.47). A light source with short coherence length is used; therefore, a more correct term would be optical low-coherence tomography. The reference mirror, located at plane R, and the object plane are imaged onto a detector that is focused onto plane R'. Planes R and R' are the same distance from the light source.

The image of an object is superimposed with a reference wave. Interference can be locally observed where the light paths from the reference and the object are approximately equal. The measurement is based on interference. Because of the short coherence length, the interference takes place only within those speckles that correspond to the surface elements close to the plane R'. These

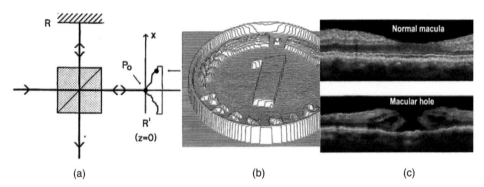

Figure 2.47 (a) Basic setup of an optical micro-radar, (b) 3D reconstruction of a coin image, and (c) tomographic images of a portion of an eye (reprinted from Ref. 14).

regions are detected and stored while the object is moved slowly along the z axis. In the same manner, the reference mirror can be movable when the object is unmoved. An example of a profile reconstructed by this technique using additional (x, y) scanning is presented in Fig. 2.47, which shows a coin.

By placing a human eye in the plane R' as an object to study and moving the plane R in the axial direction z, one can obtain the profile of backscatter from the tissues and surfaces of the eye. Because the system provides scanning in the x-y directions, a 3D image can be reconstructed [Fig. 2.47(c)]. Hundreds of publications appeared in the last decades reporting on numerous modifications of OCT and applications in different fields, especially in medicine. Ophthalmologic applications include determining the structure of the eye, and 3D imaging of blood vessels or of blood flow in the vessels. Cardiovascular applications include cranial vascularization, cardiovascular imaging, etc. Figure 2.48 shows a velocity capillary map, and Fig 2.49 shows a flow velocity distribution. It is beyond the scope of this chapter to discuss all variants of OCT and its applications, so readers are referred to encyclopedia-style books for more breadth.[172–174]

2.6.9.2 Wavefront sensing

As a response to the development of vision correction techniques, in 1996 Molebny et al. developed three types of instruments for measuring the refraction nonhomogeneity of the human eye: single-beam direct-detection ray tracing,[175] double-beam coherent ray tracing,[176] and Hartmann–Shack sensing with holographic lenslet array.[177] Single-beam ray-tracing aberrometry is based on the LiDAR principle: a laser beam is projected into the eye in parallel to the optical axis, and the position-sensitive detector measures the position of the beam projection on the retina. The license for this technique was transferred to Tracey Technologies in Texas, USA, and the iTrace instrument is now in mass production as one of the best clinical aberrometers [Fig. 2.50(a)].

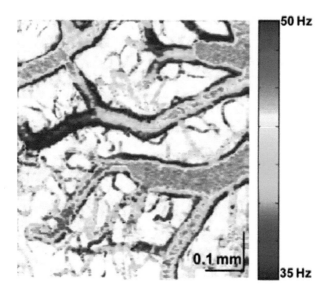

Figure 2.48 Capillary velocity map (reprinted from Ref. 14).

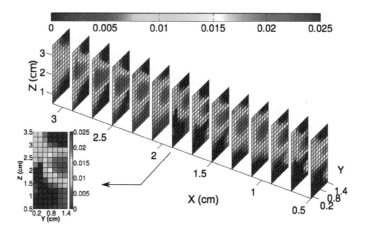

Figure 2.49 Slices of flow velocity distribution (reprinted from Ref. 14).

In 1999, Navarro and Moreno-Barriuso published their results on laser ray tracing based on electromechanical scanning with a disappointing conclusion—ray tracing does not suit live-eye studies because one measurement requires several seconds.[178] In the iTrace, the output power of the cw laser diode is about 1 mW, and the duration of the projection in one point is 1 ms or less. An acousto-optic deflector switches the beam position with a transient time of less than 10 μs. An example of the displayed data from the iTrace is presented in Fig. 2.50(b).

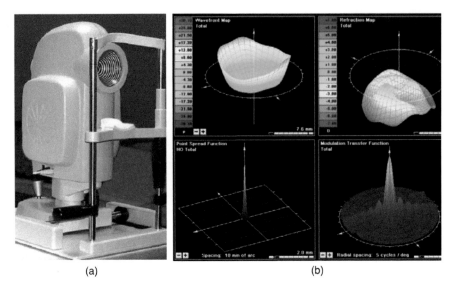

(a)   (b)

Figure 2.50 (a) The iTrace ray-tracing aberrometer and (b) the displayed data (reprinted from Ref. 14).

2.7 History Conclusion

LiDARs have passed through many development stages since the first attempts to use lasers for ranging, which resulted in broad military applications for range finding and weapon guidance, especially laser designation, a form of bistatic LiDAR. Further studies led to development of laser-imaging systems based on 2D gated viewing, and then 3D imaging, which is in the process of being fielded. Imaging systems are under intensive development, including for higher range and cross-range resolution, single-photon–sensitive arrays, and multispectral- or broad-spectrum–emitting lasers for a variety of new capabilities like better weather penetration, looking through vegetation or dense media, or target recognition.

On the civilian and dual-use side, environmental LiDARs are well established in remote sensing research of the atmosphere and the ocean, while 3D mapping LiDARs have reached operational status, being capable of 3D mapping of large areas of many countries. Lasers are becoming increasingly efficient, as well as smaller and cheaper, for use in cars or unmanned vehicles. The use of lasers in self-driving cars is likely to be the first widespread commercial application of LiDAR. This application will continue to significantly decrease the size, weight, and cost of LiDARs. It also will enable widespread use of LiDARs in UAVs as they proliferate for multiple commercial and military applications.

LiDAR technology is also finding many applications in medicine. One example is optical low-coherence tomography, which originated from LiDAR interferential reflectometry and has widespread use in ophthalmology for eye investigation with 3D reconstruction of its structures, and 3D endoscopic studies of blood vessels amplified by Doppler 3D velocimetry. Another great example of LiDAR used in medicine is refraction mapping of the human eye, which is called aberrometry.

Emerging technologies and methods to be explored in LiDAR systems include multiple and synthetic apertures, bistatic operation, multiwavelength- or broad-spectrum–emitting lasers, and photon-counting and advanced quantum techniques, as well as combined passive and active systems, and combined microwave and LiDAR systems. We expect the use of coherent LiDAR to increase as people find additional methods to utilize full-field data, including phase information. On the component side, we foresee efficient versatile laser sources, compact solid state laser scanners for nonmechanical beam steering and beam forming, and sensitive and larger FPAs for both direct and coherent detection to be combined with efficient hardware and algorithms for processing LiDAR information and high data rates.

A comparison of the achievements in LiDAR technologies that have been developed over the last 50 years across the globe shows LiDAR's pervasive attraction and its widespread application. Recently, commercial LiDAR has become more prevalent. 3D LiDARs can map large areas. All of the DARPA Grand Challenge finishers used at least one LiDAR. Google is using a LiDAR in their driverless car.

In recent history, a number of 3D solid state LiDARs have come into use. Some are scanning systems. The LOCAS sensor for munitions is one example. Teledyne Optech has a 3D LiDAR that is commercially available for mapping. Some 3D imagers are Geiger-mode APD sensors. MIT/LL has led the charge in developing these sensors. Princeton Lightwave Inc. (PLI) and Boeing SpectraLab did sell 32×32 pixel and 32×128 pixel GMAPD cameras, but Boeing SpectraLab quit selling them, and when PLI was sold to Argo, they also quit selling them. PLI had also developed an asynchronous 32×32 camera. Now Ball Aerospace has an exclusive license from PLI to sell the cameras previously sold by PLI. LMAPD-based cameras are available from Leonardo DRS, ASC, Raytheon, and Voxtel. General Atomics and Tetravue have still another version of a flash imager that uses a Pockels cell to measure time of flight.

We have had a rich half-century of LiDAR development. Many uses have evolved. It is anticipated that LiDAR is now primed to move into many more application areas. LiDAR technology has been proven as an operational system, having operated for many years in airport environments. New applications are envisioned for defense and civilian use, where better spatial and temporal resolution are required compared to what is normally available with conventional techniques. It is expected that the next half-century will see expanded use of this robust sensing technology.

References

1. G. W. Kamerman, 3D Imaging LiDAR, Lecture Course SC1103, SPIE 2016 Defense+Security Symposium.
2. V. V. Molebny, *Optical Radar Systems: Basics of Functional Layouts* [in Russian], Mashinostroenie Publishers, Moscow (1981).
3. M. S. Malashin, R. P. Kaminskiy, and Yu. B. Borisov, *Basics of LIDAR System Design* [in Russian], Vysshaya Shkola Publishers, Moscow (1983).
4. I. N. Matveev, et al., *LIDAR Technology* [in Russian], Mashinostroenie Publishers, Moscow (1984).
5. P. F. McManamon, *Field Guide to Lidar*, SPIE Press, Bellingham, Washington (2015) [doi: 10.1117/3.2186106].
6. National Academy of Sciences, *Laser Radar: Progress and Opportunities in Active Electro-Optical Sensing*, P. F. McManamon, Chair (Committee on Review of Advancements in Active Electro-Optical Systems to Avoid Technological Surprise Adverse to U.S. National Security), Study under Contract HHM402-10-D-0036-DO#10, National Academies Press, Washington, D.C. (2014).
7. V. V. Molebny, *Optical Radar* [in Russian], Kiev University Press, Kiev (1981).
8. P. A. Roos, R. R. Reibel, T. Berg, B. Kaylor, Z. W. Barber, and W. R. Babbitt, "Ultrabroadband optical chirp linearization for precision length metrology applications," *Optics Letters* **34**(23), 3692–3694 (2009).
9. J. M. Vaughan, K. O. Steinvall, C. Werner, and P. H. Flamant, "Coherent laser radar in Europe," *Proc. IEEE* **84**(2), 205–226 (1996).
10. P. McManamon, G. Kamerman, and M. Huffaker, "A history of laser radar in the United States," *Proc. SPIE*, **7684**, 76840T (2010) [doi: 10.1117/12.862562].
11. V. Molebny, P. Zarubin, and G. Kamerman, "The dawn of optical radar: a story from another side of the globe," *Proc. SPIE* **7684**, 76840B (2010) [doi: 10.1117/12.850086].
12. V. Molebny, G. Kamerman, and O. Steinvall, "Laser radar: from early history to new trends," *Proc. SPIE*, **7835**, 783502 (2010) [doi: 10.1117/12.867906].
13. T. Kobayashi, "Overview of laser remote sensing technology for industrial applications," in *Industrial Applications of Laser Remote Sensing*, T. Fukuchi and T Shiina, Eds., Bentham Science Publishers, Sharjah, UAE, pp. 3–15 (2011).
14. V. Molebny, P. F. McManamon, O. Steinvall, T. Kobayashi, and W. Chen, "Laser radar: historical prospective—from the East to the West," *Optical Engineering* **56**(3), 031220 (2016) [doi: 10.1117/1.OE.56.3.031220].

15. Optics Herald, "120th birthday of A. A. Lebedev," [in Russian] *Rozhdestvensky Optical Society Bulletin*, Issue **145**, 11–15 (2014).

16. P. A. Forrester and K. F. Hulme, "Laser range finders," *Opt. Quant. Electr.* **13**, 259–293 (1981).

17. V. P. Ponomarenko and A. M. Filachev, *Infrared Techniques and Electro-Optics in Russia: A History 1946–2006*, SPIE Press, Bellingham, Washington (2007).

18. "Weapons and Technologies in Russia," in The XXI Century Encyclopedia, Vol. 11: *Electro-Optical Systems and Laser Technologies* [in Russian] Oruzhie i tekhnologii, Moscow (2005).

19. J.-P. Fouilloy and M. B. Sirieix, "History of infrared optronics in France," *Proc. SPIE* **2552**, 804–814 (1995) [doi: 10.1117/12.218281].

20. H. Raidt and D. H. Höhn, "Instantaneous intensity distribution in a focused laser beam at 0.63 μm and 10.6 μm, propagating through the atmosphere," *Appl. Opt.* **14**, 2747–2749 (1975).

21. Yu. L. Kokurin, V. V. Kurbasov, and V. F. Lobanov, "Laser ranging of the light retroreflector installed on the" Lunokhod-1," [in Russian] *Kvantovaya Elektronika* **5**, 138 (1971).

22. N. Sugimoto, N. Koga, I. Matsui, et al., "Earth-Satellite-Earth laser long-path absorption experiment using the Retroreflector in Space (RIS) on the Advanced Earth Observing Satellite (ADEOS)," *J. Opt. A: Pure Appl. Opt.* **1**(2), 201–209 (1999).

23. D. E. Smith, M. T. Zuber, G. A. Neumann, et al., "Initial observations from the Lunar Orbiter Laser Altimeter (LOLA)," *Geophys. Res. Lett.* **37**(18), L18204 (2010).

24. T. Hashimoto, T. Kubota, and T. Mizuno, "Light weight sensors for the autonomous asteroid landing of MUSES-C mission," *Acta Astronautica* **53**, 381–388 (2003).

25. T. Mizuno, K. Tsuno, E. Okumure, and M. Nakayama, "Evaluation of LIDAR system in rendezvous and touchdown sequence of Hayabusa mission," *Trans. Japan Soc. Aeronaut. Space Sci.* **53**(179), 47–53 (2010).

26. J. T. Correll, "The emergence of smart bombs," *Air Force Magazine*, March, 2010, pp. 60–64 (2010).

27. C. Seidel, I. Schwartz, and P. Kielhorn, "Helicopter collision avoidance and brown-out recovery with HELLAS," *Proc. SPIE* **7114**, 71140G (2008) [doi: 10.1117/12.800180].

28. R. Stettner, H. Bailey, and R. D. Richmond "Eye-safe laser radar 3D imaging," *Proc. SPIE* **5412**, 111–116 (2004) [doi: 10.1117/12.553992].

29. M. A. Albota, R. M. Heinrichs, D. G. Kocher, et al., "Three-dimensional imaging laser radar with a photon-counting avalanche photodiode array and microchip laser," *Appl. Opt.* **41**(36), 7671–7678 (2002).

30. S. Kameyema, M. Imaki, A. Hirai, et al., "Development of long range, real-time, and high resolution 3D imaging ladar," *Proc. SPIE* **8192**, 819205 (2011) [doi: 10.1117/12.900108].
31. H. Tsuji, M. Takabayashi, M. Imaki, et al., "Pulsed 3D laser sensor with scanless receiver," *Proc. SPIE* **8379**, 837904 (2012) [doi: 10.1117/12. 919849].
32. N. Martin, R. Ganini, and X. Chazelle, "CLARA-A Franco-British development of airborne LiDAR," *Proc. 7th Conference on Coherent Laser Radar, Applications & Technology*, Paris, Paper FB1 (1993).
33. B. Stephan and P. Metivier, "Flight evaluation trials of a heterodyne CO_2 laser radar," *Proc. SPIE* **806**, 110–118 (1987) [doi: 10.1117/12.941404].
34. R. D. Callan, D. A. Huckridge, C. R. Nash, and J. M. Vaughan, "Active airborne linescan experiment," *Proc. SPIE* **2272**, 183–191 (1994) [doi: 10. 1117/12.191914].
35. S. W. Henderson, P. Gatt, R. Rees, and M. Huffaker, "Wind LIDAR," Chapter 7 in *Laser Remote Sensing*, T. Fujii and T. Fukuchi, Eds., CRC Press, Boca Raton, Florida, pp. 469–722 (2005).
36. J. M. Vaughan, K. O. Steinvall, C. Werner, and P. H. Flamant, "Coherent laser radar in Europe," *Proc. IEEE* **8**(2), 205–226 (1996).
37. A. I. Carswell, "Lidar measurements of clouds," in *Optical and Laser Remote Sensing*, D. K. Killinger and A. Mooradian, Eds., Springer, Berlin-Heidelberg, pp. 318–381 (1983).
38. R. M. Huffaker, A. V. Jelalian, and J. A. L. Thompson, "Laser-Doppler system for detection of aircraft trailing vortices," *Proc. IEEE* **58**, 322–326 (1970).
39. T. Sakimura, Y. Watanabe, T. Ando, et al., "1.5-μm high-gain and high-power laser amplifier using a Er, Yb:glass planar waveguide for coherent Doppler LIDAR," *Proc. 2012 Conference on Lasers and Electro-Optics (CLEO)*, 6–11 May, San Jose (2012).
40. S. Kameyama, T. Yanagisawa, T. Ando, et al., "Development of wind sensing coherent Doppler LIDAR at Mitsubishi Electric Corporation from late 1990s to 2013," *Abstracts from the International Coherent Laser Radiation Conference* (2013).
41. J. G. Hawley, R. Targ, S. W. Henderson, C. P. Hale, M. J. Kavaya, and D. Moerder, "Coherent launch-site atmospheric wind sounder: theory and experiment," *Applied Optics* **32**(24), 4557–4568 (1993).
42. T. Wilkerson, B. Bradford, A. Marchant, C. Wright, T. Apedaile, E. Fowles, A. Howard, and T. Naini, "VisibleWind™: a rapid-response system for high-resolution wind profiling," *Proc. SPIE* **7460**, 746009 (2009) [doi: 10.1117/12.826368].
43. Z.-S. Liu, D. Wu, J.-T. Liu, et al., "Low-altitude atmospheric wind measurement from the combined Mie and Rayleigh backscattering by Doppler lidar with an iodine filter," *Appl. Opt.* **41**(33), 7079–7086 (2002).

44. X. Dou, Y. Han, D. Sun, et al., "Mobile Rayleigh Doppler lidar for wind and temperature measurements in the stratosphere and lower mesosphere," *Opt. Expr.* **22**(S5), A1203–A1221 (2014).

45. H. Xia, X. Dou, M. Shangguan, et al., "Stratospheric temperature measurement with scanning Fabry–Perot interferometer for wind retrieval from mobile Rayleigh Doppler lidar," *Opt. Expr.* **22**(18), 21775–21789 (2014).

46. C. L. Korb, B. M. Gentry, S. X. Li, and C. Flesia, "Theory of the double-edge technique for Doppler lidar wind measurement," *Appl. Opt.* **37**(15), 3097–3104 (1998).

47. P. Lutzmann, B. Göhler, F. van Putten, and C. A. Hill, "Laser vibration sensing: overview and applications," *Proc. SPIE* **8186**, 818602 (2011) [doi: 10.1117/12.903671].

48. F. Hanson and M. Lasher, "Coherent laser radar at 3.6 μm," *Appl. Opt.* **41**(36), 7689–7693 (2002).

49. S. M. Beck, J. R. Buck, W. F. Buell, R. P. Dickinson, D. A. Kozlowski, N. J. Marechal, and T. J. Wright, "Synthetic aperture imaging ladar: laboratory demonstration and signal processing," *Appl. Opt.* **44**(35), 7621–7629 (2005).

50. T. S. Lewis and H. S. Hutchins, "A synthetic aperture at 10.6 microns," *Proc. IEEE* **58**(10), 1781–1782 (1970).

51. C. C. Aleksof, J. S. Accetta, L. M. Peterson, et al., "Synthetic aperture imaging with a pulsed CO_2 TEA laser," *Proc. SPIE* **783**, 29–40 (1987) [doi: 10.1117/12.940575].

52. S. Marcus, B. O. Colella, and T. J. Green, "Solid-state laser synthetic aperture radar," *Appl. Opt.* **33**, 960–964 (1994).

53. T. J. Green, S. Marcus, and B. O. Colella, "Synthetic-aperture-radar imaging with a solid-state laser," *Appl. Opt.* **34**, 6941–6949 (1995).

54. P. Park and J. H. Shapiro, "Performance analysis of optical synthetic aperture radars," *Proc. SPIE* **999**, 100–116 (1988) [doi: 10.1117/12.960229].

55. S. Yoshikado and T. Aruga, "Feasibility study of synthetic aperture infrared laser radar techniques for imaging of static and moving objects," *Appl. Opt.* **37**, 5631–5639 (1998).

56. T. G. Kyle, "High resolution laser imaging system," *Appl. Opt.* **28**, 2651–2656 (1989).

57. G. M. Revel, P. Castellini, P. Chiariotti, et al., "Laser vibrometry vibration measurements on vehicle cabins in running conditions: helicopter mock-up application," *Opt. Eng.* **50**(10), 101502 (2011) [doi: 10.1117/1.3578447].

58. J. C. van den Heuvel, F. J. M. van Putten, A. C. van Koersel, and H. M. A. Schleijpen, "Laser-induced acoustic landmine detection with experimental results on buried landmines," *Proc. SPIE* **5415**, 51–60 (2004) [doi: 10.1117/12.544074].

59. D. Pierrottet, F. Amzajerdian, L. Petway, B. Barnes, and G. Lockard, "Flight test performance of a high precision navigation Doppler lidar," *Proc. SPIE* **7323**, 732311 (2009) [doi: 10.1117/12.82190].

60. G. Fiocco and L. O. Smullin, "Detection of scattering layers in the upper atmosphere (60–140 km) by optical radar," *Nature* **199**, 1275–1276 (1963).

61. R. T. H. Collis, F. G. Fernald, and M. G. H. Ligda, "Laser radar echoes from the clear atmosphere," *Nature* **203**, 1274–1275 (1964).

62. R. M. Schotland, "The determination of the vertical profile of atmospheric gases by means of a ground based optical radar," *Proc. Third Symp. on Remote Sensing of Environment*, Ann Arbor, pp. 215–224 (1964).

63. F. Inaba, T. Kobayashi, T. Ichimura, M. Morihisa, and H. Itoh, "Fundamental study of operational performance of a ladar radar system employing A-scope representation," *Trans. IEICE(B)* **51-B**, 36–52 (1968).

64. Yu. F. Arshinov, S. M. Bobrovnikov, and S. V. Sapozhnikov, "On the technique of lidar measurement of the atmosphere temperature by signals ratio of the pure rotational spectra of N_2 and O_2," [in Russian] *Zhurnal prikladnoy spektroskopii*, **32**(4), 725–731 (1980).

65. V. E. Zuev, Yu. S. Makushkin, and V. N. Matrichev, "Laser sensing of the humidity profile of the atmosphere," [in Russian] *Doklady AN SSSR* **257**(6), 1338–1342 (1981).

66. "Weapons and Technologies in Russia," in The XXI Century Encyclopedia, Vol. 11: *Electro-Optical Systems and Laser Technologies* [in Russian], Oruzhie i tekhnologii, Moscow (2005).

67. http://niipp-moskva.ru [Russian website].

68. H. Hu and J. Qiu, "An overview of lidar study in China," *Review Laser Eng.* **23**(2), 171–174 (1995).

69. S. Svanberg, *Atomic and Molecular Spectroscopy: Basic Aspects and Applications*, Springer, Berlin, (2004).

70. E. D. Hinkley, Ed., *Laser Monitoring of the Atmosphere*, Springer, Berlin (1976).

71. T. Fujii and T Fukuchi, Eds., *Laser Remote Sensing*, CRC Press, Boca Raton (2005).

72. A. P. Ivanov, *Optics of Scattering Media* [in Russian], Nauka i tekhnika, Minsk (1969).

73. A. P. Ivanov, *Physical Fundamentals of Hydro-Optics* [in Russian], Nauka i tekhnika, Minsk (1975).

74. V. I. Tatarskiy, *Wave Propagation in Turbulent Atmosphere* [in Russian], Nauka, Moscow (1967).

75. V. E. Zuev, *Propagation of Laser Radiation in Atmosphere* [in Russian], Radio i sviaz, Moscow (1981).

76. A. S. Gurvich, A. I. Kon, V. L. Mironiv, and S. S. Hmelevcov, *Laser Radiation in the Turbulent Atmosphere* [in Russian], Nauka, Novosibirsk, (1976).

77. V. A. Banakh and V. L. Mironov, *Propagation of Lidar Radiation in the Turbulent Atmosphere* [in Russian], Nauka, Novosibirsk (1986).
78. V. M. Zakharov and O. K. Kostko, *Meteorological LIDAR* [in Russian], Gidrometeoizdat, Leningrad (1977).
79. R. R. Agishev, *LIDAR Monitoring of the Atmosphere* [in Russian], Fizmatgiz, Moscow (2009).
80. A. M. Brown, M. G. Snyder, L. Brouwer, and C. R. Philbrick, "Atmospheric aerosol characterization using multiwavelength multistatic light scattering," *Proc. SPIE* **7684**, 76840I (2010) [doi: 10.1117/12.850080].
81. C. R. Philbrick, H. Hallen, A. Wyant, T. Wright, and M. Snyder, "Optical remote sensing techniques to characterize the properties of atmospheric aerosols," *Proc. SPIE* **7684**, 76840J (2010) [doi: 10.1117/12.850453].
82. Q. S. He and J. T. Mao, "Observation of urban mixed layer at Beijing using a micro pulse LiDAR," [in Chinese] *Acta Meteorol. Sinica* **63**, 374–384 (2005).
83. Q. S. He, J. T. Mao, J. Y. Chen, and S. Y. Han, "A study of evolution and dynamics of urban atmospheric mixing-layer depth based on LiDAR data and numerical simulation," *Chinese J. Atmos. Sci.* **30**, 293–30 (2006).
84. Q. S. He, J. T. Mao, J. Y. Chen, and Y. Y. Hu, "Observational and modeling studies of urban atmospheric boundary-layer height and its evolution mechanisms," *Atmos. Environ.* **40**(6), 1064–1077 (2006).
85. S. Yuan, X. Yu, and J. Zhou, "Lidar observations of the lower atmospheric layer in Hefei," [in Chinese] *Chinese J. Atmos. Sci.* **29**, 387–395 (2005).
86. J. Zhou, D. Liu, G. Yue, et al., "Vertical distribution and temporal variation of Asian dust observed over Hefei, China by using a lidar," *J. Korean Phys. Soc.* **49**(1), 320–326 (2006).
87. T. Murayama, N. Sogimoto, I. Uno, et al., "Ground-based network observation of Asian dust events of April 1998 in east Asia," *JGR Atmospheres* **106**(D16), 18345–18360 (2001).
88. J. Zhou, G. Yu, C. Jin, et al., "Lidar observations of Asian dust over Hefei, China in spring 2000," *JGR Atmospheres* **107**(D15), 4252–4259 (2002).
89. D. Wu, J. Zhou, D. Liu, et al., "12-year LIDAR observations of tropospheric aerosol over Hefei (31.9°N, 117.2°E), China," *J. Opt. Soc. Korea* **15**(1), 90–95 (2011).
90. J. Zhou, H. Hu, and Z. Gong, "Lidar observations of Mt. Pinatubo cloud over Hefei," *Chinese Sci. Bul.* **38**(16), 1373–1376 (1993).
91. H. Hu and J. Qiu, "An overview of lidar study in China," *The Review of Laser Engineering* **23**(2), 85–88 (1994).
92. H. Hu, "LiDAR activity in China," *Paper Abstract of 17th ILRC*, Sendai, Japan (1994).

93. S.-X. Hu, H.-L. Hu, Y.-C. Zhang, and X.-Q. Liu, "Differential absorption lidar for environmental SO_2 measurements," *Chin. J. Las.* **31**, 1121–1126 (2004).

94. S. Hu, H. Hu, Y. Zhang, et al., "A new differential absorption lidar for NO_2 measurements using Raman-shifted technique," *Chin. Opt. Lett.* **1**(8), 435–437 (2003).

95. Y. Wang, X. Cao, T. He, F. Gao, D. Hua, and M. Zhao "Observation and analysis of the temperature inversion layer by Raman lidar up to the lower stratosphere," *App. Opt.* **54**(34), 10079–10088 (2015).

96. C. Xie, J. Zhou, G. Yue, F. Qi, and A. Fan, "New mobile Raman lidar for measurement of tropospheric water vapor," *Frontiers of Electrical and Electronic Engineering in China* **2**(3), 338–344 (2007).

97. J. Mao, D. Hua, Y. Wang, F. Gao, and L. Wang, "Accurate temperature profiling of the atmospheric boundary layer using an ultraviolet rotational Raman lidar," *Optics Communications* **282**(15), 3113–3118 (2009).

98. Y. Wang, F. Gao, C. Zhu, Q. Yan, and D. Hua, "Observations of atmospheric water vapor, aerosol, and cloud with a Raman lidar," *Opt. Eng.* **53**(11), 114105 (2014) [doi: 10.1117/1.OE.53.11.114105].

99. S. Gong, X. Zeng, X. Xue, W. Zheng, Z. Hu, H. Jia, H. Zhang, and Y. Liu, "First time observation of sodium layer over Wuhan, China by sodium fluorescence lidar," *Science in China A* **40**(11), 1228 (1997).

100. S. Gong, G. Yang, H. Yang, X. Cheng, F. Li, Y. Dai, and X. Li, "Lidar activity at Wuhan Institute of Physics and Mathematics, China," *Proc. SPIE* **4893**, 279–286 (2003) [doi: 10.1117/12.466596].

101. X.-K. Dou, X.-H. Xue, et al. "A statistical study of sporadic sodium layer observed by sodium lidar at Hefei (31.8°N, 117.3°E)," *Ann. Geophys.* **27**, 2247–2257 (2009).

102. F. Yi, S. Zhang, H. Zeng, et al. "Lidar observations of sporadic Na layers over Wuhan (30.5°N, 114.4°E)," *Geophys. Res. Lett.* **29**(9), 1345–1348 (2002).

103. F. Yi, S. D. Zhang, C. Yu, et al. "Simultaneous observations of sporadic Fe and Na layers by two closely collocated resonance fluorescence lidars at Wuhan (30.5°N, 114.4°E), China," *J. Geophys. Res.* **112**(D4), 303 (2007).

104. X. Hu, Z. Yan, S. Guo, Y. Cheng, and J. Gong, "Sodium fluorescence Doppler lidar to measure atmospheric temperature in the mesopause region," *Chinese Sci. Bull.* **56**(4–5), 417–423 (2011).

105. H. Inaba and T. Kobayashi, "Laser-Raman radar for chemical analysis of polluted air," *Nature* **224**, 170–172 (1969).

106. S. Nakahara, K. Ito, S. Tamura, et al., "Mobile laser Raman radar for monitoring stack effluent pollutants," *IEEE J. Quant. Electron.* **7**(6), 325–333 (1971).

107. M. R. Bowman, A. J. Gibson, and M. C. W. Sandford, "Atmospheric sodium measured by a tuned laser radar," *Nature* **221**, 456–457 (1969).

108. T. Aruga, H. Kamiyama, M. Jyumonji, T. Kobayashi, and H. Inaba, "Laser radar observation of the sodium layer in the upper atmosphere," *Report of Ionosph. and Space Res. in Japan* **28**(1–2), 65–68 (1974).

109. C. Nagasawa, M. Hirono, and M. Fujiwara, "A reliable efficient forced oscillator dye laser to measure the upper atmospheric sodium layer," *Japanese J. Appl. Phys.* **19**(1), 143–147 (1980).

110. O. Uchino, M. Maeda, T. Shibata, M. Hirono, and M. Fujiwara, "Measurement of stratospheric vertical ozone distribution with a Xe-Cl lidar: estimated influence of aerosols," *Appl. Opt.* **19**(24), 4175–4181 (1980).

111. O. Uchino, M. Tokunaga, M. Maeda, and Y. Miyazoe, "Differential-absorption-lidar measurement of tropospheric ozone with excimer–Raman hybrid laser," *Opt. Lett.* **8**(7), 347–349 (1983).

112. M. Nakazato, T. Nagai, T. Sakai, and Y. Hirose, "Tropospheric ozone differential-absorption lidar using stimulated Raman scattering in carbon dioxide," *Appl. Opt.* **46**(12), 2269–2279 (2007).

113. T. Shibata, M. Kobuchi, and M. Maeda, "Measurements of density and temperature profiles in the middle atmosphere with a XeF lidar," *Appl. Opt.* **25**(5), 685–688 (1986).

114. H. Shimizu, Y. Sasdano, H. Nakane, et al., "Large scale laser radar for measuring aerosol distribution over a wide area," *Appl. Opt.* **24**(5), 617–647 (1985).

115. Y. Sasano, H. Shimizu, and N. Takeuchi, "Convective cell structures revealed by Mie laser radar observations and image data processing," *Appl. Opt.* **21**(17), 3166–3169 (1982).

116. H. Nakane and Y. Sasano, "Structure of a sea-breeze front revealed by scanning lidar observation," *J. Meteorol. Soc. Japan.* **64**, 787–792 (1986).

117. A. B. Gschwendtner and W. E. Keicher, "Development of coherent laser radar at Lincoln Laboratory," *Lincoln Lab. J.* **12**(2), 383–396 (2000).

118. J. G. Verly and R. L. Delanoy, "Model-based automatic target recognition (ATR) system for forwardlooking groundbased and airborne imaging laser radars (LADAR)," *Proc. IEEE* **84**(2), 126–163 (1996).

119. A. V. Jelalian, *Laser Radar Systems*, Artech House, Boston (1992).

120. G. R. Osche and D. S. Young, "Imaging laser radar in the near and far infrared," *Proc. IEEE*, **84**(2), 103–125 (1996).

121. C. C. Andressen, "A 1.32 micron, long-range, solid-state imaging LADAR," *Proc. SPIE* **1694**, 121–130 (1992) [doi: 10.1117/12.138114].

122. G. Stevenson, H. R. Verdun, P. H. Stern, and W. Koechner, "Testing the helicopter obstacle avoidance system," *Proc. SPIE* **2472**, 93–103 (1995) [doi: 10.1117/12.212025].

123. J. L. Meyzonnette, B. Remy, and G. Saccomani, "Imaging CO_2 laser radar with chirp pulse compression," *Proc. SPIE*, **783**, 169–175 (1987) [doi: 10.1117/12.940593].

124. C. J. Karlsson, F. A. A. Olsson, D. Letalick, and M. Harris, "All-fiber multifunction continuous-wave coherent laser radar at 1.55 μm for range, speed, vibration, and wind measurements," *Appl. Opt.* **39**(21), 3716–3726 (2000).

125. H. Ahlberg, S. Lundqvist, D. Letalick, I. Renhorn, and O. Steinvall, "Imaging Q-switched CO_2 laser radar with heterodyne detection: design and evaluation," *Appl. Opt.* **25**(17), 2891–2898 (1986).

126. D. Letalick, I. Renhom, and O. Steinvall, "Measured signal amplitude distributions for a coherent FM-cw CO_2 laser radar," *Appl. Opt.* **25**(21), 3927–3938 (1986).

127. D. Letalick, M. Millnert, and I. Renhorn, "Terrain segmentation using laser radar range data," *Appl. Opt.* **31**(15), 2883–2890 (1992).

128. A. F. Milton, G. A. Klager, and T. R. Bowman, "Low cost sensors for UGVs," *Proc. SPIE* **4024**, 180–191 (2000) [doi: 10.1117/12.391628].

129. 129 R. L. Espinola, E. L. Jacobs, C. E. Halford, R. Vollmerhausen, and D. H. Tofsted, "Modeling the target acquisition performance of active imaging systems," *Opt. Expr.* **15**(7), 3816–3832 (2007).

130. T. R. Chevalier and O. K. Steinvall, "Laser radar modelling for simulation and performance evaluation," *Proc. SPIE* **7482**, 748206 (2009) [doi: 10.1117/12.830467].

131. P. Andersson, "Long-range three-dimensional imaging using range-gated laser radar images," *Opt. Eng.* **45**(3), 034301 (2006) [doi: 10.1117/1.2183668].

132. O. Steinvall, P. Andersson, M. Elmqvist, and M. Tulldahl, "Overview of range gated imaging at FOI," *Proc. SPIE* **6542**, 654216 (2007) [doi: 10.1117/12.719191].

133. E. Repasi, P. Lutzmann, O. Steinvall, et al., "Advanced short-wavelength infrared range- gated imaging for ground applications in monostatic and bi-static configurations," *Appl. Opt.* **48**(31), 5956–5969 (2009).

134. O. Steinvall, M. Elmqvist, K. Karlsson, H. Larsson, and M. Axlesson, "Laser imaging of small surface vessels and people at sea," *Proc. SPIE* **7684**, 768417 (2010) [doi: 10.1117/12.849388].

135. D. Bonnier, S. Lelievre, and L. Demers, "On the safe use of long-range laser active imager in the near-infrared for homeland security," *Proc. SPIE* **6206**, 62060A (2006).

136. O. David, R. Schneider, and R. Israeli, "Advance in active night vision for filling the gap in remote sensing," *Proc. SPIE* **7482**, 748203 (2009) [doi: 10.1117/12.830378].

137. P. Heckman and R. T. Hodgson, "Underwater optical range gating," *IEEE J. Quant. Electron.* **3**(11), 445–448 (1967).

138. G. R. Fournier, D. Bonnier, J. L. Forand, and P. W. Pace, "Range-gated underwater laser imaging system," *Opt. Eng.* **32**(9), 2185–2190 (1993) [doi: 10.1117/12.143954].

139. M. P. Strand, "Underwater electro-optical system for mine identification," *Proc. SPIE* **2496**, 487–497 (1995) [doi: 10.1117/12.211304].

140. J. Busck and H. Heiselberg, "High accuracy 3D lasar radar," *Proc. SPIE* **5412**, 257–264 (2004) [doi: 10.1117/12.545397].

141. H. M. Tulldahl and M. Pettersson, "Lidar for shallow underwater target detection," *Proc. SPIE* **6739**, 673906 (2007) [doi: 10.1117/12.737872].

142. S. Bailey, W. McKeag, J. Wang, M. Jack, and F. Amzajerdian, "Advances in HgCdTe APDs and LADAR receivers," *Proc. SPIE* **7660**, 766031 (2010) [doi: 10.1117/12.859186].

143. E. de Borniol, F. Guellec, J. Rothman, et al., "HgCdTe-based APD focal plane array for 2D and 3D active imaging: first results on a 320×256 with 30 μm pitch demonstrator," *Proc. SPIE* **7660**, 76603D (2010) [doi: 10.1117/12.850689].

144. A. Ashcroft and I. Baker, "Developments in HgCdTe avalanche photodiode technology and applications," *Proc. SPIE* **7660**, 76603C (2010) [doi: 10.1117/12.850133].

145. R. Heinrichs, B. F. Aull, R. M. Marono, et al., "Three-dimensional laser radar with APD arrays," *Proc. SPIE* **4377**, 106–118 (2001) [doi: 10.1117/12.440098].

146. J. P. Donnelly, K. A. McIntosh, D. C. Oakley, et al., "1-m Geiger-mode detector development," *Proc. SPIE* **5791**, 281–287 (2005) [doi: 10.1117/12.609691].

147. R. M. Marino and W. R. Davis, "Jigsaw: a foliage-penetrating 3D imaging laser radar system," *Lincoln Lab. J.* **15**(1), 23–36 (2005).

148. P. Cho, H. Anderson, R. Hatch, and P. Ramaswami, "Real-time 3D ladar imaging," *Lincoln Lab. J.* **16**(1) 147–164 (2006).

149. C. Grönwall, T. Chevalier, G. Tolt, and P. Andersson, "An approach to target detection in forested scenes," *Proc. SPIE* **6950**, 69500S (2008) [doi: 10.1117/12.777042].

150. W. Armbruster, "Exploiting range imagery: techniques and applications," *Proc. SPIE* **7382**, 738203 (2009) [doi: 10.1117/12.835136].

151. Quotes from DARPA Program Manager Jack McCrae, March 2019; see also https://www.darpa.mil/attachments/TestimonyArchived(March%2025%202004).pdf; A. K. Lal, *Transformation of the Armed Forces: 2025*, Vij Books India, New Delhi, p. 167 (2012).

152. http://www.fas.org/man/dod-101/sys/smart/docs/locaas_Industry_Day/sld013.htm.

153. C. A. Robinson, Jr. "Smack 'em flattens targets," *Signal*, AFCEA: http://www.afcea.org/signal/articles/anmviewer.asp?a=1099&print=yes.

154. C. Andressen, D. Anthony, T. DaMommio, et al., "Tower test results for an imaging LADAR seeker," *Proc. SPIE* **5791**, 70–82 (2005) [doi: 10.1117/12.605944].

155. C. M. Wong, J. E. Logan, C. Bracikowski, and B. K. Baldauf, "Automated in-track and cross-track airborne flash ladar image registration for wide-area mapping," *Proc. SPIE* **7684**, 76840S (2010) [doi: 10.1117/12.849479].

156. E. G. P. Bovenkamp and K. Schutte, "Laser gated viewing: an enabler for automatic target recognition?" *Proc. SPIE* **7684**, 76840Z (2010) [doi: 10.1117/12.849644].

157. J. Shan and C. K. Toth, Eds., *Topograhic Laser Ranging and Scanning: Principles and Processing, Second Edition*, CRC Press, Boca Raton (2018).

158. Leica AHAB DragonEye Dual Head. Oblique LiDAR System: https://w3.leica-geosystems.com/en/index.htm.

159. K. P. Corbley, "Geiger-mode LiDAR delivers fast, wide-area mapping," *Earth Imaging Journal* [online], V1 Media, June 9 (2015).

160. D. Inoue, T. Ichikawa, H. Matsubara, et al., "Highly sensitive lidar with a thumb-sized sensor-head built using an optical fiber preamplifier," *Proc. SPIE* **8037**, 80370A (2011) [doi: 10.1117/12.886478].

161. S. Yamashita and T. Okoshi, "Performance improvement and optimization of fiber amplifier with a midway isolator," *IEEE Photon. Technol. Lett.* **4**(11), 1276–1278 (1992).

162. Optech Lynx SG mobile mapper: http://www.teledyneoptech.com/index.php/product/lynx-sg1.

163. Velodyne LiDAR: "All the distance sensing data you will ever need," http://velodyneLiDAR.com/products.html.

164. J. Banic and G. Cunningham, "Airborne laser bathymetry: a tool for the next millennium," [online] EEZ Technology. http://shoals.sam.usace.army.mil/ downloads/Publications/32Banic_Cunningham_98.pdf.

165. W. J. Lillycrop, L. E. Parson, and J. L. Irish, "Development and operation of the SHOALS airborne lidar hydrographic survey system," *Proc. SPIE* **2964**, 26–37 (1996) [doi: 10.1117/12.258351].

166. A. G. Cunningham, W. J. Lillycrop, G. C. Guenther, and M. W. Brooks, "Shallow water laser bathymetry: accomplishments and applications," *Proc. Oceanol. Int'l: Glob. Ocean* **3**, 277–288 (1998).

167. G. C. Guenther, A. G. Cunningham, P. E. LaRocque, and D. J. Reid, "Meeting the accuracy challenge in airborne bathymetry," *Proc. EARSeL-SIG-Workshop LIDAR*, Dresden/FRG, June 16–17 (2000).

168. R. Nairn, "Royal Australian Navy laser airborne depth sounder, the first year of operations," *Int. Hydro. Rev., Monaco* **71**(1), 109–119 (1994).

169. W. J. Lillycrop, J. L. Irish, and L. E. Parson, "SHOALS system: three years of operation with airborne LiDAR bathymetry: experiences, capability and technology advancements," *Sea Technol.* **38**(6), 17–25 (1997).

170. O. Steinvall and K. Koppari, "Depth sounding LiDAR: an overview of Swedish activities and with future prospects," *Proc. SPIE* **2964**, 2–25 (1996) [doi: 10.1117/12.258342].

171. T. Dresel, G. Häusler, and H. Venzke, "Three-dimensional sensing of rough surfaces by coherence radar," *Appl. Opt.* **31**(7), 919–925 (1992).

172. B. E. Bouma and G. J. Tearney, Eds., *Handbook of Optical Coherence Tomography*, Marcel Dekker, Inc., Basel (2002).

173. F. Drexler and J. G. Fujimoto, Eds., *Optical Coherence Tomography: Technology and Applications*, Second Edition, Springer International Publishing AG (2015).

174. A. Girach and R. C. Sergott, Eds., *Optical Coherence Tomography*, Springer International Publishing AG (2016).

175. V. V. Molebny, I. G. Pallikaris, L. P. Noumidis, et al., "Retina ray-tracing technique for eye-refraction maping," *Proc. SPIE* **2971**, 175–183 (1997) [doi: 10.1117/12.275118].

176. V. V. Molebny, I. G. Pallikaris, L. P. Noumidis, et al., "High-precision double-frequency interferometric measurement of the cornea shape," *Proc. SPIE* **2965**, 121–126 (1996) [doi: 10.1117/12.257378].

177. V. V. Molebny, V. N. Kurashov, I. G. Pallikaris, et al., "Adaptive optics technique for measuring eye refraction distribution," *Proc. SPIE* **2930**, 147–156 (1996) [doi: 10.1117/12.260867].

178. R. Navarro and E. Moreno-Barriuso, "Laser ray-tracing method for optical testing," *Opt. Lett.* **24**(14), 951–953 (1999).

Chapter 3
LiDAR Range Equation

3.1 Introduction to the LiDAR Range Equation

In calculating the required energy, or power, to obtain a measureable return with a LiDAR, an engineer can divide the calculation into two parts. The first part is the calculated intensity of the reflected laser light from the target. This result can be stated in terms of the number of photons returned to each detector in a certain period of time. The second part calculates how much energy, in photons, is needed in each detector to measure the returned signal. For the second part, we have to compare the returned energy to the noise. We discuss the second part of this in Chapter 6, which concerns LiDAR receivers. For a given noise, we can set a threshold that results in a certain probability of both detection and false alarm. The first part of this calculation—how many photons are returned—is usually called a link budget calculation. The second part of the calculation is associated with the receiver sensitivity. There are many different types of receivers, as discussed in Chapter 6. This chapter focuses primarily on the link budget portion of the calculation. For the calculations in this chapter, it is assumed that a certain number of received photons is required in a given period for the receiver to detect the return signal. Chapter 6 provides more detail on receiver sensitivity and probabilities, which depend on the type of receiver being considered.

3.2 Illuminator Beam

To calculate the number of photons returned to each detector, an engineer can start with the transmitted laser power. The transmitter can have a beam shaped as a Gaussian beam or a super-Gaussian, or it can be a flat-top beam. For the computationally simple flat-top beam, the irradiance (watts/cm^2) is the laser power divided by the area of the beam footprint. This is a significant gain over radiating throughout a sphere, since lasers have small beam divergence. Beam divergence can be smaller for large-aperture transmitters and for shorter-wavelength transmitters because the diffraction limit is a smaller angle. The full–angular-width, half-maximum diffraction limit ϑ is given by

$$\vartheta \approx \frac{\lambda}{D}, \tag{3.1}$$

where λ is the wavelength, and D is the beam diameter. For transmitters, the beam diameter has to be smaller than the transmit aperture diameter to avoid significant clipping. Figure 3.1 shows a Gaussian beam compared to a super-Gaussian beam with $N = 5$.

The equation for the Gaussian beam is

$$f(x) = ae^{-\frac{(x-b)^2}{2c^2}}, \tag{3.2}$$

whereas a super-Gaussian beam shape is given by

$$f(x) = ae^{-\frac{(x-b)^N}{2c^2}}. \tag{3.3}$$

In both Eqs. (3.2) and (3.3), a is the amplitude of the Gaussian beam, c is a measure of the width, and b is an offset. The term b does not matter if we do not care about the exact location of the Gaussian beam, so it can be ignored in many applications. With a super-Gaussian beam, N is larger than 2. In Fig. 3.1 (which is the same as Fig. 1.9), N is 5 for the super-Gaussian case.

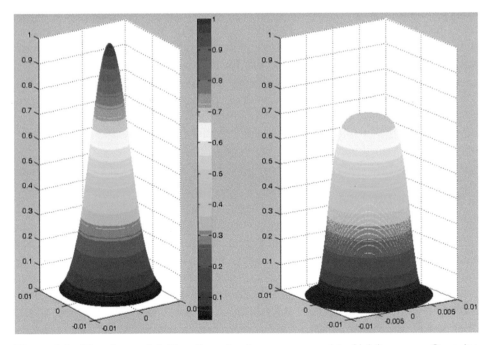

Figure 3.1 The shape of (left) a Gaussian beam compared to (right) a super-Gaussian beam (right) with $N = 5$ (reprinted from Ref. 1).

Beams that are more flat topped have less clipping at the apertures. Clipping of the Gaussian beam is shown in Fig. 3.2. This 18.5% clipping loss was calculated using a MATLAB® function, which is given below.

```
function [Loss w]=Gaussa(D,sigma)
% Gaussa calculates the near field pattern for a Gaussian
beam emitted
% through an aperture of radius D, and the shape of a clipped
near field,
% and the loss
w=zeros(101);
x=linspace(-D*.7,D*.7,101);
y=linspace(-D*.7,D*.7,101);
for r=1:numel(x);
  for c=1:numel(y);
  w(r,c)=exp(-(x(r)^2/(sigma^2)+(y(c)^2/(sigma^2))));
  end
end
figure(3)
surf(w)
value1=sum(sum(w));%value1 should be integral under the 2D
Gaussian
apert=ones(101);
radius=36;
for r=1:101;
  for c=1:101;
  if (r-51)^2+(c-51)^2>radius^2;
  apert(r,c)=0;
  w(r,c)=w(r,c)*apert(r,c);
  end
  end
end
value2=sum(sum(w));% value2 is integral under the Gaussian
after zeroing the part which is cut off
Loss =100*(1-(value2/value1));
figure(4)
surf(w)
```

We then look at a 2D profile of a Gaussian beam and super-Gaussian beams up to $N = 5$ compared to a 3D profile as shown in Fig. 3.3, which shows 2D Gaussian and super-Gaussian profiles for $N = 2$ through $N = 5$. This chart was generated in Excel. We can see from this 2D representation that higher-order super-Gaussian beams have less clipping for a given beam width because the sides of the distribution fall faster. The ratio of beam width

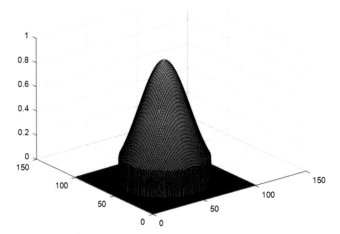

Figure 3.2 Clipped Gaussian beam with $c = 1$ and aperture diameter $= 2.5$.

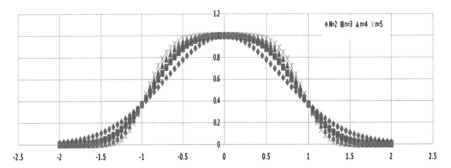

Figure 3.3 Gaussian and super-Gaussian distributions for $N = 2$ through $N = 5$.

to diameter can be larger for a given amount of clipping when using a super-Gaussian beam distribution for object illumination. Many LiDAR systems avoid this problem by using a holographic optical element (HOE) to shape the beam profile and give a flat profile that matches the FOV of the receiver.

3.3 LiDAR Cross-Section

Next we create a fictitious area we call the cross-section. This is not a physical area but is related to the physical area. For area targets, if you have a flash-imaging LiDAR, by using many detectors, you can only count the cross-section seen by each detector when calculating cross-section because we are calculating how much power, or energy, is being reflected into a given detector. Flash imaging means using an array of detectors, with the whole area viewed by the array illuminated at the same time. Higher ground sample distance means that each detector sees a smaller area and a smaller cross-section. If the target illumination area is fixed, an increased imaging resolution (i.e., increased number of detector pixels) results in decreased signal per detector because the cross-section viewed by each detector is smaller, so less power is reflected into each detector. The signal returned can be increased by increasing the transmitter flux/power to compensate for illuminating more pixels. Surfaces with high reflectivity in the backward direction (toward the LiDAR receiver) also have a higher cross-section.

For a target whose illumination area is larger than the detector angular subtense (DAS), we can assume that the cross-section viewed by a receiver pixel is proportional to the projected area of a pixel times the reflectance. For square receiver pixels, we have

$$\sigma = \rho_t A_p = \rho_t d^2, \tag{3.4}$$

where d is length of one side of the projected area of a pixel on the target, ρ_t is the reflectance of the area, and A_p is the area of the pixel, which for a square pixel is d^2. Surfaces with high reflectivity have a higher cross-section. If the LiDAR is looking straight down from nadir, the cross-range resolution is also called the ground sample distance (GSD). The area seen by the detector grows with range according to

$$\sigma = \rho_g \times (DAS)^2 \times R^2, \tag{3.5}$$

where ρ_g is the ground reflectivity, and $DAS \times R = d$ in Eq. (3.4). For a point target or line target with dimensions smaller than the pixel size at the target location, the cross-section will be smaller due to a smaller area that is reflecting laser light.

Figure 3.4 shows a point target that is much smaller than the illuminating beam (1), a linear target that is larger than the illumination beam in one dimension (2), and an area target that is larger than the illumination beam in both directions (3). Cross-section times illumination flux is constant versus

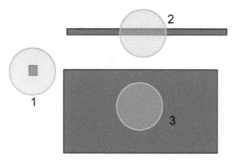

Figure 3.4 Area reflecting laser light for a reflector smaller than the illuminated area (1), for a linear reflector that is larger than the GSD in one dimension (2), and for an area that is larger than the GSD (3) (reprinted from Ref. 1).

range for an area larger than the illumination area, but varies with range or range squared for the first two cases. Some LiDAR range equations use target reflectivity rather than cross-section and therefore can have R^3 or R^4 in the denominator.

High range resolution also reduces the effective cross-section if there are scatterers at multiple ranges within a DAS. The return within a given pixel may come from multiple distinct surfaces. Let us say that we have reflections from three different ranges within a single GSD. At any given range, we do not receive reflection from the whole area of the pixel. We only receive reflection from that portion of the pixel at the measured range, meaning that the cross-section for three equal reflectivity areas at different ranges would be $\rho_t \times 0.33 \, (DAS)^2 R^2$.

The accepted definition of cross-section is different for LiDARs than for microwave radars. For LiDAR cross-section, engineers usually assume that scattering is Lambertian, and reflected light is equivalent to being reflected into π steradians (sr). We arrive at π sr as the equivalent solid angle of reflected light by assuming a cosine-distribution reflected light over a hemisphere (2π sr) for Lambertion scattering of light from a rough surface. While light is scattered over 2π sr, the regions near the surface do not receive much scattered light because the projected area of the scattered surface becomes smaller by the cosine factor. There is a cosine decrease in the scattered light as you move away from normal due to the projected area of the reflecting surface, as shown in Fig. 3.5. This means that the total scattered light is the same as in the case where the reflected light is reflected uniformly over an angular region of π sr. With Lambertian scattering, for simplicity in calculations, we assume that light is scattered into π sr, even though it is really scattered over 2π sr with a cosine distribution.

In microwave radar, the cross-section definition usually involves scattering light over 4π sr from a small, round, gold ball (Fig. 3.6). This makes sense for radar, where the radar wavelength is often longer than the diameter of the round, gold ball. For LiDAR this does not make as much sense because the ball would be much larger than the wavelength, so it would

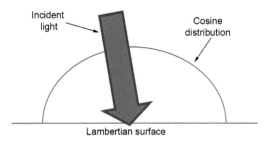

Figure 3.5 Lambertian scattering over 2π sr with a cosine distribution of scattered light, resulting from the projected area of the scattered surface (reprinted from Ref. 1).

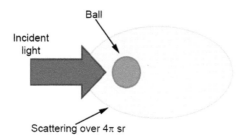

Figure 3.6 Scattering from a small, round, gold ball as used in microwave systems to define cross-section (reprinted from Ref. 1).

block forward radiation. The cross-section in either case is proportional to the area of the target being illuminated inside of one DAS and surface reflectivity. The radar cross-section is therefore one-quarter as large as the LiDAR cross-section based on the difference in definition between scattering reflected EM radiation into 4π sr versus π sr.

3.3.1 Cross-section of a corner cube

The cross-section of a corner cube is much larger than the cross-section of other surfaces because light hitting the corner cube is reflected back toward the source. It is like an antenna with gain. A 2D representation of a corner cube is shown in Fig. 3.7. Light hits the corner cube, bounces twice, and then reflects back toward the place from which the light originated. However, the corner cube is only so large, so the reflected light is limited in angular divergence by the diffraction limit given by Eq. (3.1). We said earlier that a Lambertian scatterer has an average solid angle return of π. A corner cube has an angular return that is about λ/a in each dimension, with a in this case being the diameter of the corner cube in each dimension. The increase in effective cross-section compared to a flat plate is the ratio of these solid angles to each other. In one case, the return is spread over a wide angle, whereas a corner cube concentrates the return beam, making the radiance of the return beam larger. Of course, a corner cube can be dihedral, having two planes that meet, so concentrating light in only one dimension. Alternatively, a corner cube can be trihedral, concentrating light in two dimensions. People usually

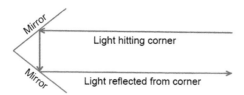

Figure 3.7 Schematic diagram of a 2D corner cube showing reflection back toward the illumination source (reprinted from Ref. 1).

mean a trihedral when they speak of a corner cube. A trihedral's cross-section would be

$$\sigma_t \approx \frac{4a^4}{\lambda^2}.$$

(3.6)

3.4 Link Budget Range Equation

The power received in a given detector can be calculated using

$$P_R = P_T \frac{\sigma}{A_{illum}} \frac{A_{rec}}{\pi R^2} \eta_{atm}^2 \eta_{sys},$$

(3.7)

where P_R is the power received, P_T is the power transmitted, σ is the cross-section in square meters, A_{illum} is the area illuminated, A_{rec} is the area of the receiver, R is the range, η_{atm} is the transmission efficiency through the atmosphere, and η_{sys} is the receiver system's optical efficiency. You can see that the power received is the power transmitted times two ratios of areas, times appropriate efficiency terms. The first ratio of areas is the cross-section divided by the illuminated area at the object plane. The second ratio of areas is the receiver aperture area divided by the effective average area illuminated by the reflection from the target. For example, we could use π sr for light reflected from a Lambertian scattering target. The cross-range resolution of a pixel cannot be better than the diffraction limit, or

$$d \geq \frac{R\lambda}{D_{rec}}.$$

(3.8)

Equation (3.8) is derived from Eq. (3.1), but for the receiver, we obviously use the full receiver aperture diameter D_{rec} as the diameter of the receiver. Equations (3.1) and (3.8) use the full width, half power, diffraction-limit definition of beam divergence. Often other values are used, such has half-width, or the width at zero power. For a circular aperture with A_{rec} as the area of the receiver,

$$A_{rec} = \pi \left(\frac{D_{rec}}{2}\right)^2,$$

(3.9)

which is for a single receive aperture. If you have multiple subapertures, then calculating the receive area can be more complex.

The area illuminated can be no smaller than the diffraction limit as given by

$$A_{illum} \approx \pi \left(\frac{\lambda R}{2D}\right)^2.$$

(3.10)

The area illuminated can be larger than is given in Eq. (3.10) if the transmit beam is spoiled, or if the transmit beam is not diffraction limited. We can then invert Eq. (3.7) to obtain the required laser transmit power for a given LiDAR range:

$$P_\text{T} = \frac{\pi P_\text{R} A_\text{illum} R^2}{\sigma A_\text{rec} \eta_\text{atm}^2 \eta_\text{sys}}.$$ (3.11)

The terms in Eqs. (3.7) and (3.11) can be expanded if desired. When using these equations, care must be taken to properly evaluate the cross-section per detector pixel for a given target, such as area targets, line targets, or point targets. Assuming that enough is known about the target's scattering properties and its shape, calculation of the effective cross-section is straightforward.

If we multiply both sides of Eq. (3.7) by laser pulse width in units of time, the result will be energy received and energy transmitted. The time multiplier can be derived from one over the receiver bandwidth, assuming that the laser pulse width is at least that long in time. We can then solve for required transmitted energy using

$$E_\text{T} = E_\text{R} \frac{A_\text{illum}}{\sigma} \frac{\pi R^2}{A_\text{rec}} \frac{1}{\eta_\text{atm}^2} \frac{1}{\eta_\text{sys}},$$ (3.12)

where

$$\eta_\text{atm} = \exp(-R\beta).$$ (3.13)

The atmospheric attenuation has exponential decay, which is usually called Beer's law. Visibility is defined at 550 nm. To adjust to a different wavelength, we use[2]

$$\beta(\lambda) \propto \left(\frac{\lambda_0}{\lambda}\right)^q,$$ (3.14)

where parameter q is defined in Ref. 2 and adjusts the β parameter in Eq. (3.14) for wavelength:[3]

$$q(\lambda) = 0.1428\lambda - 0.0947.$$ (3.15)

Therefore, the value of q for 1.55 μm using this formula is 0.12664.

Human visibility can be used as a factor in calculating atmospheric transmission. In Europe, visible range (or visibility) is usually defined when contrast is reduced to 5% of its original value. In the U.S. the contrast threshold for visible range is 2% contrast, or 0.02. Engineers usually consider β in terms of decay in a 1-km path, but for cloud or fog distances, it may be advantageous to use decay per meter instead of decay per kilometer. Parameter β is defined based on the definition of visible range (visibility). The visibility in clouds is very low. You probably have personal experience trying to drive in heavy fog, where visibility is short range, so you have an idea of LiDAR range in heavy fog. A typical human eye responds to wavelengths that range from about 390 to 700 nm. An eye-safe LiDAR may have a longer wavelength, say at 1550 nm, but the difference in wavelength will not be enough to significantly increase visibility.

This chapter develops the q and β parameters for fog based on experimental results from a 5.5-m–deep fog tank. It is anticipated that this model adjustment is good for fog but may not be as good for snow or rain. It is also anticipated that LiDAR imaging will do better through rain and snow. We use 0.55 μm as λ_0. Changing wavelength from 0.55 μm to 1.55 μm decreases β to 0.877 of its value at 0.55 μm. This means that we can see 1.14 times as far using the same laser power, considering only attenuation due to scattering, not R^2 or other loses.

The required received energy E_R can be specified as the energy in N photons. The energy is a single photon is given by

$$E_p = h\nu = \frac{hc}{\lambda}. \tag{3.16}$$

Therefore, it is possible to replace E_R with $N \times$ the energy in each photon. Then the link budget equation in energy is

$$E_T = \frac{Nhc}{\lambda} \frac{A_{\text{illum}}}{\sigma} \frac{\pi R^2}{A_{\text{rec}}} \frac{1}{\eta_{\text{atm}}^2} \frac{1}{\eta_{\text{sys}}}. \tag{3.17}$$

This equation can be solved in the other direction—for the number of photons received per detector—given a certain transmitted energy. The ratio of area illuminated to cross-section is

$$\frac{A_{\text{illum}}}{\sigma} = \frac{1.1\, N_{\text{det}}}{\eta_{\text{ref}}}, \tag{3.18}$$

where N_{det} is the number of detectors. Based on Eq. (3.18), we can solve for the required transmitter energy per pulse:

$$E_T = \frac{4.4Nhc}{\lambda} \frac{N_{\text{det}}}{\eta_{\text{ref}}} \left(\frac{R}{D}\right)^2 \frac{e^{2\beta R}}{\eta_{\text{sys}}}. \tag{3.19}$$

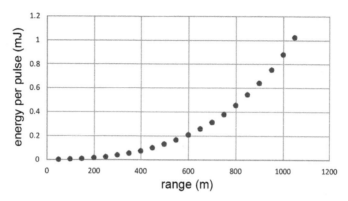

Figure 3.8 Required energy per pulse versus range for 5-km visibility, 20 photons per detector, a 128 × 128 detector array, and 10% reflectivity.

One way to solve this equation is by using an Excel spreadsheet. If we assume that 20 photons are required per pulse, a 128 × 128 detector array, 5-mm visibility, 10% reflectivity from a large area target, and a system's efficiency of 60%, then we can see from Fig. 3.8, which shows required energy versus range, that 0.878 mJ is required from the transmitter at a range of 1000 m.

3.5 Atmospheric Effects

The discussion in the previous section talked about visibility and Beer's law. Here it will be useful to further elaborate on the atmospheric loss mechanism. The atmosphere has nitrogen, oxygen, CO_2, water vapor, and many other components, although those listed are the major ones. All of these components may absorb light. Unlike atmospheric scattering, atmospheric absorption tends to be very wavelength specific for a given molecule, so small shifts in laser wavelength can make a significant difference in the absorption component to the exponential decay of laser radiation while traversing through the atmosphere.

Figure 3.9 shows atmospheric transmission versus wavelength at the Wright-Patterson Air Force Base (WPAFB). The plot in Fig. 3.9(a) represents 5 km to space in the winter, and the plot in Fig. 3.9(b) represents ground to space in the summer. Note that scales in the horizontal axes use a 10-μm wavelength for the value of 1. You can see a significant change in the longer-wavelength transmission, but not as much change in the shorter wavelengths. For most LiDARs being built today, the difference between the two cases would not be significant because most of these systems are in the shorter wavelengths, where little difference is seen between the two plots.

Higher humidity in the summer, with transmission all the way to the ground, is a likely cause for many of the differences between the two plots. We

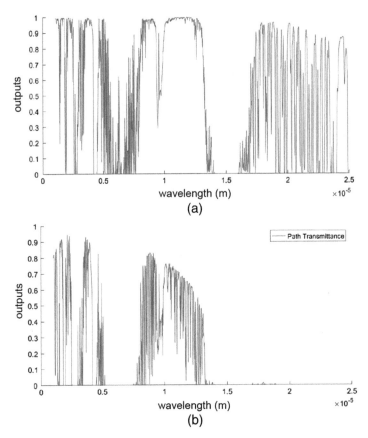

Figure 3.9 Atmospheric transmission at WPAFB: (a) winter atmosphere with path geometry of 5 km to space; (b) summer atmosphere with path geometry of surface to space (courtesy of Air Force Institute of Technology and Dr. Steven Fiorino[4]).

see that the summer (from the ground) has much less transmission in the LWIR regions, consistent with higher absorption in the LWIR due to high absolute humidity.

Popular LiDAR wavelengths are at 1.064 μm due to Nd:YAG availability and at 1.55 μm due to eye safety and the availability of telecom components at this wavelength. While absorption is not zero, both of those regions avoid any significant atmospheric absorption. More recently, there has also been some interest in wavelengths at ~2 μm and in the MWIR region. Atmospheric absorption at 2 μm and in the MWIR is also not significant, but scattering is less at longer wavelengths. Real beam resolution of course is better at the shorter wavelengths. When deciding on a LiDAR wavelength, the designer will choose an atmospheric line that has little absorption; however, many of these lines are spread across a fairly wide range from the visible to the LWIR. It should be noted that Fig. 3.9 does not assume a large presence of water vapor in the air. At high absolute humidity, the LWIR wavelength region has significant absorption, so it should be avoided for those situations. Many other parameters will, however, help determine the exact wavelength to be chosen, including atmospheric parameters and available lasers.

3.5.1 Atmospheric scattering

One of the main loss factors in the LiDAR equation is atmospheric loss, and much of that loss comes from scattering. We have three main regimes of scattering: where the particle is much smaller than the EM wavelength, where the particle is on the order of the same size as the EM wavelength, and where the particle is much larger than the EM wavelength. Rayleigh scattering (Fig. 3.10) is the elastic scattering of light by molecules and particulate matter that are much smaller than the wavelength of the incident light. Rayleigh scattering intensity is proportional the sixth power of the diameter of the scattering particle and inversely proportional to the fourth power of the wavelength of light. This means that the shorter wavelengths in visible light (violet and blue) are scattered more strongly than the longer wavelengths toward the red end of the visible spectrum. This scattering is responsible for the blue color of the sky during the day and the orange colors during sunrise and sunset.

Mie scattering theory is used for particles that are on the order of the same size as the wavelength of the EM radiation. Mie scattering (Fig. 3.11) has more forward scattering than in any other direction. Raleigh scattering is in

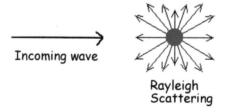

Figure 3.10 Diagram of Rayleigh scattering (reprinted from Ref. 1).

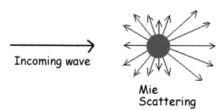

Figure 3.11 Diagram of Mie scattering (reprinted from Ref. 1).

all direction. Particles much larger than the wavelength of light have mostly forward scattering. Each wavelength will have a different exponential decay. Scattering is only one component of that decay. Aerosols are one particle to consider, but dust, fog, sand, and other particles can also contribute to atmospheric scattering.

3.5.2 Atmospheric turbulence

Because of small variations in the index of refraction of the air, there is a maximum effective aperture size on receive when a LiDAR is looking through the atmosphere, and on transmit through the atmosphere. This will limit the diffraction-limited resolution of a LiDAR unless turbulence is compensated. While these index-of-refraction variations are small (on the order of 0.0001 or less) they cause phase changes across an aperture. The largest diameter that an aperture can have without corruption due to phase variations is called the Fried parameter r_0. Because Fried named this parameter r_0, it is sometimes called the Fried radius, but that is incorrect; it is a diameter, not a radius. Figure 3.12 shows a beam transmitting through atmosphere that has turbulence variations.

When looking through turbulence, the effects are not symmetric. If you image from the ground looking up through turbulence, the angular deflections will be large, but while looking down, the turbulence will have little effect.

Lastly, we should mention the frozen-flow atmospheric turbulence assumption, which states that atmospheric turbulence does not change but simply moves from one place to another at a certain velocity; the movement can be due to wind or vehicle motion. The higher the velocity the faster the atmospheric turbulence changes at a given location. The atmospheric coherence time at a given location is given by

$$\tau_0 = \frac{r_0}{v}, \qquad (3.20)$$

where v is the velocity of the atmosphere past a certain location.

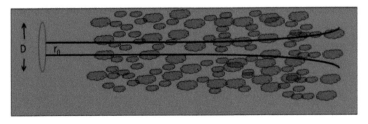

Figure 3.12 A representation of a beam transmitting through atmospheric turbulence (reprinted from Ref. 1).

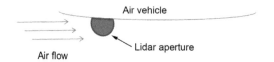

Figure 3.13   Diagram of aero-optical air flow (reprinted from Ref. 1).

3.5.3 Aero-optical effects on LiDAR

When a LiDAR is on an air vehicle, there may be a rapid variation in the index of refraction of air near the surface of the vehicle due to the aero-optical effect. This is especially true when the LiDAR aperture juts out into the wind stream, as shown in Fig. 3.13. Often a gimbal is used to point a LiDAR. In this case, a round half-ball juts into the wind stream, creating additional turbulence. Even a smooth surface will create some turbulence. Recall the discussion about a frozen atmosphere, and how coherence time varies with the velocity of the wind. In this case, the important velocity is the velocity of the air vehicle traveling through the air, which can be much faster than the wind velocity. It is not surprising that aero-optical effects contain much higher frequency variations than normal turbulence. Obviously, the frozen atmosphere assumption is valid until the air vehicle disturbs the air flying through it, creating turbulence. For the aero-optical effect, we can have velocities from 50 m/s up to near Mach 1, say 300 m/s, and some vehicles can even move at supersonic speeds, creating a significant disturbance in the air.

When sensing on receive through an aero-optical layer, the refractive index variations are close to the air vehicle, so they are essentially in the pupil plane of the sensor. However, when emitting a laser beam to illuminate an object, the beam goes through the aero-optical turbulent layer immediately after being emitted, so the aero-optical turbulent layer is positioned to cause a significant impact on the illuminator beam. Fortunately, for many LiDARs, the illuminator beam can be fairly large, simultaneously illuminating a large area to be imaged by an array of detectors. This can mitigate the effects on the outgoing beam because for flash illumination we can illuminate a larger area compared to a diffraction-limited angular region.

3.5.4 Extended (deep) turbulence

It is relatively easy to compensate for turbulence in the pupil plane of a sensor. Unfortunately, often turbulence is distributed between the target and the sensor, especially if the path is horizontal. Turbulence in the pupil plane creates a phase change, and an adaptive optics mirror also creates a phase in the pupil plane. Therefore, an adaptive optics mirror with opposite phase change will add to the pupil plane atmospheric phase change to cause zero effective phase change. When the turbulence is spread throughout the laser path length, phase-only compensation does not completely correct for the turbulence because, as light travels from the far field to the pupil plane, some

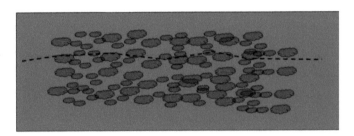

Figure 3.14 Diagram of extended (deep) turbulence (reprinted from Ref. 1).

phase information is converted to amplitude information. In theory, if we had an adaptive optics aperture that could impose plus and minus phase and amplitude variations, we could exactly compensate for extended, or deep, turbulence (Fig. 3.14). If the only amplitude variation that we can impose is a reduction in amplitude, this is a loss. So, a deformable mirror (DM), even with amplitude reduction added, cannot completely compensate for deep turbulence. While a DM cannot compensate for amplitude variations in the pupil plane, a number of people have shown that the major effect on an image is the phase, not the amplitude. Still, phase correction alone is not a complete correction.

A second way to compensate for deep turbulence is to use a computer. The nice thing about using a computer is that you can insert a virtual phase screen at multiple locations in the path to compensate for the real turbulence that occurs at multiple locations.

If the turbulence is not too strong, the lack of precision in compensating for extended turbulence is not a significant issue; therefore, for mild extended turbulence, adaptive optics will compensate for the atmosphere.

3.5.5 Speckle

When you illuminate an object with a narrowband laser beam, you see light and dark areas, which are called speckle. Speckle is interference that occurs due to scattering from one part of an object interfering with scattering from another part of the object (Fig. 3.15). We do not see speckle when sunlight illuminates an object because there are many different wavelengths in the illuminating light, and the interference from each wavelength averages out when added to all of the other wavelengths. If we use a laser with a broad spectral band, we can eliminate speckle in a direct-detection laser for the same reason that sunlight does not have speckle.

To obtain a new speckle realization, a 2π phase difference between the two wavelengths illuminating the target (with a certain surface roughness) is assumed:

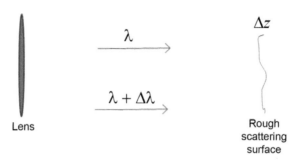

Figure 3.15 The effect of surface roughness on speckle (reprinted from Ref. 1).

$$\Delta\lambda = \frac{\lambda^2}{\Delta z - \lambda}. \qquad (3.21)$$

For just two surface locations, Eq. (3.21) shows the required change in wavelength to obtain a different speckle realization. Speckle depends on surface roughness and coherence length. The coherence length must be longer than the depth of the surface roughness in order to have speckle. The second way to average out speckle is to have enough movement such that subsequent laser pulses provide angular diversity in view angle, which creates different interference and therefore different speckle.

The size of an illuminated object or the area being illuminated—whichever is smaller—determines the extent of speckle lobes in the same way that the size of an aperture determines the diffraction limit. This is diffraction of the scattered return, just like the diffraction of an emitted beam. We can only have interference over the illuminated area of an object. This means that we can tell how big an object is by measuring the angular extent of speckle lobes. Being able to measure the size of an object based on the smallest speckle returned from an object is one of the ways speckle can be used in a LiDAR sensor. The smallest area of interference is set by the resolution of the sensor and is limited either by sampling or by diffraction from the aperture of the sensor.

The isoplanatic patch is the physical area at any location over which the speckle patterns are the same. The isoplanatic angle is the angular region over which the speckle patterns are the same. These two terms will appear in further discussions of LiDAR technologies, so their meanings should be understood.

Problems and Solutions

3-1. Assume a 10-cm–diameter transmitter beam at a wavelength of 1550 nm. Assume a diffraction-limited beam. Calculate the angular beam divergence and the effective antenna gain, where antenna gain is defined as the ratio of the solid angle covered by the beam to 4π sr.

3-1 Solution: Use $\vartheta \approx \lambda/D$ with the parameters chosen.
Generally, the solid angle is

$$\Omega \approx \frac{A}{R^2},$$

where A is the illuminated area, and R is the distance to the origin. The solid angle is calculated as

$$\Omega \approx \iint \sin\theta d\theta d\phi,$$

where θ is the divergence angle, which can be calculated as

$$\theta \approx \frac{\lambda}{D},$$

where D is the transmitter diameter, and λ is the wavelength. Therefore, performing the integral, the solid angle becomes

$$\Omega = \int_0^{2\pi} d\phi \int_\theta^0 \sin\theta d\theta = 2\pi(1 - \cos\vartheta) \rightarrow \Omega = 2\pi(1 - \cos\theta).$$

We can approximate $\cos\theta$ with the Taylor series as

$$\cos\theta = 1 - \frac{\theta^2}{2}.$$

Therefore, the solid angle approximately equals

$$\Omega = 2\pi(1 - \cos\vartheta) = 2\pi\left(1 - 1 + \frac{\theta^2}{2}\right) = 2\pi \times \frac{\theta^2}{2} = \pi\theta^2,$$

where $\theta \approx \lambda/D$. Therefore, we have

$$\Omega = 2\pi(1 - \cos\theta) = \pi\theta^2 = \pi\left(\frac{\lambda}{D}\right)^2.$$

Thus, the illumination area is

$$A = \Omega R^2 = \pi\left(\frac{\lambda}{D}\right)^2 R^2 = \pi\left(\frac{R\lambda}{D}\right)^2.$$

The beam divergence is 15.5 μrad:

$$\vartheta \approx \frac{\lambda}{D} = \frac{1.550 \times 10^{-6}\,\text{m}}{0.1\,\text{m}} = 15.5\,\mu\text{rad}.$$

$$\Omega = 2\pi(1 - \cos\vartheta) = 2\pi[1 - \cos(15.5\,\mu\text{rad})] = 7.548 \times 10^{-10}\,\text{sr}.$$

$$Gain = \frac{4\pi}{\Omega} = \frac{4\pi}{7.548E - 10} = 1.665 \times 10^{10}.$$

The antenna gain is 1.665×10^{10}.

3-2. Assume a range of 1000 m and a square DAS of 0.1 mrad × 0.1 mrad. Assume that we have returns from three different ranges within the pixel. The first return is from 10% of the pixel area and has a reflectivity of 30%. The second return is from 10% of the area and has a reflectivity of 25%. The last return is from the remaining area and has a reflectivity of 10%. Assume that there is no shadowing due to light going in and out at an angle. What is the cross-section at each range?

3-2 Solution: The GSD will be 10 cm × 10 cm. We get this by multiplying range by DAS. For the first range return, we multiple 0.01 sq. m by 0.1 and 0.3, obtaining 0.0003 sq. m. For the second range return, we multiple 0.01 sq. m by 0.1 and 0.25, resulting in 0.00025 sq m. For the last range return, we have $0.01 \times 0.8 \times 0.1 = 0.0008$.

3-3. Assume the same 1000-m range and a square DAS of 0.1 mrad × 0.1 mrad. We have a highly reflecting wire passing through the pixel. The wire has a reflectivity of 60%. It is long in one dimension and 1 cm in the other dimension. What is its cross-section? If we decrease the range to 500 m, what is its cross-section?

3-3 Solution: Each side of the GSD is 10 cm for the 1000-m range. The cross-section at 1000 m is therefore $0.1 \times 0.01 \times 0.6 = 0.0006$ sq m. At 500-m range, the width of the wire stays the same, but the GSD doubles to 20 cm × 20 cm. Therefore, the cross-section is now twice as large: $0.2 \times 0.01 \times 0.6 = 0.0012$ sq m.

The important point of this problem is that the wire is linear with respect to the pixel. Its width is smaller than the pixel and does not depend on range.

3-4. For a trihedral corner cube that has 95% reflectivity and area of 1 cm × 1 cm, and reflects 1550-nm–wavelength light, calculate the beam spread of the return beam in angle and the effective cross-sectional area.

3-4 Solution: The beam spread is determined using Eq. (3.1) to be 155 μrad. Equation (3.6) can be used for the cross-section:

$$\sigma_t \approx \frac{4a^4}{\lambda^2} = \frac{4(0.95)\,(1\,\text{cm})^4}{(1550\,\text{nm})^2} = 15816.85\,\text{m}^2.$$

3-5. Calculate the required energy per pulse for a 128×32 detector array, 10% reflectivity, 0.8 received photons per detector, a visibility of 10 km, a receive aperture size of 10 cm diameter, and a range of 2000 m.

3-5 Solution: 0.035 mj/pulse, using the appropriate numbers in the spreadsheet below. The equation in the last term is

$4.4 \times (Nhc/\lambda) \times$ (# detectors in one dimension $\times$ # in other direction/ reflectivity) $\times (R/(D_r)^2 \times (1/Qe) \times$ (1/two-way transmission).

Number of photons received from a 10% reflective object

Human visibility (m)	$\beta_{0.55}$	$\beta_{1.55}$	Imaging range	One-way transmission	Two-way transmission	Design # of photons	Receiver aperture diameter (mm)	Required transmitter energy (mj) for 10% reflectivity
5000	0.0006	0.0005	15	99.2%	98.4%	20	50	0.0003
5000	0.0006	0.0005	30	98.4%	96.9%	20	50	0.0011
5000	0.0006	0.0005	45	97.7%	95.4%	20	50	0.0026
5000	0.0006	0.0005	60	96.9%	93.9%	20	50	0.0047
5000	0.0006	0.0005	75	96.1%	92.4%	20	50	0.0075
5000	0.0006	0.0005	90	95.4%	91.0%	20	50	0.0109
5000	0.0006	0.0005	105	94.6%	89.6%	20	50	0.0151
5000	0.0006	0.0005	120	93.9%	88.2%	20	50	0.0201
5000	0.0006	0.0005	135	93.2%	86.8%	20	50	0.0258
5000	0.0006	0.0005	150	92.4%	85.4%	20	50	0.0323
5000	0.0006	0.0005	165	91.7%	84.1%	20	50	0.0398
5000	0.0006	0.0005	180	91.0%	82.8%	20	50	0.0481
5000	0.0006	0.0005	195	90.3%	81.5%	20	50	0.0573
5000	0.0006	0.0005	200	90.0%	81.0%	75	50	0.2273
5000	0.0006	0.0005	210	89.6%	80.2%	20	50	0.0675
5000	0.0006	0.0005	225	88.8%	78.9%	20	50	0.0787
5000	0.0006	0.0005	240	88.2%	77.7%	20	50	0.091
5000	0.0006	0.0005	255	87.5%	76.5%	20	50	0.1044
5000	0.0006	0.0005	270	86.8%	75.3%	20	50	0.1189
5000	0.0006	0.0005	285	86.1%	74.1%	20	50	0.1346
5000	0.0006	0.0005	300	85.4%	73.0%	20	50	0.1515

3-6. For Rayleigh scattering

a) What is the scattering coefficient that is proportional to the wavelength? What power is used with the wavelength?

b) Why is Rayleigh scattering responsible for blue sky?

3-6 Solution:

a) Rayleigh scattering is caused by air and haze molecules that are small compared to the wavelength of light. The scattering coefficient is proportional to λ^{-4}, which describes the Rayleigh law.

b) At $\lambda < 1$ μm, Rayleigh scattering produces the blue color of the sky because blue light is scattered much more than red light. Rayleigh scattering is caused by air molecules and haze that are smaller than the wavelength of the radiation.

3-7. For Mie scattering

a) What is Mie scattering?
b) Why is the sunset red?

3-7 Solution:

a) Mie scattering is the scattering of particles comparable in size to the wavelength of the light being scattered.

b) Mie scattering losses decrease rapidly with increasing wavelength, eventually approaching the Rayleigh scattering case. That is why Mie scattering is responsible for sunsets appearing red.

3-8. If 25 photons per pulse are required and we are using a 128 × 128 detector array, the receiver aperture radius is 75 mm, visibility is 3 km, and reflectivity is 10% from a large-area target, plot the required energy per pulse versus the range.

3-8 Solution:

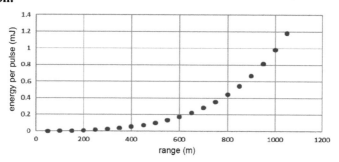

3-9. Use the data provided in the last problem to answer the following:

a) For a systems efficiency of 40%, how much energy per pulse is required from the transmitter at a range of 1050 m?

b) If the maximum energy that may be provided per pulse is 1 mJ, how much does the range need to be decreased?

3-9 Solution: The following spreadsheet is used to calculate required energy versus range.

a) 7.1 mJ/pulse
b) The range should be decreased from 1050 m to 600 m.

| | | | | | | | Required |
Human visibility (m)	$\beta_{.55}$	$\beta_{1.55}$	Imaging range	One-way transmission	Two-way transmission	Design # of photons	Receiver aperture diameter (mm)	transmitter energy (mj) for 10% reflectivity
3000	0.0010	0.0009	50	95.7%	91.6%	25	150	0.0005
3000	0.0010	0.0009	100	91.6%	83.9%	25	150	0.002
3000	0.0010	0.0009	150	87.7%	76.9%	25	150	0.005
3000	0.0010	0.0009	200	83.9%	70.4%	25	150	0.0097
3000	0.0010	0.0009	250	80.3%	64.5%	25	150	0.0165
3000	0.0010	0.0009	300	76.9%	59.1%	25	150	0.026
3000	0.0010	0.0009	350	73.6%	54.2%	25	150	0.0386
3000	0.0010	0.0009	400	70.4%	49.6%	25	150	0.055
3000	0.0010	0.0009	450	67.4%	45.5%	25	150	0.076
3000	0.0010	0.0009	500	64.5%	41.7%	25	150	0.1024
3000	0.0010	0.0009	550	61.8%	38.2%	25	150	0.1352
3000	0.0010	0.0009	600	59.1%	35.0%	25	150	0.1756
3000	0.0010	0.0009	650	56.6%	32.0%	25	150	0.225
3000	0.0010	0.0009	700	54.2%	29.3%	25	150	0.2848
3000	0.0010	0.0009	750	51.8%	26.9%	25	150	0.3568
3000	0.0010	0.0009	800	49.6%	24.6%	25	150	0.4431
3000	0.0010	0.0009	850	47.5%	22.6%	25	150	0.5461
3000	0.0010	0.0009	900	45.5%	20.7%	25	150	0.6682
3000	0.0010	0.0009	950	43.5%	18.9%	25	150	0.8127
3000	0.0010	0.0009	1000	41.7%	17.4%	25	150	0.9829
3000	0.0010	0.0009	1050	39.9%	15.9%	25	150	1.1828

Number of photons received from a 10% reflective object

3-10. What is the diffraction-limited cross-range of a pixel for a LiDAR at a range of 500 m, using 1550-nm wavelength, a circular aperture, and a receiver area of 15 cm^2?

3-10 Solution:

$$A_{\rm rec} = \pi\left(\frac{D_{\rm rec}}{2}\right)^2, \quad D_{\rm rec} = 2\sqrt{\pi A_{\rm rec}} = 2\sqrt{\pi(15\,{\rm cm}^2)} = 13.73\,{\rm cm},$$

$$d \geq \frac{R\lambda}{D_{\rm rec}} = \frac{(500\,{\rm m})(1500\,{\rm nm})}{13.73\,{\rm cm}} = 56.45\,\mu{\rm m}.$$

3-11. What is the relative power received in a given DAS of 0.1 mrad × 0.1 mrad, using a range of 1000 m, a 128 × 128 receiver array, reflectivity of 0.3, receiver area of 15 cm^2, and wavelength of 1550 nm? Do not consider system efficiency.

3-11 Solution:

$$d^2 = (DAS)^2 \times R^2 = (0.1\,{\rm mrad})^2(1000\,{\rm m})^2 = 0.001\,{\rm m}^2,$$

$$\sigma = \rho_t A_p = \rho_t d^2 = (0.001\,{\rm m}^2)(0.3) = 0.003\,{\rm m}^2.$$

Since we're covering 128 × 128 detectors in an array,

$$\sigma = 128^2(0.003\,\text{m}^2) = 49.152\,\text{m}^2,$$

$$A_{\text{rec}} = \pi\left(\frac{D_{\text{rec}}}{2}\right)^2,$$

$$D_{\text{rec}} = 2\sqrt{\pi A_{\text{rec}}} = 2\sqrt{\pi(15\,\text{cm}^2)} = 13.73\,\text{cm},$$

$$A_{\text{illum}} \approx \pi\left(\frac{\lambda R}{2D}\right)^2 = \pi\left[\frac{(1550\,\text{nm})(1000\,\text{m})}{2(13.73)\,\text{cm}}\right]^2 = 355.7\,\mu\text{m},$$

$$\eta_{\text{atm}} = e^{-\beta R} = e^{-0.877} = 0.4160,$$

$$\frac{P_T}{P_R} = \frac{\pi A_{\text{illum}}R^2}{\sigma A_{\text{rec}}\eta_{\text{atm}}^2} = \frac{\pi(355.7\,\mu\text{m}^2)(1000\,\text{m})^2}{(0.0003\,\text{m}^2)(15\,\text{cm}^2)(0.4160)^2} = 875.82.$$

In this scenario, 143.49 million times more power is transmitted than received!

3-12. Using the information in Problem 3-11, plot relative power versus range over 1000 m. At what range will the power transmitted be less than 100 times greater than the power received?

3-12 Solution:

```
aill = 355.7*1e-6;
cs = 49.152;
atm = 0.416;
arec = 0.15;
range = linspace(1,1000,1000);
p = pi*aill*range.^2/(cs*arec*atm^2);
plot(range,1+p)
```

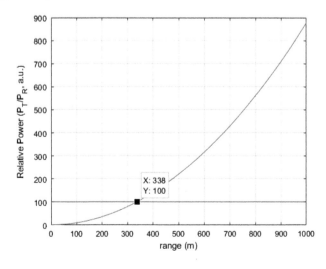

3-13. Given a visibility of 7 km, defined using the U.S. standard of $\lambda_0 = 0.55$ μm, and 2% contrast, calculate the atmospheric attenuation at a range of 3 km and a wavelength of 1064 nm.

3-13 Solution:

$$q(\lambda) = 0.1428(1.064) - 0.0947 = 0.0572,$$

$$\beta(\lambda) \approx \left(\frac{\lambda_0}{\lambda}\right)^q = \left(\frac{0.55}{1.064}\right)^{0.0572} = 0.9630,$$

$$\eta_{\text{atm}} = e^{-\beta R} = e^{-(0.9630)(3)} = 0.556.$$

3-14. What is the surface roughness required for a new speckle realization with a wavelength of 1064 nm and a bandwidth of 0.25 nm?

3-14 Solution:

$$\Delta\lambda = \frac{\lambda^2}{\Delta z - \lambda},$$

$$\Delta z - \lambda = \frac{\lambda^2}{\Delta\lambda},$$

$$\Delta z = \frac{\lambda^2}{\Delta\lambda} + \lambda = \frac{(1064\,\text{nm})^2}{(0.25\,\text{nm})^2} + 1064\,\text{nm} = 4.5\,\text{mm}.$$

This surface roughness would be large enough to visually see with aid (however, it falls into the near-IR spectrum).

Notes and References

1. P. McManamon, *Field Guide to Lidar*, SPIE Press, Bellingham, Washington (2015) [doi: 10.1117/3.2186106].
2. I. I. Kim, B. McArthur, and E. J. Korevaar, "Comparison of laser beam propagation at 785 nm and 1550 nm in fog and haze for optical wireless communications," *Proc. SPIE* **4214**, 26 (2001) [doi: 10.1117/12.417512].
3. M. Ijaz, Z. Ghassemlooy, J. Pesek, O. Fiser, H. L. Minh, and E. Bentley, "Modeling of fog and smoke attenuation in free space optical communications link under controlled laboratory conditions," *Journal of Lightwave Technology* **31**(11), 1720–1726 (2013).
4. LEEDR transmission plots courtesy of the Air Force Institute of Technology and LEEDR technical developer Dr. Steven T. Fiorino.

Chapter 4
Types of LiDAR

As stated in the introduction, LiDAR has a large number of available phenomenologies. There are therefore many types of LiDAR. In this section we briefly discuss a variety of LiDAR types. A traditional method of sorting LiDARs has been direct detection compared to coherent detection. Coherent detection uses a local oscillator and can measure both phase and intensity, whereas direct detection only measures the intensity of the returned signal because no detector can respond to a frequency as high as that of optical signals. This book uses the direct detection versus coherent sorting method but then further divides coherent detection into real beam versus aperture-synthesis–based LiDAR. Direct-detection LiDAR is always restricted to the real beam because, in order to synthesize an aperture larger than the real aperture, it is necessary to capture both phase and amplitude. There are techniques in which the phase is estimated without measuring it;[1] in this case, we could synthesize an aperture larger than the real aperture without actually measuring phase. Aperture synthesis is included the coherent LiDAR section.

4.1 Direct-Detection LiDAR

1D, 2D, or 3D LiDAR can also be made in a coherent embodiment, but these LiDAR types usually use direct detection because phase information is not needed for these LiDAR modalities. The only major exception to this rule is when very high range resolution is required. A technique that allows higher range resolution using a coherent LiDAR is available in which the local oscillator and the signal can both be chirped, allowing a lower-bandwidth detector to act in the LiDAR system like a high-bandwidth detector. Most discussion of 1D, 2D, and 3D LiDAR in this book are for the direct-detection LiDAR embodiments.

4.1.1 1D range-only LiDAR

A 1D image is a range profile of an object, with the only dimension being range. The range profile does not depend on the size of the receiving optical apertures, except for signal-to-noise considerations, so it can be used at long range if sufficient signal is available. This can be very useful in environments that are not cluttered and for objects that are far away, where the transverse dimensions of the object are not resolved by an optical system. A range profile gives the size of an object in one dimension and the distribution of its scattering centers along the orientation of that particular range profile.

A high-range-resolution 1D profile is most useful when the orientation of the object is known. For example, one generally assumes that an airplane is oriented such that the nose is pointed in the direction that the plane is traveling. Starting from the time when the detector is first triggered, the dimensions of the rest of the airplane along the vector of the range profile can be calculated by the relative time delay. Figure 4.1 shows the range profile of an airplane viewed in 1D from the front, although the image is shown from the side. A recent paper showing multiple 1D range images was given at the 2018 Defense and Commercial Sensing conference in Orlando.[2]

4.1.2 Tomographic imaging LiDAR

If we have multiple high-range-resolution 1D images from different directions, we can develop cross-range information as well as range information. The wider the angular distribution of the high-range-resolution images the better

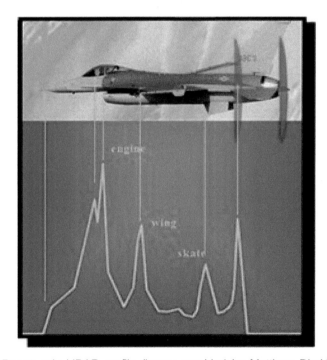

Figure 4.1 Range-only LiDAR profile (image provided by Matthew Dierking, Air Force Research Lab).

the cross-range imaging information. This is similar, but opposite, to the idea of stereo vision, where angle resolution is measured, but range information is inferred. In tomographic imaging, range information is measured from multiple angles, making it possible to infer cross-range information in the plane of the angle measurements. A time-dependent return signal can be collected by a single detector to provide a high-range-resolution image. This provides a 1D slice of the 3D spatial profile of the object. When the object rotates, or the LiDAR moves, different range slices of the 3D profile can be obtained. If a sufficient number of 1D slices is collected, a 2D range/angle image can be reconstructed in the plane of the various slices. The eye sees the azimuth/elevation in two dimensions, but in this case, we would have a range/azimuth image in dimensions similar to those that occur with a synthetic-aperture LiDAR, as is discussed later in this chapter. Various algorithmic approaches can be used to combine multiple range profiles to make a 2D range/angle image from multiple 1D range profile images taken at different angles. The more complex the object, and the more cluttered the background, the greater the need to obtain range profile information over wider angles and possibly with finer angular sampling before the object will be recognized. Figure 4.2 shows an airplane being viewed from multiple angular locations. This type of angular diversity can be used to develop a range/cross-range image.

When a coherent detection system is employed and data are collected from multiple views, an object's Doppler spectrum can be used to assist in determining the angle from which a given 1D image has been collected. Also, with Doppler techniques, the rotation speed of an object can be precisely measured. If the object rotates or the sensor moves, different range projections are formed, so knowing the exact rotation speed can make it easier to generate tomographic images from multiple 1D images. The Doppler shift can also provide other discriminants, such as vibration characteristics. This will be further discussed in Section 4.2 on coherent LiDAR.

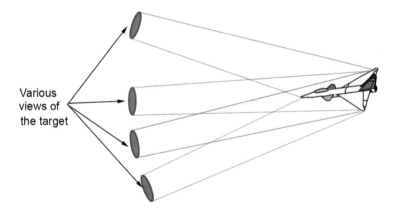

Various views of the target

Figure 4.2 Multiple range-only images create a 2D image (reprinted from Ref. 39).

4.1.3 Range-gated active imaging (2D LiDAR)

Range-gated active imaging (2D LiDAR) is similar to passive imaging, but with illumination. The view is similar to what is generated by the eye, which also is a 2D sensor. For active gated imagers, the receiver is only open when the gate is open, so noise can only enter during that period. Active gated imagers increase SNR by decreasing the noise allowed into the receiver.

One benefit of 2D LiDAR compared to passive sensing is that the camera can see at night while using a shorter-wavelength sensor. To gather sufficient signal at night with a passive sensor, engineers use MWIR or LWIR thermal imagers. We can use 1.06- or 1.55-μm laser radiation for illumination. This means that the diffraction-limited resolution is better than that of passive sensing, as indicated in Eq. (3.1). With better diffraction-limited resolution, a design engineer is free to choose a longer focal length, with a detector angular subtense (DAS) that matches the improved diffraction-limited angular resolution.

Also, since you bring your own illumination, there are no thermal crossover issues like you have with thermal imagers. At night, passive imagers tend to use MWIR (3- to 5-μm) wavelengths or LWIR (8- to 12-μm) wavelengths, whereas a 2D gated imager will probably be at a wavelength of 1 or 1.5 mm. The choice between 1 and 1.5 μm for the 2D gated LiDAR is determined primarily based on eye safety considerations. To enhance SNR, a gated 2D camera is preferred, which will only gather noise over a short time period, as compared to continuously gathering noise. This type of LiDAR will likely frame at low repetition rates, consistent with the repetition rates of framing cameras; 10–30 Hz would be a typical repetition rate. While nanosecond-class laser pulses are not a requirement, it is likely that Q-switched lasers will be used, resulting in 5- to 15-ns pulse widths. While the pulse widths are likely to be nanosecond class, the detector bandwidth does not have to be that fast because we are not accurately measuring range. Framing cameras at 30 or 60 Hz can be used, but with the addition of gating circuitry, which might be microsecond class. This means that the gate is open to receive signal for some time value, such as one microsecond. Gated 2D LiDAR has a simpler readout circuitry than a 3D imaging LiDAR camera.

Another benefit of gated 2D LiDAR imaging compared to passive imaging is the ability to gate out obscurants that occur before the object of interest, as well as clutter that occurs before or after the obscurants. This concept is shown in Fig. 4.3. You can, for example, have a range-gated receiver that is closed when return from the fog, or foreground clutter, hits it and only opens around the approximate target range. Gating out backscatter from ranges that are not of interest also occurs for 3D LiDARs. Figure 4.4 shows a diagram of the most popular 2D gated SWIR imaging camera, Intevac's LIVAR® M506.

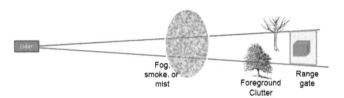

Figure 4.3 Diagram of 2D LiDAR range-gating out backscatter that occurs at ranges other than that of the object of interest (reprinted from Ref. 39).

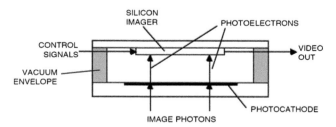

Figure 4.4 Diagram of Intevac's 2D gated camera system (reprinted from Ref. 3).

This form of imaging started with an electron-bombarded active-pixel sensor (EBAPS) based on a GaAs or InGaAs photocathode in proximity focus with a high-resolution, CMOS active-pixel sensor (APS) imager anode. The SWIR version of this thermoelectrically cooled sensor has a spectral response that ranges from 950 to 1650 nm. Electrons emitted by the photocathode are directly injected in the electron-bombarded mode into the anode. Low noise gain is achieved in the CMOS anode by converting the high-energy photoelectrons (resulting from 1- to 2-keV bias applied between the photocathode and CMOS anode) to electron–hole pairs in the anode via the electron-bombarded semiconductor (EBS) gain process. The electrons are collected in the anode pixel and subsequently read out. The EBS gain process is inherently low noise with an excess noise factor of less than 1.1.[3,4]

A second, more-recent gated imaging system performs both gated 2D active functions and gated 2D passive functions in two spectral bands. It uses an MWIR HgCdTe electron-initiated avalanche photodiode (e-APD) FPA[5] with 640 × 480 pixel format and 25-μm–pixel pitch. The MWIR e-APD FPA is fabricated using the HgCdTe technology described above. In the MWIR band, the noise-equivalent temperature difference (NETD) at gain is ~18 mK (with *f*/4 optics and 16-ms integration), while in the SWIR band used for gated imaging, the noise-equivalent photons (NEPh) is ~2 photons for an integration time of 100 ns.

4.1.4 3D scanning LiDAR

Scanned LiDAR is one approach for collecting 3D images. This used to be the main approach until 2D high-bandwidth detector arrays became available.

Over time, we have moved away from scanned 3D LiDAR, although for low-cost applications such as auto LiDAR, it will still be the most prevalent approach. A single detector, or a limited number of individual detectors, is scanned at a rapid rate to make a 3D image. Each detector has a sufficiently high bandwidth to measure the time of the return laser pulse accurately enough to provide range information to the precision desired. The need for a high-bandwidth detector response and readout in 3D imaging is the reason that scanning LiDAR has been popular, especially for commercial applications where cost is a significant issue. Detectors for scanning LiDARs can be obtained more easily and with less cost than for flash 3D LiDARs, but they require higher-speed scanners and improved stabilization. A scanning LiDAR must be stabilized to a small portion of an individual DAS, and the scanner has to move, or scan, a small number of detectors over the full image space to create each frame. A scanning LiDAR is similar to a TV raster scan, but it creates a 3D image.

A couple of significant, current, scanning LiDAR applications are 3D mapping and driverless cars, which are two of the largest applications of commercial LiDAR. Figure 4.5 shows one implementation of a scanning LiDAR, in which a limited number of detectors are scanned in azimuth and then dropped down in elevation and scanned again. This scanning approach must scan quickly in azimuth, and then scan slower in elevation. Unless the elevation scan is a step in elevation angle, the fast azimuth scan will be at a slant angle across the horizontal FOV, starting at one elevation and ending at a lower or higher elevation. Most auto LiDARs have a 1D detector array that covers the full required elevation, so they do not need to scan in elevation. This eliminates the need to scan in both directions.

3D scanning LiDARs have all of the same benefits as 2D LiDARs, but in addition provide 3D shape information. 3D shape is a great discriminant for recognizing objects because shape does not change as much as the illumination features associated with most 2D imagery. 3D mapping is becoming widespread as one commercial application of LiDAR because it is possible to map a downtown area, or an area with flood issues, in three dimensions. Figure 4.6 shows a 3D image of lower Manhattan, including the

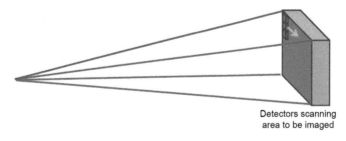

Detectors scanning
area to be imaged

Figure 4.5 Scanning LiDAR representation (reprinted from Ref. 39).

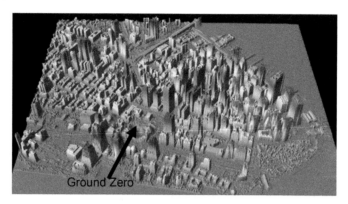

Figure 4.6 A 3D image of lower Manhattan taken using an Optech scanning LiDAR (image derived from Teledyne Technologies ALTM™ data with permission).

rubble pile resulting from the collapse of the twin towers on Sept. 11, 2001. This image was taken because of concern at the time of a potential shift in the rubble; rescue personnel were very concerned about the safety of personnel who had to walk on the rubble pile to rescue victims.

4.1.5 3D flash imaging

Flash imagers use a large array of detectors (e.g., a 128×128 pixel array) to develop the image. One of the first methods of large-array flash imaging used polarization change to measure range.[6] This was done because at the time FPAs with large format were not available with high-bandwidth detectors. More recently, high-bandwidth FPAs with appropriate ROICs have been developed. For flash imaging, the return laser power is imaged onto an array of detectors; therefore, we do not need to scan as fast, saving size, weight, and power. Also, for flash imagers, we only have to stabilize to a fraction of the angular size of the array, rather than to a fraction of the size of one DAS. This reduction in required stabilization precision can be a significant systems benefit, whereas the cost of large-format FPAs can be a disadvantage. For flash imagers, the energy returning to the receiver is divided among a large number of detectors. An alternative way to look at this is that the cross-section of the target in each pixel is reduced for an area target that is divided into more pixels. Thus, imaging of targets using flash imaging in angle/angle space requires high peak-illumination power and/or very sensitive detectors compared to single-pixel cameras, or cameras with a limited number of detectors.

One method to increase sensitivity of a direct-detection receiver is to amplify the return signal so that the signal level is above the noise sources introduced downstream from the initial detection. Avalanche photodiodes (APDs) are often used to increase sensitivity. Linear-mode APD detectors have a gain that provides a linear relationship between the number of photons received and the amplitude of the signal output from the detector. Geiger-mode APDs (GMAPDs) have a maximum avalanche amplitude regardless of the number of photons input. These two gain approaches will be further discussed in Chapter 6. Fiber pre-amplification before detection is also possible but to date has not been used with arrays of LiDAR detectors, although it is used with single detectors to increase sensitivity. Figure 4.7 shows a flash imaging system that illuminates an area and then overlays an imaging array on that area. This is meant to depict a flash imager; in reality, the array is in the receiver, but it is correct that the area viewed by the array is illuminated by a single laser "pulse" or sample time.

Polarization-intensity-encoded imaging addresses a different issue. Instead of having high-bandwidth readout circuits, intensity-encoded imagery uses polarization rotation to measure range, which means that simple framing cameras can be used to do 3D flash imaging. This is further discussed in Chapter 6. The disadvantage of this type of imaging is the requirement for a Pockels cell to rotate the polarization. Pockels cells can require high voltage and can have a narrow field of view.

4.1.6 3D mapping applications

3D mapping systems have three variants: linear-mode scanning LiDAR, Geiger-mode LiDAR (GML), and single-photon LiDAR (SPL). The design and comparative performance of these LiDAR systems are discussed in Chapter 11. Efficient 3D mapping requires careful consideration of various factors affecting the data collection and quality of data. This includes the local topography, vegetation, scene content, data accuracy requirements, spatial data reference information, desired acquisition window, weather conditions, unique conditions/restrictions, ground survey data acquisition, flight plans, tiling scheme, output products, and application of rigorous quality control measures throughout the process. Mission, flight, and ground control plans are typically generated to support LiDAR data acquisition, and some of the collection parameters are used as metadata for processing the acquired raw data. Quality checks need to be performed preflight, inflight, and postflight to

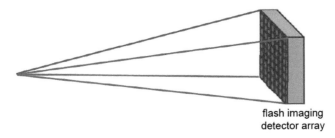

flash imaging
detector array

Figure 4.7 Flash imaging LiDAR, showing an area being viewed rather than a single pixel being viewed (reprinted from Ref. 39).

ensure that collected data pass quality requirements [i.e., no voids, cloud obstructions, shadows, or edge mismatch; point precision; and quality of data (e.g. precision dilution of precision (DOP) of the Global Positioning System (GPS) / Global Navigation Satellite System (GNSS)].

To meet the data accuracy for standard elevation map products as defined in the U.S. Geological Survey's (USGS) *LiDAR Base Specifications* (version 1.3) and the American Society for Photogrammetry Engineering and Remote Sensing's "ASPRS Accuracy Standards for Digital Geospatial Data" (version 1.0), collection of the following ancillary data is required: (1) a GPS/GNSS data network providing data at ≤1-s epochs and operating at a specified distance from the aircraft; (2) establishment and collection of ground control point (GCP) data to improve the positional accuracy of the data products; and (3) establishment and collection of error check points (ECPs) in the project area for ground truthing and assessing the quality (i.e., accuracy) of the data products.[7,20]

4.1.6.1 GNSS position data collection

A network of ground-based GPS/GNSS base stations within the collection area will be required to provide positional accuracy for the aircraft and to ensure that the vertical and horizontal accuracy of the data products can be determined. A temporary GNSS data network should be established if no Continuously Operating Reference Station (CORS) networks are available in the collection area of interest. A network of GPS/GNSS base stations with 100% overlapping coverage within the project area will be required to achieve uniform, high vertical and horizontal data product accuracies. In the United States, the National Geodetic Survey (NGS) agency within the National Oceanic and Atmospheric Administration (NOAA) operates and maintains a nationwide network of ground-based CORSs, providing continuous positional data for LiDAR processing.[8] In locations where a CORS network is not available or there are gaps in coverage, a temporary base-station network must be established.

4.1.6.2 Ground surveys

Well-distributed, GPS-surveyed ground control points are established, and the *x, y, z* positional data of the points are collected to supplement the onboard, platform GPS/GNSS position measurements. Field procedures consistent with the U.S. National Geodetic Survey's "Guidelines for Real Time GNSS Networks" should be followed for the ground surveys.[8] These procedures include making redundant occupations under different satellite configurations and field conditions for each point. The points must be distributed throughout the project area and must be LiDAR identifiable (LID). The points should be in open and flat terrain where there is a high probability that the sensor will have detected the ground surface without interference from surrounding vegetation and/or objects. LID points must be visible in the LiDAR data for the x, y, and z axes. Point accuracy should satisfy a local network accuracy of 5 cm or better at a 95% confidence level both horizontally and vertically to meet the 10-cm overall product accuracy for USGS QL1-2 products. Accuracy should be tested using the National Standard for Spatial Data Accuracy (NSSDA) guidelines.[9]

The GCPs are used to facilitate calibration of the LiDAR data points and perform position adjustments (i.e., the airborne LiDAR time of flight and the associated metadata recorded on board are processed and re-projected to an earth-referenced coordinate system and adjusted in x, y, and z using the GCPs.) Another set of survey points is used as ECPs to support processing and product quality assessment. The ECPs are used for testing the final root-mean-square error vertical ($RMSE_z$) accuracy and horizontal ($RMSE_{x,y}$) accuracy of the final data products. Two sets of ECPs are collected: nonvegetated vertical accuracy (NVA) check points and vegetated vertical accuracy (VVA) check points; the number and distribution of these check points should follow the ASPRS-recommended guidelines for digital elevation data quality assessment.[9]

4.1.6.3 Data acquisition conditions

The preferred data-acquisition window may be based on tidal, weather, ground, and vegetation conditions. LiDAR data collections are preferred when the weather conditions are the best (i.e., low probability of cloud cover—no low clouds between the aircraft and ground—and rain in the collection area) during aerial surveys. Some areas can have persistent cloud cover for most months such that the window of opportunity for LiDAR collection may be limited to few months in the year. Other factors considered are the snow and rainwater accumulation on the ground. Vegetation conditions are specified as leaf-off or leaf-on for the collections. The leaf-off condition is when all leaves fall off of trees before the winter and before the first new leaves grow in spring. The leaf-on condition is when trees have fresh leaves or are full leaf in the summer; in some cases, year-round evergreens are considered as leaf-on. Most U.S. Federal civilian collections prefer leaf-off, but in northern latitudes where snow is prevalent in many months, this results in short windows for preferred collections. For lower clouds, lower-altitude 3D mappers have an advantage.

Tidal water bodies are defined as oceans, seas, gulfs, bays, inlets, salt marshes, and very large lakes that are affected by tidal waves. Tidal variations over the course of a collection or between different collections will result in lateral and vertical discontinuities along the shorelines. Coastal area collections typically have LiDAR collections constrained to certain time windows around the low tide. The USGS and NOAA recommend data acquisitions to be constrained to ±2 hours about the low-tide occurrence on any day of collections of tidal water bodies. If possible, reasonable planning efforts should be made to minimize tidal deviations.

4.1.6.4 Data processing and dissemination

Wide-area-mapping LiDAR collection techniques—especially for Geiger-mode LiDAR (GML)—at high point density requires large data storage and can pose significant challenges for data processing, storage, management, and dissemination. These LiDAR systems can produce raw data sizes on the order of a terabyte per hour of data collection, requiring large storage volumes (both onboard aircraft and on-the-ground processing systems), processing power, and infrastructure to support the data production. Data compression methods may be considered to alleviate the problem to some extent. Further processing of data in a shorter time requires high throughput and high-capacity, distributed-processing hardware, storage, access, and dissemination techniques. A dedicated production center with a parallel-grid computing infrastructure (having large core servers, huge memory, fast input/output (I/O) transfer devices and interfaces, and petabyte-class disk-storage spaces) may be required. One such system that is in use for commercial GML data production for fully autonomous point-cloud processing and product creation, as well as for data management, is described in the 2014 National Academy of Sciences study on laser radar.[16] As discussed in Chapter 6, GMAPD-based LiDARs have more-complicated processing due to the required coincidence of multiple pulses, but they have simpler readout circuitry. Historically, the first 3D mapping LiDARs built for the military required a long time to create 3D point-cloud images; however, over time, that limitation has been mitigated.

Multiresolution processing. Linear-mode systems collect data at a specific resolution and are therefore direct measurement systems. Geiger-mode and single-photon LiDAR systems oversample the scene, so sampling is derived through processing rather than collection. As a result, multiple resolutions can be derived from the same dataset based on user requirements. Data are processed by aggregation of multiple data frames through consensus modeling (Fig. 4.8) and processed into pre-defined voxels, providing a more uniform data product (Fig. 4.9). Note that multiple resolutions can be produced from GML data via voxel-based processing.

Spatial reference system. LiDAR data products are created in a spatial reference system that is defined by a project. The spatial reference system defines the coordinate systems, the horizontal projection and datum, the vertical datum, the geoid model, and the unit of reference. Reference information concerning the available spatial reference system used in North America can be found on the NGS/National Spatial Reference System (NSRS) website.[10] The coordinate system typically used in North America is either the State Plane Coordinate System (SPCS) or the Universal Transverse

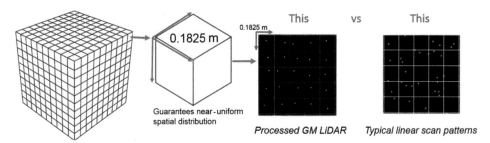

Figure 4.8 Voxel-based processing example to create a 30-point per square meter calculation: $1/\sqrt{30} \approx 0.1825$-m voxel dimension (reprinted with permission from Harris Corp.).

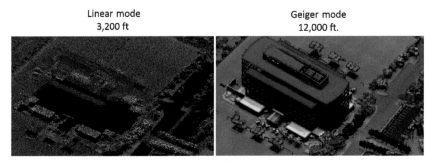

Figure 4.9 Linear-mode and Geiger-mode LiDAR image comparison (reprinted with permission from Harris Corp.).

Mercator (UTM) with appropriate local zone number for data projection. Countries around the world use other coordinate reference systems, such as the World Geodetic System (WGS84)-based spatial referencing. In the United States, the horizontal data for latitude, longitude, and ellipsoidal heights will be the North American Datum of 1983 (NAD 83), using the most recently published adjustment of the NGS (currently, NAD 83 epoch 2010.00). The vertical datum for orthometric heights is the North American Vertical Datum of 1988 (NAVD 88), and the geoid model used to convert between ellipsoidal heights and orthometric heights is the latest hybrid geoid model of the NGS (currently, the GEOID12B). Horizontal and vertical units are defined in meters, U.S. feet, U.S. Survey Feet, or International Feet.

Data products. Table 4.1 illustrates data products, accuracies, and formats that are typically produced for topographic mapping and other Federal/civilian applications. Definitions of each product are indicated in the table. The base product is an unclassified, calibrated, point-cloud dataset that is derived from raw LiDAR data and transformed in an Earth-referenced coordinate frame. Other value-added products often include base and extended classifications of point data, and extraction of features for asset management (utility poles, wires, vegetation, and transportation features), utility vegetation management, planimetry, and forestry management.

Table 4.1 LiDAR-derived products and specifications.

Point Density	$\geq$ 2–100 points per square meter (ppsm) area (first return); Lower ppsm at ground level below vegetation canopy, depending on sensor and canopy type.
Accuracy	Vertical absolute accuracy: $\leq$10 cm (RMSE$_z$) in nonvegetated environments for USGS QL1–QL2 products; Horizontal absolute accuracy: $\leq$30–100 cm RMSE$_{x,y}$ horizontal, depending on application; Relative accuracy: $\leq$5 cm RMSE$_{x,y,z}$; Precision: $\leq$5 cm RMSE$_{x,y,z}$; Classification level of accuracy: defined in % accuracy of the classified points.

Data Products

Product Name	Description	Standard Formats and Tiling Scheme
Calibrated Point Cloud	This is base-processed, unclassified elevation point-cloud data used for applications such as topography, geology, environment, flood, utility, and urban planning/modeling applications.	Produced in industry-standard ASPRS LAS formats (e.g., version 1.2 to 1.4; tiled per project need.[13]
Classified Point Cloud: Base and Enhanced Classification	A base-classified point cloud product is calibrated elevation data-classified terrain (ground points) cloud data. An enhanced classified point cloud is calibrated elevation data-classified terrain (ground points), vegetation, or infrastructure (buildings, utility features, transportation features etc.). Applications include topography, geology, environmental, forestry biomass, utilities, 3D building models, planimetry, and transportation.	Produced in industry-standard ASPRS LAS formats (e.g., version 1.2 to 1.4); tiled per project need.
Digital Elevation Model (DEM): Bare Earth and Hydro-flattening	This is a digital representation of continuous elevation values at regularly spaced or gridded intervals in x and y directions. DEMs are typically created using ground (bare earth) classified data. Hydro-flattening is a process applied to the bare-Earth DEM using breaklines so that water bodies (rivers, streams, reservoirs) are flat.	Produced in industry-standard, GIS-compatible raster format (such as Geotiff or Erdas.img); cell size is per project need.

(continued)

Table 4.1 *(Continued)*

Data Products

Product Name	Description	Standard Formats and Tiling Scheme
Digital Surface Model	This is a numerical model of the elevations of points on the Earth's surface. Digital records of the terrain elevation for all nonground positions (i.e., buildings, manmade structures, tree canopy, etc.) based on first and intermediate signal returns. Data are produced at regularly spaced intervals.	Produced in industry-standard, GIS-compatible raster format (such as Geotiff or Erdas .img); gridded with cell size and tiled per project need.
Relative Intensity Image	This is a raster image of regularly spaced points, with each point represented by a relative or absolute reflectance (i.e., intensity) as derived from the LiDAR signal from reflective surfaces.	Produced in industry-standard, GIS-compatible raster format (such as Geotiff or Erdas .img); gridded with cell size and tiled per project need
Breaklines	A breakline is a linear feature that describes a change in smoothness or continuity of a topographic surface. Breakline defines interruptions in surface smoothness to delineate water bodies such as rivers, streams, shorelines, dams, building footprints and other locations).	Produced in industry-standard, GIS-compatible formats (e.g., Esri shapefile or Geodatabase); non-tiled Esri feature class for the entire AOI in polygon.
Elevation Contour	This is a map of contour lines that trace areas of equal elevation at a specified interval (e.g., 1 ft). It is an imaginary line on the ground, all points of which are at the same elevation above or below a specified vertical datum. Contours are created from the bare-Earth data or DEM. Contours are used to produce flood plain delineations.	Produced in industry-standard, GIS-compatible formats (e.g., Esri shapefile or Geodatabase); tiled per project need.
Custom Ultrahigh-Density Products	Custom data products are high-point-density (up to 100 ppsm) vector products that capture fine details of utility poles, structures, wires or traffic control structures, and pavement markings to support asset and inventory management, as well as monitor vegetation management. These data products also support high-density collection with high level of foliage penetration to detect ground level, vegetation height, shape, and density.	Produced in industry-standard, vector GIS-compatible formats; tiled per project need.

The USGS requires a minimum set of classification for topographic mapping applications, which include classification of points into bare earth, water, bridge deck, ignored ground, snow (if present and identifiable), and low noise.[8] The typical value-added products used for topographic mapping are: classified point clouds, digital elevation models (DEMs), digital surface models (DSMs), relative reflectance or intensity images, contours, and hydro-flattened breakline data. A variety of commercial-off-the-shelf (COTS) software tools are available that are used to produce such products. The production of derived products is mostly labor intensive, as the software tools utilized are based on manually assisted processing to produce quality data. In the United States, topographic data products and metadata typically conform to USGS, ASPRS, Open Geospatial Consortium (OGC), U.S. National Digital Elevation Program (NDEP), and U.S. Federal Geographic Data Committee (FGDC) metadata standards.

Data accuracy standards. As indicated in Table 4.1, three sets of standardized accuracy parameters are defined for the data products. Precision is a measure of the repeatability of a LiDAR point, i.e., the closeness to which measurements agree with each other, even though they may all contain a systematic bias. Precision accounts for inherent instrument measurement error when the data are collected multiple times. Accuracy is stated as the closeness of a measured value to a standard or accepted (true) value of a LiDAR point. Typically, the accepted value is determined from an accurately (GPS) surveyed (x, y, z) position of a ground truth like an ECP. Absolute accuracy accounts for all systematic and random errors (or the so-called bias) in the dataset. Absolute accuracy is stated with respect to a defined datum or reference system. Another metric that is collected is the relative accuracy, which is a measure of the variation in point-to-point accuracy in a dataset. The relative accuracy metric is also specified as the positional agreement between points within a swath, adjacent swaths within a lift, adjacent lifts within a project, or between adjacent projects.

Root-mean-square error (RSME) or 1σ error is a way of reporting the accuracy for any of the above three parameters. Alternatively, a 95% confidence level, which equates to 2σ error that in turn is 1.96 times the RMSE value, is also specified. As an example, the absolute nonvegetated vertical accuracy (NVA) for USGS Quality Level 1 (QL1 or 8 ppsm) products must meet $\leq$10 cm $RMSE_z$ (or equivalently, $\leq$19.6 cm at 95% confidence level). The absolute accuracy in the horizontal direction may vary by the application, i.e., nominal GPS accuracy (approx. 1 m) for utilities and $\leq$30 cm for typical topographic applications. Point cloud classification may also include additional levels of accuracy for the classified points.

Data dissemination. High-density LiDAR data require special handling of data management and delivery methods. The following delivery methods are in use today: (1) shipping on data hard drives (typically limited to 2 or 3 TB); (2) FTP (file transfer protocol), or (3) via data-hosting services. Higher–point-density data and/or larger projects may need the option to access and download the data holdings, or simply visualize them, through data-hosting services instead of receiving through normal deliveries shipped via data hard drives. One such implementation of data deliveries and longer-term data-hosting services is highly secure cloud-based services, such as Amazon Web Services (AWS). Data-hosting services offer a secure cloud environment from which users and authorized third-party providers can download the project's deliverable data to their local environment on demand.

Applications. LiDAR elevation data have become an essential tool to helping people gain a general understanding of the physical landscape for planning and infrastructure development. LiDAR data are collected at various densities to support various end-user applications. Figure 4.10 shows the general alignment between LiDAR point density data and the applications that are most commonly supported at these resolutions.

Often individual cities and counties collect LiDAR data on a smaller-area basis, constrained by available funding. To overcome this issue, a common approach is to combine the collective resources from multiple different agencies or entities for more-cost-effective, larger-area collections, processing, and dissemination. Events like erosion and flooding, new road construction, and

Base-Mapping LiDAR Data

LiDAR processed at nominal pulse spacing of 2 ppsm to 4 ppsm provides base mapping information for wide-area 2D mapping, urban planning, conceptual analysis, and emergency planning.

Engineering-grade LiDAR Data

LiDAR processed at nominal pulse spacing of 8 ppsm or greater with accuracy to achieve 1-ft contours for civil earthwork and grading plans, planimetric products for pavement and building asset tracking, and 3D digital terrain and surface water models for waterflows, wetlands and slope design.

Ultrahigh-Density LiDAR Data

LiDAR processed at higher density nominal pulse spacing, greater than 20 ppsm, provides highly detailed data for utility mapping, asset inventory management, roadway and bridge design, detailed as-built documentation.

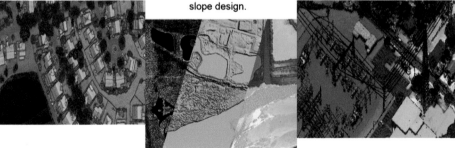

Figure 4.10 General-application density requirements (reprinted with permission from Harris Corp.).

expanding community footprints physically change the Earth's surface. Critical decisions about city expansions, emergency preparedness, utilities and transportation asset management, and other long-range planning activities often require a comprehensive look at broader areas to properly assess the impacts. New technology developments such as the introduction of photon-counting LiDAR systems to the commercial market has enabled a paradigm shift in the approach used by local and state governments by enabling data processing at multiple resolutions (i.e., point densities) and multiple-use applications from a single, broad-area collection project, making it a more attractive, cost-effective, and viable solution for many stakeholders. The remainder of this subsection discusses some representative applications of LiDAR data.

Mapping land topography. LiDAR provides high-resolution elevation data of the physical locations of natural land-surface features like rivers, trees, and hills, as well as manmade features like buildings and roadways, giving a precise, 3D position of every point collected. LiDAR data are used to produce digital elevation models of the terrain at ground level (i.e., DEM bare earth) based on the last LiDAR return signal as well as at the highest surface (buildings, manmade structures, tree canopy, etc.) based on first and intermediate return signals (e.g., DSM). LiDAR-derived DEMs are produced at much higher resolution (30 to 50 cm) and higher precision than commercially derived DEMs from optical or SAR airborne or satellite sources. Other topographic products include relative or absolute (reflectance) intensity images derived from LiDAR signals from reflective surfaces and iso-elevation contour lines.

Potential stakeholders who can benefit from high-resolution elevation data include government planning departments, land managers, infrastructure (roads, rails, buildings) planners/developers, farmers, and engineering firms. LiDAR-derived high-resolution contours and elevation models can be used for transmission corridor and pipeline planning, storm water modeling, drainage analysis, master planning, grading, erosion control, and roadway design. Figure 4.11 shows a sample of a high-definition point cloud of a cityscape produced using GML data.

Figure 4.11 High-definition point cloud of the cityscape of Chicago (reprinted with permission from Harris Corp.).

Flood plain mapping, flood hazard prediction, and water management. Accurate digital elevation data are a prerequisite for flood plain mapping and flood inundation modeling (from river flooding or storm surge). LiDAR-derived bare-earth DEMs, with or without hydro-flattening, greatly improve the accuracy of flood mapping. LiDAR elevation data from land topographic and hydrographic surveys fused with aerial imagery can provide synoptic 3D views of the areas of vulnerability, assess potential damage spots, and evaluate changes occurring in near real time during heavy hurricane- or monsoon-related flooding in known flood zones. The data can be used to provide actionable intelligence for first responders, using flood modeling and prediction tools integrated with real-time weather and flood gauges for accurate risk analysis and disaster response. Figures 4.12 and 4.13 show the contrast of non–hydro-flattened versus hydro-flattened DEMs that were collected with GML for the State of North Carolina in support of North Carolina Emergency Management's flood inundation mapping and early flood warning program. The new LiDAR information is allowing the State to calculate potential flooding in parking lots, on roadways, and around

Figure 4.12 Non–hydro-flattened LiDAR DEM (reprinted with permission from ESP Associates, Inc.).

Figure 4.13 Hydro-flattened LiDAR DEM (reprinted with permission from ESP Associates, Inc.).

buildings in residential neighborhoods. The State can now make plans to mitigate impacts that affect individuals directly before the hazard strikes and during flood events in near real time.

Note the surface roughness in the river and lake areas in Fig. 4.12, which illustrates the necessity of hydro-flattening to produce accurate water flow and subsequent modeling as shown in Fig. 4.13. High-quality DEMs and DSMs derived from LiDAR data are used by the Federal Emergency Management Agency (FEMA) to conduct research and formulate regulations related to mapping flood hazards, changes to surface topography from flood event, and flood loss estimation analysis. Flood hazard maps produced by FEMA provide spatial variation of flood velocity and depth. FEMA's flood loss estimation analysis helps to estimate the structural damage risks to buildings and infrastructure based on vulnerability. FEMA publishes an official Flood Insurance Rate Map (FIRM) to delineate the flood hazard areas and the risk premium zones, which are then used for flood insurance rates.

Engineering and survey-grade LiDAR can be used for many different aspects of water management and distribution. The highly accurate LiDAR data provide valuable information for pipeline routing, including ground elevation to minimize pumping requirements, as well as surface corridor mapping for access. LiDAR-derived DEM products provide detailed surface water information for reservoirs (storage) and drainage.

Mapping utility assets and vegetation encroachment. Utility companies are challenged with managing a wide range of assets (poles, wires, and third-party attachments) and vegetation management across their large service areas. Until recently, identification, tracking, and cataloging the assets were conducted in 2D through GPS location capture by walk downs, feature coding, and then subsequent updating of the geographic information system (GIS) base map. LiDAR is revolutionizing utility/corridor mapping, vegetation encroachment over power lines, emergency response, line re-rating, and asset monitoring (e.g., aging infrastructure) by providing high-density data with high accuracy over large areas in a timely, safe, and cost-effective manner.

Figure 4.14 illustrates several utility products and analytics being produced on network-wide projects that cover tens of thousands of square miles. Optionally, additional geodata from aerial cameras and a wide assortment of commercial satellites are also often acquired to supplement the LiDAR data. Combined with automated data-processing capabilities, the high-density LiDAR approach produces actionable information used to comply with regulations and standards (such as the U.S. NERC standard FAC-003) to increase reliability and to supply affordable energy.

Mapping transportation assets. LiDAR data provide high-precision DSM and intensity data that can be used in urban traffic and transportation

Figure 4.14 Utility product examples (reprinted with permission from Harris Corp.).

management. Transportation corridor mapping with LiDAR supports engineering planning and highway design, corridor development, traffic flow, highway safety (grading and cross-slope) and critical infrastructure protection, utility wiring and obstructions, and urban feature surveying. Surface terrain data like DSM can be used to economically site or relocate existing infrastructure facilities or design new highway infrastructure. A study report by the National Cooperative Highway Research Program (NCHRP) discusses the application of LiDAR and the required accuracy to support various transportation applications.[11] The Departments of Transportation (DOTs) in many states have identified the need for defining 3D workflows to support data collection and maintenance operations for various transportation stakeholders, such as traffic engineering, bridges employees, plant maintenance staff, environmental monitors, and others.

LiDAR mapping is used for transportation asset and inventory mapping of highway/street furniture and extraction of 2D and 3D asset features such as road markings, signs, streetlights, barriers, guardrails, signals, and others. High-density aerial LIDAR data collection supports detailed point classification and feature extraction, and compliments mobile LiDAR collections commonly used by various DOTs. High-density aerial LiDAR also provides critical data for designing highways in long and difficult, densely wooded area with slopes and ravines. These sensors support efficient and cost-effective transportation network data collection in dense urban and suburban environments.

Mapping 3D urban infrastructure and modeling. High-density LiDAR collection provides DSM height maps, intensity data, and other crucial data such as natural (e.g., trees) and manmade (e.g., buildings, rooftop structures, roads, road lane markings, railroads, water towers, electric transmission lines) land features through enhanced classification of point data. Large cities (like Chicago, Seattle, and others) have been mapped quickly and efficiently at

Figure 4.15 Point cloud with 3D planimetric extraction of transportation features and lighting (reprinted with permission from Harris Corp.).

high point densities of 8–30 ppsm from altitudes of 12,000–27,000 ft above mean sea level using GML technology. Figure 4.15 shows a point cloud with 3D planimetric extraction of transportation features and streetlights using GML at an altitude of 12,000 ft. The LiDAR perspective shown in Fig. 4.11 illustrates the quality and homogeneous nature of data that is collected with the GML sensor, mapping dense city infrastructures such as tall buildings, stadiums, freeways, city multistory structures, rooftop building structures, parking spaces, etc., with minimal shadows and voids. This type of data supports urban/city governments for planning, emergency management, route planning and security, and comprehensive, 3D fly-through situational awareness supporting major events.

Additional processing of LiDAR data also provides valuable 2D/3D planimetric data to extract highly accurate building footprints, pavements, and impervious road surfaces. 3D modeling of building and rooftop extraction is also performed to support assessment of rooftop geometry for insurance and solar panel site assessments.

Mapping hazards and emergency management. LiDAR was used for disaster management and evacuation during the 2010 Haiti earthquake relief mission conducted by the U.S. Government. An MIT/LL 3D mapping LiDAR called the ALIRT system was deployed in Haiti and was used to quickly map critical areas. From the dense point-cloud data, DSM, bare-earth DTM, and change detection products were produced to support damage analyses, infrastructure assessments (e.g., bridges and government buildings), trafficability and evacuation route assessments, watershed analyses, and shelter placements for displaced residents and for distribution of medical supplies and food.

Forest inventory management. Natural resource analysis and forest inventory management requires LiDAR data collection at high point densities, over large areas. The data collection needs to be fast and economical so that the data can be used for performing detailed analytics in evaluating the condition and maturity of the trees and vegetation. Analytical data of interest include tree segmentation, canopy height modeling, species identification, timber volume estimations, and slope/area calculations. This requires sensors that can penetrate the canopy while flying at a relatively higher altitude for faster data collection compared to a low-flying system. Traditionally, linear-mode sensors are used for this purpose. Recent advancements in GML and SPL technologies present valuable new opportunities for the forest industry to enable large areas to be collected cost effectively and efficiently with high point-density data. Figure 4.16 shows classified LiDAR point-cloud data, color coded by species, of a mixed-species forested area.

Mapping ancient archeological sites. LiDAR is also one of the newer techniques used by archeologists to study landscapes and map ancient structures hidden under tropical canopies.[12] Highly dense vegetation covering most ancient remains such as Mayan sites in Mexico, Guatemala, Belize, etc., preclude archeologists from discovery and documentation of the extent of the archeological sites using traditional approaches, which are very slow and use a laborious process of clearing the canopy areas. Foliage penetration using high-density LiDAR enables faster and enhanced detection of archeological landscapes, analysis of the size and distribution of their urban centers, and extraction of the details of the monument architecture, roads, industrial and residential neighborhoods, houses, reservoirs, and agricultural terraces.

Higher point-density data: improving efficiency and cost. The 3D mapping LiDAR market has been growing over the last decade as the technology has proven the survey mapping accuracy to be in the 5- to 10-cm vertical accuracy range. With ever-increasing collection rates, especially with the newer photon-counting LiDAR systems, various government users at local, state, and

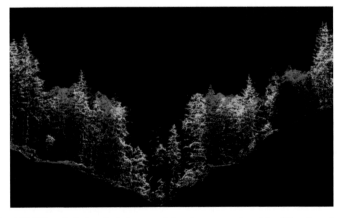

Figure 4.16 LiDAR point-cloud data used for tree species identification (reprinted with permission from Object Raku Technology).

national levels have an emerging mapping need for LiDAR data collection at higher point densities and over large areas. Collection of elevation data at higher point densities together with mapping and locating assets and other derivative products is highly desired by various federal and civil agencies, including the USGS, NOAA, FEMA, U.S. Department of Agriculture (USDA), U.S. Forestry Service (USFS), U.S. Army Corp of Engineers (USACE), Federal and State DOTs, and county/city governments. The commercial LiDAR industry is primarily driven by the following markets: private energy/utility companies, oil/gas and mining, engineering projects, forest industry, and other stakeholders. Many of these stakeholders need data collections to update their data holdings periodically at some regular interval that may vary from every few days or weeks for engineering or mining site-specific locations to every few years for base-mapping applications.

In 2012, a consortium of U.S. federal agencies led by the USGS jointly conducted a National Enhanced Elevation Assessment (NEEA) study to address the need for higher-resolution elevation data. The NEEA study included base-recommended requirements, business use cases, and the cost benefit analysis for a nationwide mapping program across the United States.[13,14] USGS then established the 3D Elevation Program (3DEP) with the objective of accelerated 3D elevation data generation and updates over an eight-year cycle. The core recommendation over the conterminous United States (Lower 48) was to deliver a minimum of a Quality Level 2 (QL2) product that comprised a calibrated and classified LiDAR point cloud (at least 2 ppsm) and a 1-m hydro-enforced DEM with <10-cm vertical accuracy. The study also concluded that the greatest value for multi-agency re-use was for higher-resolution LiDAR data (>8 ppsm); however, given the fact that only linear-mode LiDAR technology was assessed at the time, the greatest cost/benefit ratio was found in the QL2 product. But because new SPL and GML systems have become commercialized, resulting in possible lower costs for large-area, higher-density collections, NOAA and USGS are in the process of undertaking an updated nation-wide survey called 3D Nation.

A similar type of assessment study was conducted in 2017 by the State of Florida's Department of Environmental Protection for topographic LiDAR, bathymetric LiDAR, and/or topo-bathymetric LiDAR to document the business uses, and identify the need and frequency, of LiDAR collections. This survey indicated a greater need for higher–point-density data (up to ~30 ppsm to support water management districts, utility clients, and DOTs) but a minimum base requirement of a QL1 product (>8 ppsm) with a goal to update the elevation data on a three-year cycle for the State of Florida.[15] Similarly, utility companies would like to perform their vegetation and asset management analysis on three- to five-year cycles. As such, collection and production of higher-density LiDAR data for stakeholders and end users, at reduced cost and efficiency, and in greater re-visit cycles, is highly desired.

4.1.7 Laser-induced breakdown spectroscopy

Laser-induced breakdown spectroscopy (LIBS) uses intense, very short, laser pulses to induce optical breakdown in a material, and then detects the emission spectrum from the breakdown. For nanosecond-duration pulses, the initial breakdown creates a plasma that then absorbs a portion of the energy in the pulse. Shorter pulses, in the femtosecond to picosecond range, are over before the plasma absorbs more of the pulse energy. Recent advances in LIBS technology employ two pulses, spaced apart by 1–10 μs, where the second pulse acts to heat the plasma created by the first pulse. The detection system then observes the emission from the plasma, with time gating of the detector to optimize the desired emission spectrum signal. If the pulse, or pulses, are powerful enough, the emission contains components from the electronic transitions of individual, ionized atomic constituents of the irradiated material. Through spectroscopic analysis of the emission wavelengths, typically in the 170- to 1100-nm range, one can determine the atomic constituents of the material, using the characteristic wavelengths of their emission lines. From the strength of the signals at each line, it is possible to determine their relative concentration, and from this, to gain knowledge of the chemical makeup of the irradiated material. This is a form of differential absorption LiDAR. Figure 4.17 shows a diagram of a LIBS LiDAR.[16]

4.1.8 Laser-induced fluorescence

Laser-induced fluorescence (LIF) LiDARs detect material in the atmosphere or on the surface of a target of interest, creating transitions by exciting material to higher electronic levels. In some cases, the material returns to the ground electronic level by fluorescing, which can then be detected by a receiver. Figure 4.18 shows a diagram of a LIF LiDAR,[17] where range is shown as a

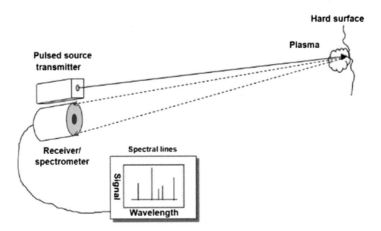

Figure 4.17 Laser-induced breakdown spectroscopy, a form of differential absorption LiDAR (reprinted from Ref. 16).

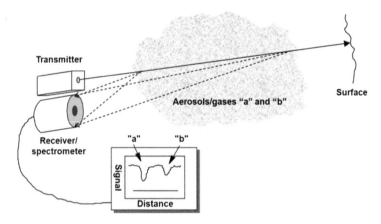

Figure 4.18 Laser-induced fluorescence LiDAR (reprinted from Ref. 17).

time response in the chart in the lower part of the figure. A LIF LiDAR system employs a laser to generate fluorescence and a receiver tuned to the wavelength of the fluorescence. Through time analysis of the fluorescence, it is possible to localize the detected material in the atmosphere, thus providing range-resolved data. By employing a tunable receiver, or multiple receivers, one can detect multiple materials ("a" and "b" in the figure) as a function of range because different materials fluoresce at different wavelengths.

Because fluorescence is a rapid process (fluorescence times are on the order of a picosecond to a femtosecond), the fluorescence time does not influence any but the most precise range measurements. It is also possible to detect fluorescence from materials on a surface. Typical systems employ pulsed UV sources for the excitation of the electronic levels required to create the fluorescence. You can obtain range distribution for each species. The decay time of the returned signal, as well as the fluorescence spectrum itself, can also provide some ability to distinguish different materials.

4.1.9 Active multispectral LiDAR

Color can be a very useful discriminant. Color distinctions for an active sensor like a LiDAR are based on the differences in reflectivity variation with wavelength. Active multispectral imaging for targets can have an advantage over passive imaging of targets because with the former, it is possible to control the illumination source. Therefore, even at nighttime, near-IR wavelengths can be used, whereas there would not be enough signal to use passive multispectral at these wavelengths at night. There is more reflectivity variation with wavelength in the near-IR than in the MWIR, so being able to operate a spectral-based classifier in the near-IR during the night can be very useful. We would prefer to stay above the eye-safer spectral limits of about 1.5 μm because it will be easier to meet eye safety considerations, but it is possible to keep a LiDAR eye safe on the ground by appropriate design, even using wavelengths below 1.5 μm. The LiDAR has control over both the spectrum and direction of the light source (as well as enabling range resolution for obscurant and foliage penetration).

An active EO multispectral sensor combines the benefits of conventional LiDAR and multispectral wavelengths in a band that has significant color variation in its reflectance. Conventional LiDARs and some passive imaging sensors utilize the shape of a target for detection and/or identification purposes. This can work well when the object is clearly visible. In cases where the target is partially obscured, active multispectral sensing can be a great adjunct sensing mode for a LiDAR. Color can highlight an object even when the shape is obscured. Active multispectral sensing is similar to laser vibration as an active EO adjunct sensing modality because neither of them depends on object shape. They both have the potential to be used in applications beyond the diffraction limit of a LiDAR in cases where the object is partially obscured and enough signal is used to create sufficient signal to noise.

Two different approaches have been used to deploy active multispectral sensors against hard targets. One approach is to use the laser wavelengths that are easy to generate, e.g., 1.064 μm and around 1.5 μm, and take whatever recognition benefit one can gain. The second approach is to determine what wavelengths make the best active multispectral recognition probabilities, and make the lasers appropriate to these discriminants. In the second case, a class separability metric is formulated and optimal wavelengths are selected.[18,19] Figure 4.19 shows reflectivity versus wavelength for a number of materials in the 1- to 5-μm region. This is an excellent wavelength region for active

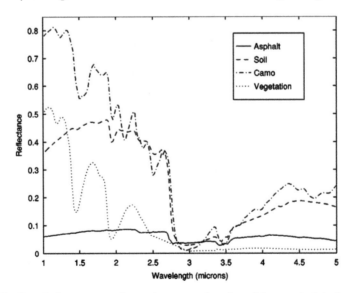

Figure 4.19 Spectral responses for various wavelengths of four materials (reprinted from Ref. 18).

multispectral sensing. Obviously, customizing the selection of the wavelengths to discriminate one material from another will be more powerful than just selecting a couple of readily available wavelengths, but this will put more of a load on supplying laser sources.

Another issue to address concerning active multispectral LiDAR is whether discreet laser lines or a laser continuum will be used. Obviously, having an available continuum may be more difficult from the laser source point of view. Even with a number of discrete wavelengths, it should be possible to achieve high probabilities of material identification. Another tradeoff is spectral line width. Ideally, spectral lines should be broad enough to average out speckle, but narrow enough to be a good discriminate. A recognition algorithm must distinguish between L different possible hypotheses. The only difference in the observations from different classes should result from the different object reflectivity spectra. Atmospheric absorption must, however, be taken into account as a possible confuser. The material classes considered include both natural (vegetation and soil) and manmade (camouflage cloth and tar or asphalt). Spectral discrimination can be an adjunct sensing mechanism for which probabilities of target identification do not need to be as high. If spectral discrimination is used in conjunction with other discriminants, it must only separate among classes that are not discriminated by the primary discriminant.

4.1.10 LiDARs using polarization as a discriminant

Light is a transverse electromagnetic wave (i.e., the oscillations are perpendicular to the direction of propagation), similar in some respects to the waves that propagate down a taunt string. The wave may oscillate horizontally, vertically, in a right- or left-handed spiral, or in some combination of the four. Mathematically, the four components of the polarization are represented by a set of quadruples known as the Stokes vector. With active polarization LiDAR, one controls the polarization and direction of the illumination, enabling, in principle, full Mueller matrix characterization of the target reflectance. A Mueller matrix is a 4×4 matrix describing the effect an optical element has on the polarization state of light. Multiple measurements with polarizers rotated with respect to each other can identify the degree of polarization (DOP) of the light received. Surface roughness seems to be one of the parameters that influences the ratio of light reflected in one polarization to that reflected in the opposite polarization. Smooth surfaces tend to have more of a polarizing effect than rough surfaces. Rough surfaces have many facets, so reflected polarization is diverse in nature. People often wear polarized sunglasses to reduce glare from smooth surfaces.

In a similar fashion, a LiDAR that detects the polarization of reflected light can provide an indication of surface roughness. Passive DOP measurements have been limited by the uncertainty in the initial polarization state of the ambient illumination and the angle of incidence of the light. In addition, if more than one material is present within the FOV of a single pixel (e.g., when leaves or camouflage are in front of a vehicle), the intensity and the DOP measured will be the result of the summation of all of the surfaces within the FOV. Since these proportions are not known, the so-called mixed-pixel effect further adds to the uncertainty. Active EO polarimetric measurements can eliminate this uncertainty by controling the direction of illumination and the initial state of its polarization. Active polarization LiDARs can also separate returns in range, dramatically reducing the mixed-pixel effects. The signal generated by foliage or camouflage can be separated in range from the signal generated by whatever lies behind, or in front of, the object of interest. This minimizes the mixed-pixel effect unless the two surfaces are separated by less than the range resolution of the LiDAR. Manmade objects are often smoother than natural objects, allowing a LiDAR that discriminates polarization to more easily identify manmade objects compared to objects in nature.

4.2 Coherent LiDAR

4.2.1 Laser vibration detection

The Doppler shift changes the frequency of the return laser signal based on

$$\Delta f = \frac{2V}{\lambda}, \qquad (4.1)$$

where V is velocity, and λ is wavelength. The Doppler shift is used to measure velocity. It can provide a very accurate measure of velocity and as a result is often used to measure the small back-and-forth velocity caused by vibration. Table 4.2 provides the Doppler frequencies resulting from various velocities for a wavelength of 1.55 μm. A 10-μm/s velocity yields a frequency shift of 12.9 Hz, which can be measured in a fraction of a second. As an object vibrates, it creates velocities toward and away from the sensors. For vibration, we see a frequency-modulated return signal, with the frequency of the modulation as the vibration frequency and the magnitude of the frequency shifts equal to the varying Doppler shifts. The following equation shows the variation in position resulting from a vibration:

Table 4.2 Doppler frequency resulting from a certain velocity when using 1550-nm illumination.

Velocity (μm/s)	Frequency (Hz)
1	1.29
10	12.9
100	129.03
1000	1290.32

$$x = A \sin(2\pi f t), \tag{4.2}$$

where A is the amplitude of the vibration, and f is its frequency. A Doppler vibrometer measures velocity to and from the LiDAR. The velocity is the derivative of Eq. (4.2) for the surface position of a vibrating object:

$$V = \frac{dx}{dt} = 2\pi f A \cos(2\pi f t). \tag{4.3}$$

Equation (4.3), along with Eq. (4.1), can be used to generate a table showing the frequencies resulting from illuminating a vibrating object with light at a wavelength of 1.55 μm. In Table 4.3, the required (minimum) sample time is twice one over the maximum Doppler shift frequency. Even for a 1-μm magnitude of vibration at 10 Hz, there will be a maximum frequency for 81 Hz, which should be measurable.

For high Doppler sensitivity, detection is usually performed using temporal heterodyning, where the return signal is combined on the detector with a local oscillator to create a frequency-downshifted signal, called a beat frequency. Laser vibrometers can be used to characterize many types of vibration. The type of engine in a vehicle can be identified; it is easy to tell the difference between a turbine engine and a piston engine. It should be easy to determine how many cylinders a piston engine has. Almost any type of vibration can be identified. The vibrational image of an internal combustion engine can be used to identify combustion pressure pulses and inertial acceleration of the pistons and drive trains. It can also help identify mechanical imbalances and misfires. Power through a transformer, or liquid through a pipe, can be characterized. If the return Doppler shift is sent to headsets, we will hear noise. A diesel engine will sound like a diesel engine, and the same is true for other common vibrations.

Flash illumination for vibration detection would be powerful because it allows simultaneous vibration measurement over an area, but it requires a 2D array of photodetectors operating at very high frame rates. Such imaging

Table 4.3 Doppler frequencies resulting from surface vibration.

Vibration frequency (Hz)	Vibration amplitude (μm)	Max. velocity (μm/s)	Max. Doppler shift frequency (Hz)	Required sample time (ms)
10	1	63	81	24.7
50	1	314	405	4.9
100	1	628	810	2.5
200	1	1257	1622	1.2
10	0.1	6	8	250.0
50	0.1	31	40	50.0
100	0.1	63	81	24.7
200	0.1	126	163	12.3

cameras exist at low frame rates in the range of a few hundred hertz. Area vibrometers are just now being developed. Point vibrometers such as Polytec's laser vibrometer are currently commercially available.

The use of area laser vibrometers will enable a significant number of new application areas. For example, if you look at a bridge while traffic is going over it, you will see spatial vibration modes.[20] As the bridge decays, it will begin to crack. Those cracks will change the vibration modes. Also, area laser vibrometers will be ideal for mapping vibration sources across any machinery or underground seismic source.[21] These applications can be attempted with point laser vibrometers, but you cannot see a flash image of the vibration modes like you can with an area-based laser vibrometer.

4.2.2 Range-Doppler imaging LiDAR

A microwave radar can make a 2D image using only one detector. A LiDAR can do this as well, but the field of view is much smaller because the diffraction limited spot size is much smaller. We have previously talked about range resolution being available from a single detector based on the speed of light and the time for a laser pulse to go to the target and return. Assuming that the LiDAR is closing on the target, the velocity toward the target is

$$V_T \geq V \cos(\vartheta) \cos(\varphi), \tag{4.4}$$

where ϑ is the elevation angle, and φ is the azimuth angle.

If the LiDAR is traveling in a certain direction at a certain velocity, the Doppler shift varies with angle. The LiDAR is looking down at an angle; therefore, the Doppler shift is spread in angle, as conceptually illustrated in Fig. 4.20. The solid green arrow pointing at the military vehicle has a beam width. The closer portion of the beam hitting the ground sees a smaller Doppler shift than the farther portion of the beam width hitting the ground because the down angle on the closer portion is larger. This variation with angle can provide better cross-range information than the diffraction limit of the LiDAR. Even within a single detector, we can have a distribution in

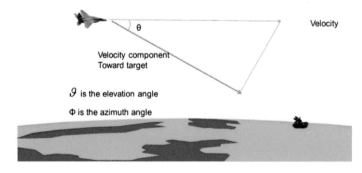

Figure 4.20   Range-Doppler imaging LiDAR (reprinted from Ref. 39).

Doppler shift frequencies that allows a resolution beyond the diffraction limit. This is called range-Doppler imaging because a 2D image can be made using range and Doppler information obtained from a single detector. This works well for radar because a radar beam width is so large compared to a LiDAR beam width, making it easy for the spread in Doppler shift to provide better cross-range resolution than the diffraction limit. In order to make the range-Doppler image larger than a single pixel, a larger area must be illuminated and a detector array must be used; otherwise, the FOV of the LiDAR is the angular subtense of the diffraction limit, using a DAS equal to the diffraction limit. Doppler imaging can provide elevation information that is beyond the diffraction limit and is also of higher resolution than the DAS. For long-range imaging, when looking down as an air vehicle moves forward, this can be a very useful feature.

4.2.3 Speckle imaging LiDAR

This discussion on speckle imaging LiDAR does not necessarily need to appear in the coherent LiDAR section of this chapter, but coherent LiDARs must have a narrow linewidth, making them more susceptible to speckle. A direct-detection LiDAR can have a broad linewidth, which averages speckle effects, although it does not have to have a broad linewidth. Therefore, although it could have easily gone before the coherent LiDAR section, this subsection is part of the coherent LiDAR discussion.

When a beam of light from a laser transmitter illuminates a target, phase irregularities occur in the backscattered light due to surface roughness on the scale of a wavelength of light. Interference between the various contributions to the optical field produces a speckle pattern of bright and dark intensity regions in the receiver. This surprised early laser researchers because we do not see speckle in our normal life. Our eyes see broadband light after the interfering returns are averaged such that we do not see light and dark areas. In many optical applications, speckle is considered a nuisance (e.g., it degrades target images for narrowband LiDARs). However, researchers since the 1960s have known that the speckle pattern carries information about the physical properties of a target. The interference we call speckle only comes from the illuminated region of a target, so the finest speckle return we will see in angle space is based on the either the size of the target or the size of the illumination area, whichever is smaller. For an area target, the speckle size is limited by the illumination pattern. For a small target, it is limited by the target size. Figure 4.21 illustrates information about speckle size.

Our full-LiDAR-aperture diameter on receive then limits the finest speckle angular resolution that can be seen, although the DAS could limit the finest observable angular resolution if the LiDAR is not diffraction limited. The achievable resolution in a speckle imaging system is limited by the spatial

Figure 4.21 Speckle LiDAR speckle size (reprinted from Ref. 39).

extent of the pupil plane sampling (which can be thought of as the diffraction limit of the speckle imager).

4.2.4 Aperture-synthesis–based LiDAR

Weight and space constraints are frequently placed on active EO systems, including LiDAR systems. For real-beam-aperture LiDARs, high-resolution imaging in cross-range requires large optics due the diffraction limit. Large optics are frequently impractical. One reason for this is that a large, monolithic-aperture imaging system has to be deeper than a smaller lens or reflector to maintain the same $f/\#$, contributing to large optics being heavy and expensive. To achieve high resolution at lighter weight and smaller size, methods have been developed to synthesize large apertures. This can be done by using either the motion of a smaller aperture or an array of smaller apertures.[22,23]

Using multiple transmitter apertures and multiple receive apertures allows design flexibility not present in monolithic apertures. Synthetic-aperture radar (SAR) comprises a single moving aperture, so the transmit and receive apertures both move.[24–27] Synthetic-aperture imaging was developed for radar decades before it was developed for LiDAR because, when used with radar, (1) it is more resolution limited and (2) it is not as limited in range by atmospheric considerations; therefore, radars are often very long-range sensors. A synthetic-aperture LiDAR (SAL) uses the motion of the single aperture to synthesize a larger effective aperture. For both SAR and SAL, the synthetic aperture is almost twice as large as the actual flown distance due to the angle of incidence equaling the angle of reflection. This has been experimentally demonstrated over decades in the microwave region, and recently in the optical regime, when SAL was demonstrated.[22,28]

Duncan and Dierking[29] provide a cross-range resolution equation for spotlight-mode SAL:

$$\delta = \frac{R\lambda}{2L + D_{\text{real}}}, \qquad (4.5)$$

where D_{real} is the size of the aperture, R is the distance to the object, λ is the wavelength, and L is the distance moved. As a result, the effective size of a synthetic aperture based on motion from a diffraction point of view is

$$D_{\text{eff}} = 2L + D_{\text{real}}. \tag{4.6}$$

For radiofrequency (RF) systems, the size of the real aperture is neglected, making the effective aperture twice as large as a monolithic aperture of width equal to the distance flown. The real aperture in RF systems is often on the order of a meter compared to the distance moved, which is multiple kilometers for a SAL. In an EO system, the real aperture can be a significant fraction of the size of the moved aperture, so the D_{real} term is retained. For example, a SAL might have a real beam size of 15 cm and a flight distance of 1 m.

In order to synthesize a large aperture, the returned field, including both amplitude and phase, must be obtained either by measuring it or by estimating it. Optical carrier frequencies cannot be measured directly by detectors. A 1.5-μm wavelength has a carrier frequency of 200 THz, which is orders of magnitude higher than even the highest-bandwidth detectors. Visible lasers have even higher frequencies. In order to measure phase as well as amplitude, it is necessary to use a local oscillator (LO) to beat against the return signal.

For temporal heterodyne detection, we can choose to have a LO that aligns with the return signal but uses a different frequency than the illuminator. The two return signals beat against each other to create an intermediate frequency (IF) return. If the LO and the transmitted signal are stable, the IF return will have the same phase difference as the carrier return. This allows us to measure phase of the returned carrier signal.

Alternatively, we can use spatial heterodyne detection—which falls under the broader category of digital holography—to measure the return phase of the signal. In spatial heterodyne detection, we use an LO that is the same frequency as the illuminator signal, but is offset in angle from the return signal. This creates a spatial beat that generates fringes across the receive pupil plane, allowing us to measure spatial variations in phase. Both temporal and spatial heterodyne methods can use either pupil-plane or image-plane imaging to measure the return fields. Once the field is measured, it is possible to convert from image plane to pupil plane, and back, as well as to digitally adjust the focus or digitally correct for aberrations. Once the field of the pupil plane is stored digitally, a Fourier transform will take us to the image plane. A parabolic optical path delay will be a lens.

All LiDARs that synthesize larger apertures than the real aperture generate the larger synthetic aperture in the same way, as seen from a physics point of view. Multiple pupil-plane samples of the received fields are measured. Each sample of the pupil-plane field is at a somewhat different location. The pupil-plane fields can then be added to include a larger pupil-plane area than a single real aperture. If the added pupil-plane fields overlap, some normalization may be required. If there are three overlapping measurements in a given pupil-plane region, it may be necessary to divide the intensities by a factor of three in that pupil-plane region, using what is called aperture weighting. Aperture weighting will influence the final image results. Once a larger pupil-plane field is sampled, it is possible to Fourier transform that large pupil-field sample to obtain a higher-resolution image of an object.

The rest of the discussion on SAL presents engineering details. A SAL moves a single aperture, including both the transmitter and the receiver, to make the synthesized pupil-plane field sample larger. A multiple-input, multiple-output (MIMO) LiDAR has multiple physical transmitters and receiver apertures used to make a larger synthetic-aperture pupil-plane field sample. It may also make use of motion, or may not. Any motion must be compensated before we add the pupil-plane field samples. Also, in order to translate the receive aperture back to its real location, it is necessary to perform coordinate transformations whenever the transmitter is moved. It is possible to guess the phase and use a merit function to judge the quality of the estimation. The estimation can be optimized by trying multiple guesses and comparing against the merit function.[30]

4.2.4.1 Synthetic-aperture LiDAR (SAL)

Synthetic-aperture LiDAR uses the motion of the LiDAR to develop a synthetic pupil-plane aperture that is larger than the real aperture. This technique is derived from the 1951 work of Carl Wiley,[31] who proposed the technique we call synthetic-aperture radar (SAR) to overcome the diffraction limit of an imaging radar's aperture. This technique collected data from multiple aperture locations and then computationally built an effective aperture that was twice the size of the aperture swept out by the motion. This technique allowed radars, despite their long wavelength, to achieve unprecedented resolution. At a given location, the LiDAR emits a pulse and measures or estimates the returned field from the target. If the return field can be measured at many different locations, those fields can be stitched together in the pupil plane to develop a larger pupil-plane image in the direction of the motion. Normalization, or aperture weighting, may be required where more than one pupil-plane field overlaps. This larger pupil-plane image can be Fourier transformed to obtain a high-resolution image.

Engineers involved in SAR or SAL refer to slow time and fast time. Range is measured in fast time. Light is sent out, bounces off of an object, and returns to the receive aperture. This happens at the speed of light, so it is fast. Azimuth is measured in slow time. In order to develop a full synthetic aperture, the SAL has to move the required distance. This is done at the speed of the vehicle carrying the SAL, so it is in slow time. SAL is a high-resolution 2D sensor, but the two dimensions are range and azimuth, unlike the eye, which also develops an image in two dimensions, but these dimensions are azimuth and elevation. A SAL can also image in elevation, but this is done

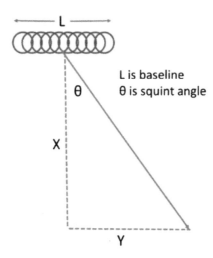

Figure 4.22 Diagram illustrating a squint angle.

with the resolution of the SAL's real beam apertures, not the synthetic aperture. A squint angle is the angle between the forward motion of the SAL and the view angle to the target, as shown in Fig. 4.22.

It is not surprising that synthetic-aperture imagers were developed first at microwave frequencies, where a typical wavelength for a tracking radar will be 3 cm or 10 GHz. With a carrier wavelength of 3 cm, the real-beam-aperture angular resolution for a 60-cm–diameter aperture is 50 mrad. At a range of 100 km, this would be a spatial resolution of 5 km. No wonder back in the 1960s people referred to radars as blob detectors. There was significant incentive to find ways to increase radar resolution compared to the diffraction limit. Also, one challenge for synthetic-aperture sensors is the need to align the pupil-plane images before adding them. Any motion or atmospheric disturbance between pupil-plane images has to be corrected. It is easier at longer wavelengths to align pupil-plane images to a fraction of a wavelength; 5% of 3 cm is 1.5 mm, whereas 5% of 1.5 μm is 75 nm. Aligning multiple pupil-plane field measurements to 1.5 mm is obviously easier than aligning them to 75-nm precision. In addition, the real-aperture angular resolution of LiDAR is very good, so the need for SAL is less. Radar apertures are usually larger; let's compare the angular resolution of a 60-cm–diameter, 10-GHz radar to the angular resolution of a 15-cm–diameter, 1500-nm LiDAR. The radar will have a real beam resolution of 50 mrad, as stated earlier. The LiDAR will have a real beam resolution of 10 μrad. Usually, optical systems have a shorter range than radar systems because of atmospheric attenuation, but for comparison, let's take a range of 40 km in both cases. The 60-cm–diameter radar will have a real-beam cross-range spatial resolution of 2,000 m, and the LiDAR will have a real-beam cross-range resolution of 0.4 m. With optical systems tending to have shorter ranges anyway, as well as great angle resolution, it is easy to see why there was a huge incentive to develop a SAR many decades ago, but less incentive in the optical regime to develop a SAL.

Synthetic-aperture radar has existed for decades. Usually in SAR, the point of closest approach, called the phase center, is set as zero for the reference system. Because of the reference frame adjustment and, more fundamentally, because both the transmitter and the receiver move, the resolution in the along-track direction for SAR is given by

$$\vartheta \approx \frac{\lambda}{2L},$$ (4.7)

where L is the distance flown. This is twice the resolution of a real aperture with a diameter the same as the distance flown. The transmit aperture is always transformed to a location at the reference center. When we transform the location of the transmit aperture, we move the receiver aperture by the same amount, but in the opposite direction. This expands the size of the generated synthetic pupil-plane aperture. For a 1.5-μm SAL, the required distance flown will be 20,000 times shorter than at 10 GHz to obtain the same angular resolution, but the motion compensation requires 20,000 times more precision. SAL should be able to provide long-range, high-resolution imaging in two dimensions, one being range and the other being angle. The range resolution is

$$\Delta R = \frac{c}{2B}.$$ (4.8)

For spotlight SAL, we look at the same target, slewing our look angle, as shown in Fig. 4.23. However, from a phase perspective, the figure looks different, as shown in Fig. 4.24. As we move in slow time while viewing the target, we start out with the target farther away, thereby having more phase difference. We then come to the point of closest approach with a minimum phase difference, and then proceed to the farthest distance with maximum phase difference again. There is a quadratic phase difference as we look at the

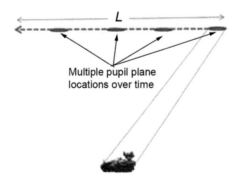

Figure 4.23 Geometry of a SAL (reprinted from Ref. 39).

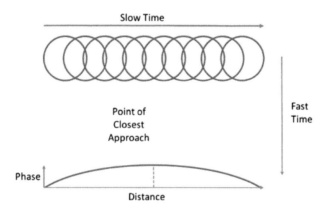

Figure 4.24 Conceptual diagram of a SAL flight path.

target, so the phase forms a parabola. Figure 4.24 is shown inverted in the sense that it shows the closest approach as the maximum phase, but that is because the reference is actually the maximum phase at both ends, and the minimum phase is the highest point in the middle of the figure.

Another thing to consider with SAL is the squint angle. The best azimuth resolution occurs perpendicular to the line of flight. If the object being viewed is forward at some angle then it reduces the effective flight path by a factor of the cosine of the angle. A SAL does not have enhanced resolution directly in front of the vehicle carrying the SAL because there is no effective motion. Holding the 2D resolution constant, SAL systems have power requirements that scale as $1/R^3$.

4.2.4.2 Inverse SAL

Inverse synthetic-aperture LiDAR (inverse SAL) or inverse synthetic-aperture radar (inverse SAR) is the same as SAL or SAR, respectively, except that the angular diversity is obtained by the motion of the object being viewed instead of by the motion of the sensor (see Fig. 4.25). The same physics is at play as in SAL.

In SAL, the target stays oriented the same way, but the LiDAR moves, creating an angular rotation that allows you to obtain angular diversity. The sensor moves a distance L, as described above, and is a distance R away from the target. If the target is a distance R away, then the resolution obtained is $R \times \Delta \vartheta$. If, instead of the LiDAR moving, the target rotates, then

Target rotation can provide angular diversity

Figure 4.25 Inverse-SAL conceptual diagram (reprinted from Ref. 39).

$$\Delta\varphi \approx \frac{L}{R}. \tag{4.9}$$

Now we have the same angular diversity and can obtain the same resolution. Historically, radar scientists test SARs by looking at targets placed on a rotating platform. The physics is again the same. The pupil-plane field must be captured multiple times with the target in multiple orientations. There is a reference frame adjustment, and then a larger pupil-plane image is created, which is Fourier transformed to obtain the object plane image of the target. In SAL, the purpose of reference frame adjustment is to adjust each pupil plane image as though the illuminator remained in the same place. In inverse SAL, the reference frame adjustment takes place as though the illuminator remains at the same angle with respect to the target. The target is rotating, but we consider the motion as if the rotation of the target with respect to the LiDAR were coming from the motion of the LiDAR with respect to the target.

4.3 Multiple-Input, Multiple-Output Active EO Sensing

Instead of using motion to synthesize a larger effective aperture, multiple-input, multiple-output (MIMO) active EO sensing uses multiple physical subapertures to create a larger synthetic pupil-plane aperture. An array of receive-only subapertures can synthesize a larger aperture, as large as the receive array, assuming that the field across the array can be measured or estimated. Multiple transmit apertures further expand the size of the synthetic pupil-plane aperture. In a SAL, both the transmit and receive apertures move. In MIMO, both apertures can be fixed in space, but distributed in angle using multiple subapertures.

Multiple transmit and receive subapertures increase the angular resolution, similar to the angular resolution increase from motion-based SALs. Instead of motion, an array of n subapertures that transmit and receive can be used. It is also possible to have separate arrays for transmitting and for receiving. For nine transmit/receive subapertures in a row, the array will have a diffraction limit consistent with a monolithic aperture that is 1.89 times as large as the length of the array in that dimension. Eight subapertures are equivalent to the distance moved [L in Eq. (4.5)], while one subaperture is equivalent to the real aperture diameter D_{real}. If transmission occurs from one subaperture in the middle of an array, the receive aperture array is effectively in its normal location. If the transmit beam is moved up one subaperture, it is as though the receive aperture were moved down one subaperture. If this process continues, the result is something like what is shown in Fig. 4.26, where the lighter-color linear arrays indicate the perceived location of the linear receive array, depending on which transmit subaperture is used. The dark-colored column shows the actual location of the arrays. The full extent

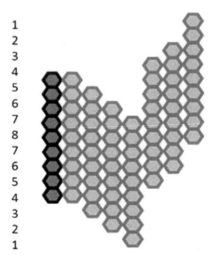

Figure 4.26 Effective receiver aperture placement based on the transmit subaperture utilized. Dark color shows actual location of the arrays. Light color shows perceived location. The number on the left shows how many received subapertures are perceived to be at that location (reprinted from Ref. 39).

of the linear arrays in 1.89 times larger, but it is sampled more near the middle of the synthetic aperture, as represented by eight samples across in the middle, down to one across on either end. This is similar to an apodized aperture. An alternative view of the increased resolution is provided in Fig. 4.27.

The main thrust of Ref. 32 by G. A. Tyler is to argue that, due to the multiple receive and transmit apertures, there are multiple ways to obtain the same optical distances. This allows a scientist to solve for the difference in path lengths between the subapertures and to calculate out that optical path difference without any iteration.

Rabb et al. use only one receive aperture, but multiple transmit apertures,[33] which can also increase resolution. This configuration can be

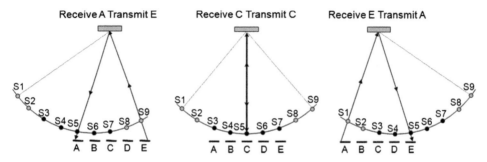

Figure 4.27 Speckle measurements create a synthetic aperture that is almost twice as large as the array. These diagrams illustrate the nature of the speckle return from the object (reprinted from Ref. 32 with permission from OSA).

very useful if the receive apertures are expensive. If tagged transmitters are spaced at distances less than the diameter of a receive subaperture, the resulting sampling can provide a closed-form solution to atmospheric path differences across the subaperture array.[34] An example of the resulting synthetic receiver pupil using illuminator diversity is shown in Fig. 4.28. Because of the overlap, an engineer or scientist can solve for the phase between transmit subapertures, allowing its use to compensate for atmospheric phase disturbances in the pupil plane.

MIMO optical imaging systems are new, and all of their uses have most likely not yet been discovered. Multiple transmitters of course can allow more rapid compensation for speckle because more realizations of speckle can be gathered quickly.

Rabb et al.[33] show that by using multiple closely spaced illuminators, with one receive aperture, curve fitting can be used on the phase differences of the overlapped wavefront fields. In this method, the authors used bivariate expansion to model intra-aperture phase aberrations. Gunturk et al. showed that intra-aperture phase aberrations can be solved for a single receive aperture using a linear systems of equation with three illuminators.[35] The wavefront difference in this research was used to calculate the coefficients up to seventh-order Zernike polynomials to correct for the phase errors. Both of these methods deal with intra-aperture aberrations but do not address the need for image phasing in a multiple–receive-aperture imaging system.

Similar to the mentioned studies of Rabb et al. and Gunturk et al., Krazcek et al. showed that intra- and inter-aperture phase aberrations can be solved in a closed-form manner. These aberrations are then removed from the field.[36]

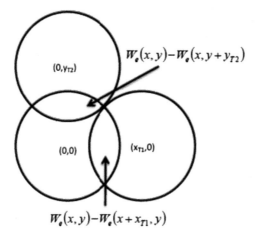

Figure 4.28 Multiple-subaperture overlap using three transmitters spaced one-half of a receive subaperture diameter apart. When images are adjusted to account for illuminator location, the receiver subaperture pupils overlap as shown (reprinted from Ref. 33 with permission from OSA).

At this point, the wavefront difference is all in one plane across a single receive aperture, with just piston, tip, and tilt present (measured with respect to a common plane for all of the receive apertures). These phases are not removable using just one aperture because they are constant across the aperture. Using the wavefront difference of the overlapped fields from the combination of a different aperture and illuminator pair, the tip and tilt can be removed by using the gradient of the wavefront difference. With the removal of tip and tilt, the remaining average phase in the wavefront difference is the piston.

In addition to phasing multiple receive apertures using three illuminators, Krazcek et al.[36] also considered a fourth illuminator widely separated from the three-illuminator cluster. This can greatly extend the size of the synthetic pupil-plane aperture and will approximately double the frequency response cutoff of the system because the fourth illuminator aperture is placed about the width of the aperture array away from the set of three illuminators.

Before performing inter-aperture, closed-form phasing, intra-aperture phase errors must be removed. The method to remove intra-aperture phase errors closely follows Gunturk's approach using three illuminators to remove the intra-aperture phase errors on a given subaperture. The left side of Fig. 4.29 shows an example of a multi-illuminator, single-receive-aperture system. The receive aperture is the large circle on the right, and the three-illuminator cluster is on the left. The cluster is set to the vertices of an equilateral triangle with side lengths equal to the radius of the receive apertures. The right side of Fig. 4.29 shows the measured received fields referenced to a common coordinate system. This is similar to Fig. 4.28, taken from Rabb et al. The overlap regions of the received fields in the pupil plane are due to the different illuminator positions, as shown. The outer edge of this image is the shape of the synthetic pupil. The synthetic pupil results from the different fields being coordinate transformed such that they can be referenced

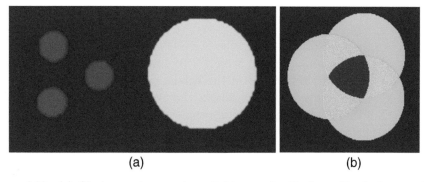

(a)   (b)

Figure 4.29 (a) Single receive aperture (light green) with three illuminator apertures (magenta). (b) Field overlap due to a single receive aperture and a three-illuminator cluster.

to a common coordinate system. This larger synthetic pupil gives rise to a high-resolution image.

As described in Ref. 37, a shift in the illuminator location has an equal and opposite result on the movement of the placement of the received field in the synthetic pupil. This necessitates a coordinate change to compensate for different illuminator locations before the pupil-plane fields can be coherently added. The synthetic pupil is built from the individual pupil fields being registered in the correct location in postprocessing. Pupil fields were assembled by Krazcek et al. using known illuminator positions, since this work was done in simulation.

The fields incident on the receive aperture can be described by

$$U_i(x,y) = P(x,y)\exp[j2\pi W_e(x,y)]U_b(x - x_i, y - y_i), \quad i = 1,2, \qquad (4.10)$$

where $P(x, y)$ is the pupil function of the receive aperture, $W_e(x, y)$ is the wavefront error over the aperture, and $U_b(x - x_i, y - y_i)$ is the received field with the subscript i referring to the ith illuminator. Defining $W_b(x, y)$ as the wavefront of $U_b(x, y)$, the received wavefront is

$$W_i(x,y) = W_e(x,y) + W_b(x - x_i, y - y_i), \quad i = 1,2. \qquad (4.11)$$

The object wavefront is not well defined, and it is desirable to calculate the wavefront aberrations independent of the intended target. It is necessary to register the different fields based on the field shifts caused by illuminator positions. In this way, the numerical dependence on the object wavefront can be removed. Taking the difference of the overlapping areas in the received wavefronts gives

$$
\begin{aligned}
\Delta W(x,y) &= W_1(x - x_1, y - y_1) - W_2(x - x_2, y - y_2) \\
&= [W_e(x + x_1, y + y_1) + W_b(x,y)] - [W_e(x + x_2, y + y_2) + W_b(x,y)] \\
&= W_e(x + x_1, y + y_1) - W_e(x + x_2, y + y_2).
\end{aligned}
$$

$$(4.12)$$

Due to the coordinate shift, the wavefront difference is a difference between the wavefront errors, which can be calculated and removed. Figure 4.30 shows the difference of the wavefront of two overlapped pupil-plane fields. The fringed area is where the two fields overlap. Notice in Fig. 4.30 that there is no speckle within the region of overlap. In that region, the speckle has been subtracted out.

Using the wavefront difference, Gunturk et al.[35] are able to remove the intra-aperture phase errors using a set of overdetermined linear equations, by solving for the Zernike polynomials describing the phase errors across a circular pupil. However, their method does not work for inter-aperture phase

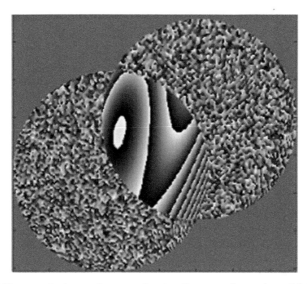

Figure 4.30 Difference between the wavefronts of two overlapped pupil fields. The fringed section in the middle is where the two fields overlap.

errors because the Zernike coefficients are not necessarily the same between two apertures. The zeroth- and first-order, piston, tip, and tilt cannot be solved for in the method described above because they do not show up in the wavefront difference when using the same receive aperture. Piston is a flat phase, and there is no difference in that term across the aperture. Tip and tilt have constant slopes across the aperture and therefore also cancel out when taking the difference of the two wavefronts in the same aperture.

A different method must be developed to correct for the inter-aperture tip-tilt-piston aberrations. The top of Fig. 4.31 shows a three-illuminator, five-receive-aperture array. This is an example of a multi-illuminator, multi-aperture (which is a MIMO) system that can be used for inter-aperture closed-form phasing. The bottom of Figure 4.31 shows the pupil-plane overlap pattern for the system shown in the top of the figure. Each individual receive aperture provides coverage in the same pattern as shown in Fig. 4.29(b). There is significantly less overlap between each aperture, but some overlap still exists. After coordinate transformation, the receive apertures should be located as close together as is physically permitted to allow for greater inter-aperture overlap of the synthetic pupil. This greater overlap is beneficial in the registration process. The wavefront difference is the wavefront error across the overlapped apertures. Assuming that the intra-aperture phase errors have been removed, the wavefront difference is now defined as

$$\Delta W(x,y) = W_{e,i1,j1}(x,y) - W_{e,i2,j2}(x,y) \quad i1 \neq i2, j1 \neq j2, \qquad (4.13)$$

where $i1$ and $i2$ are different illuminators, and $j1$ and $j2$ are different receive apertures.

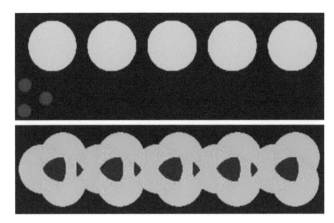

Figure 4.31 (Top) Example of a three-illuminator, five-aperture array. The light green circles are filled receive apertures, and the pink circles are illuminators. (Bottom) Example of the field overlaps from the array shown on top. The light-blue areas are where only one field is present, the yellow are areas of two fields overlapping, and the maroon is where three fields overlap.

Figure 4.32 is an example of the wavefront difference between two pupils coming from two different apertures. The wavefront difference shows only the relative difference between the phase errors on the two apertures. Figure 4.32(a) shows the wavefront difference after intra-aperture corrections have been made, and Fig. 4.32(b) shows the same wavefront difference after tip-tilt corrections have been made. The gradient can be used on the area between phase wraps to gain an estimate of the tip and tilt between the two apertures. This estimate can then be used to correct tip and tilt. At this point, all that is left to correct is piston.

Figure 4.32(b) shows an example of the wavefront difference after all other phase errors are removed, resulting in only piston phase error remaining. The

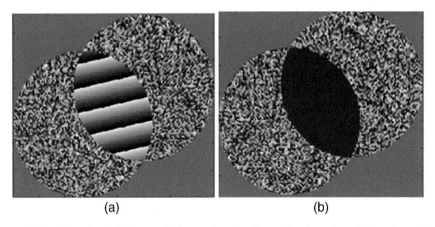

(a)   (b)

Figure 4.32 Wavefront difference between two pupils coming from two different apertures: (a) Piston, tip, and tilt errors remain. (b) Piston error remains (reprinted from Ref. 40).

value of the dark section is the piston phase. In a noise-free environment, both the piston and the tip and tilt corrections can be calculated with very few pixels. In a noisy system, more pixels will be needed, but as phase noise is generally zero-mean Gaussian, using a small number of pixels should gain a reasonably accurate estimate. When correcting for these inter-aperture phase errors, one receive aperture is set as a reference to which all others are corrected. The locations of the different transmitters require this correction. The corrections are daisy chained from the reference field to the field that is farthest away. As seen in Fig. 4.32, only two illuminators are required for inter-aperture phasing. Three illuminators are still used for intra-aperture phasing and to facilitate inter-aperture phasing with 2D receive aperture arrays.

Adding a fourth illuminator can significantly increase the size of the synthetic pupil, depending on the location of the fourth illuminator. The top of Fig. 4.33 shows a four-illuminator, five-receive-aperture, MIMO LiDAR. The bottom of Fig. 4.33 shows the shape of the synthetic pupil. The set of five circles on the left-hand side comes from the fourth, widely separated illuminator. Those fields do not need to overlap since they come from the same apertures that already had the intra- and inter-aperture phase aberrations removed. The exception is that there needs to be overlap between at least one pupil field from the fourth illuminator with one of the pupil fields from the initial three-illuminator cluster. In this example, the farthest right field from the fourth illuminator and the farthest left field from the illuminator cluster overlap. This overlap is used to obtain an additional piston correction for all pupil fields from the fourth illuminator and allow for field registration. Subpixel registration on the speckle in the pupil plane is used. It is assumed that the exact positions of all illuminators are known but overlap is maintained. An additional piston correction will be necessary to account for the different position of the fourth illuminator fields, from where the initial inter-aperture measurements were made. Movement of the pupil fields causes

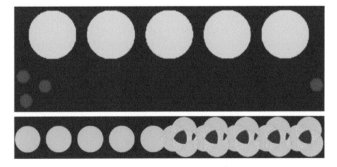

Figure 4.33 Top: Example of a four-illuminator, five-aperture array. The light-green circles are filled receive apertures, and the magenta circles are illuminators. Bottom: Example of the field overlaps from the array shown on top. Colors denote the same areas as in Fig. 4.31 (reprinted from Ref. 40).

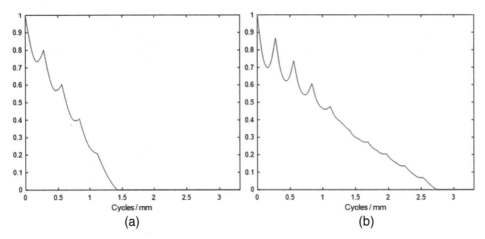

Figure 4.34 MTF plots for the arrays shown in Fig. 4.33: (a) vertical MTF for the array and (b) horizontal MTF for the array (reprinted from Ref. 40).

an additional piston term in the wavefront difference, due to tip-tilt, that needs to be corrected.

Figure 4.34(a) shows the theoretical modulation transfer function (MTF) in the vertical dimension, and Fig. 4.34(b) shows the horizontal dimension. Notice in the horizontal direction that the cutoff frequency for the three-illuminator, five-aperture array is about one-half of the cutoff of the array with four illuminators and five apertures. This is because the synthetic pupil size is about half as large in that dimension.

MIMO techniques, as described here, can be implemented either using temporal heterodyne or spatial heterodyne (digital holography) techniques. However, it will be much easier to tag the emitted transmitter signals (allowing simultaneous transmission) if a high-bandwidth temporal hetero-dyne technique is used, since high-bandwidth tagging schemes can then be used. Radiofrequency MIMO techniques that use multiple, simultaneous phase centers have been developed.[38]

The angular resolution of an array of subapertures can be almost twice that of the diffraction limit for a monolithic aperture. In addition, atmospheric turbulence between the imaged object and the sensor can be very quickly and accurately compensated because we do not need to use an iterative approach to find the required optical path correction to compensate for the turbulence. This compensation is relatively straightforward for turbulence in the pupil plane, but will be more difficult for volume turbulence. These are narrowband sensors; however, using multiple transmitters allows speckle mitigation. A significant limitation is that the received signal is captured in a smaller receive aperture area. The required laser power will increase by the ratio of the area of the monolithic aperture to the area of the received aperture array. Also, if the angular resolution is greater than that of a monolithic subaperture, less laser return will come from each image voxel.

Arrays of high–temporal-bandwidth detectors will be very helpful in implementing MIMO techniques for EO imaging. Rabb et al.[33] and Krazcek et al.[36] sequenced through the multiple transmitters because they implemented MIMO using a digital holography / spatial heterodyne approach to imaging, using low-bandwidth framing detector arrays. If high-bandwidth detector arrays and a temporal heterodyne approach to imaging are used, it should be possible to simultaneously emit multiple tagged transmitter beams and to have each receiver be able to distinguish the transmit aperture any photon came from. High-bandwidth detectors will allow high-bandwidth modulations to be imposed on each transmitted beam and to be sorted on receive. Temporal heterodyne arrays may need to be AC coupled, or have high dynamic range, or have high sensitivity such that a temporal heterodyne technique can be implemented using a relatively weak LO, while still having sufficient signal to noise.

A technical hurdle to overcome is implementing MIMO in volume turbulence. Techniques for calculating and compensating for volume turbulence need to be developed.

MIMO technology allows imaging with high angular resolution using much-lighter and more-compact aperture arrays than a monolithic aperture. An array of small subapertures can be much thinner and lighter than a monolithic aperture. Also, a MIMO approach can achieve almost twice the diffraction-limited angular resolution of a monolithic aperture. Speckle averaging using multiple transmitters will be another advantage.

To freeze the atmosphere, MIMO should be implemented using high-bandwidth detector arrays, which still need to be further developed. A digital implementation will be needed for conducting the required calculations. Many different optical trains will be needed as well, complicating the optical system. Also, narrow-linewidth lasers will be needed in order to perform either spatial or temporal heterodyne sensing. Higher-power lasers will be required to image a given area using a MIMO array compared with imaging with a monolithic aperture.

This technology is well suited for long-range imaging applications from air or space. MIMO could also be used in the cross-range dimension along with motion-based synthetic-aperture imaging.

MIMO approaches for active EO sensing can, at a minimum, increase the effective diameter of an aperture array by almost a factor of two, and can allow multiple subapertures on receive to be phased using a closed-loop calculation of the phase difference between subapertures. This can compensate for atmospheric turbulence at least at some locations between the sensor and the imaged object.

Practical implementation of this approach will require narrowband lasers and coherent imaging, either spatial or temporal. To implement this approach in a changing atmosphere will require gathering all of the data for phasing during a single instantiation of the atmosphere. If the atmosphere changes while various measurements are made, it will be impossible to use this closed-form phase compensation approach.

Appendix 4.1: MATLAB® program showing synthetic-aperture pupil planes and MTFs

```
close all; clear all; clc;
%% Multi_trans_ap_array_and_MTFv2.m
% This simulation allows the user to define a pupil array
% with multiple illuminators by clicking on the displayed
% figure. It then displays the effective pupil with and
% without overlap and several MTF plots. Finally it
% displays the MTF of a circular pupil of equal area to the
% effective pupil.

% Instructions are given in the command window and in the
% title of the figures. Left click on the figure that matlab
% brings to the front when and how you are directed.

% Author: Jeffrey Kraczek
% University of Dayton
% 3/20/15, Last modified 4/17/15
%% Setup
% These are the initial parameters. dx D_phys and ill_scan be
  changed
dx = 100e-6; % Grid increment
D_phys = 22.9e-3; % Physical subaperture diameter [m]
ds = 5e-3; % Dead space
cs = ds + D_phys; %center to center spacing
ill_s = 3e-3; % Clearence for illuminators from edge of
apertures.
p_cs = cs/dx; % Center to center spacing in pixels
%% Hex Center Finder
% Finds the centers for a hex array
Num=5; % Must be odd integer. Number of apertures in middle
row. Number of rows.
Nfinal=(Num+1)/2; % Number of apertures in the shortest row
NTotal=Num+2*sum(Nfinal:(Num-1)); %Total number of apertures

a1=cs*[1 0]; %Transfer vector from triangular coords.
a2=cs*[.5 (3^.5)/2]; %Transfer vector from triangular
coords.
center=zeros(1,2);
```

```
subap_x_cen=zeros(NTotal,1);
subap_y_cen=zeros(NTotal,1);
q=1; %Index for center
for i=1:Num
   v=(Num-1)/2-(i-1); %Initial (i=1) Brings to top of hex
   if i<(Num+1)/2 %top half of hex array
     for k=-Nfinal+1:i-1
       u=k;
       center(1,:)=a1*u+a2*v; %Vector transformation from
       triangular coordinates
       subap_x_cen(q)=round(center(1,1)/dx); %x center
       subap_y_cen(q)=round(center(1,2)/dx); %y center
       q=q+1; %array index
     end
   else %bottom half of hex array
     for k=-Num+i:Nfinal-1
       u=k;
       center(1,:)=a1*u+a2*v;
       subap_x_cen(q)=round(center(1,1)/dx);
       subap_y_cen(q)=round(center(1,2)/dx);
       q=q+1;
     end
   end
end
%% Space builder
x_max = max(subap_x_cen)+p_cs; % Largest x pixel value
x_min = min(subap_x_cen)-p_cs; % Smallest x pixel value
y_max = max(subap_y_cen)+p_cs; % Largest y pixel value
y_min = min(subap_y_cen)-p_cs; % Smallest y pixel value

n = x_min:1:x_max; % Pixel numbers in x centered
m = y_min:1:y_max; % Pixel numbers in y centered
Nx = length(n); % X pixels
Ny = length(m); % Y pixels
% [nx ny] = meshgrid(n,m); % Cartesian grid
R_pix = round(D_phys/(2*dx)); % Radius of subaperture in
pixels
[nxs, nys] = meshgrid(-R_pix:R_pix); % Small Cartesian
grid for single subaperture
sub_ap = nxs.^2+nys.^2 < R_pix^2; % fills subaperture, ones
ill_pix = ill_s / dx; %Space for illumunator in pixels
[nxill, nyill] = meshgrid(-ill_pix:ill_pix); % Small
Cartesian grid for illuminators
```

```matlab
sub_ill = nxill.^2 + nyill.^2 < ill_pix^2; % Fills illumi-
nator, ones

%% Build Hex 19
Aperture = zeros(Ny,Nx); %Array that will contain full
aperture
for count = 1 : NTotal %Fills Aperture with subapertures
  cxi = subap_x_cen(count) + 1 - x_min; %Center x pixel
  cyi = subap_y_cen(count) + 1 - y_min; %Center y pixel
  Aperture(cyi - R_pix:cyi + R_pix, cxi - R_pix:cxi + R_pix) =
  sub_ap;%->
  % Makes a portion of the Aperture array = to sub_ap which is
  % a single subaperture
end
figure
hex19 = gcf; %gcf is a function that grabs the current figure
for figure handles
imagesc(n,m,Aperture);
colormap jet
drawnow; % Forces graphing
%% Choose sub_apertures
commandwindow; % Brings command window to the front
n_aps = input('How many subaperture will you choose? Enter
a number 1 - 19 then press the enter key.');
if isempty(n_aps) || n_aps <1 || n_aps >19 %Makes  sure
numbers within range
  n_aps = 2;
  fprintf('You chose poorly. Number of sub_aperture will be
  %d.\n', n_aps)
end

cxsa = zeros(n_aps,1); % Center in x of pick sub_apertures
cysa = zeros(n_aps,1); % Center in x of pick sub_apertures
sapi = zeros(n_aps,1); % Indicies of chosen subapertures
bool = zeros(n_aps,1); % logical array for to check valid
placement

message = 'Please click on desired subaperture';

CM = [0 0 0.563; 0.5 0 0; 0.563 1 0.438]; % array for new color
map

for count = 1 : n_aps
  logical = 0;
  while logical == 0
    figure(hex19)
```

```
imagesc(n,m,Aperture);
title(message)
[cxsa(count),cysa(count)]=ginput(1);   %Gets   pixel
values from click
cxsa(count) = round(cxsa(count));  %Rounds pixels  to
whole value
cysa(count) = round(cysa(count));
logical = sqrt((cxsa(count)-subap_x_cen).^2 + (cysa
(count)-subap_y_cen).^2) < R_pix;%->
%Indicates if chosen pixel is within a subaperture's
area
if sum(logical) == 0 %checks to see if click off of subap
    message = 'Failed!!!  Please   click   on   desired
    subaperture';
    beep
end
ind = find(logical); %finds index of chosen
if sum(logical) ~= 0 %Records filled subap
for p= 1:n_aps
    bool(p)=sapi(p)==ind;
end
if sum (bool) %Checks to makes sure didn't choose subap
twice
    logical = 0;
    message = 'Failed!!!  Please  click  on  a  different
    subaperture';
    beep
end
end
end
sapi(count) = ind; %Records filled subap
cxi = subap_x_cen(sapi(count))+1-x_min;   %center   x
pixel within array
cyi = subap_y_cen(sapi(count))+1-y_min;   %center   y
pixel within array
Aperture(cyi-R_pix:cyi+R_pix,cxi-R_pix:cxi+R_pix)...
    = Aperture(cyi-R_pix:cyi+R_pix,cxi-R_pix:cxi+R_pix)
    + sub_ap;
% Mrks choosen subap
colormap(CM)
message = 'Please choose next subaperture';
end
```

```
imagesc(n,m,Aperture);

%% Choose transmitters
commandwindow; % Brings command window to the front
% This section determines how many illuminators are desired
oneorthree = input('Do you want a cluster of three closely
spaced illuminators? Enter y or n then press enter','s');
Yes = 'y'; No = 'n';
if strncmpi(oneorthree,Yes,1)
  cluster = 3;
  message = 'Please choose the location for three illumina-
  tor cluster';
elseif strncmpi(oneorthree,No,1)
  cluster = 1;
  message = 'Please choose the location for your first
  illuminator';
else
  fprintf('You chose poorly. You get a cluster of three
  closely spaced illuminators\n')
  cluster = 3;
  message = 'Please choose the location for three illumina-
  tor cluster';
end
n_ill = input('How many additional illuminators will you
choose? Enter a number 0 - 4 then press the enter key.');
if isempty(n_ill) || n_ill <0 || n_ill >4
  n_ill = 0;
  fprintf('You chose poorly. Number of subaperture will be
  %d.\n', n_ill)
end

cxill = zeros(n_ill+cluster,1); % Illuminator x coord
cyill = zeros(n_ill+cluster,1); % illuminator y coord
CM = [0 0 0.563; 0.5 0 0; 0.563 1 0.438; 1 0 1]; % new colormap
for count = 1 : n_ill+1 %Illuminator placement
  logical = 1;
  while sum(logical) ~= 0 %Checks to make sure illuminators
  in valid location
  figure(hex19)
  imagesc(n,m,Aperture);
  title(message)
  [cxill(count),cyill(count)]=ginput(1); %gets pixel from
  clicking on figure
```

```
cxill(count) = round(cxill(count)); %Rounds to whole
pixel
cyill(count) = round(cyill(count)); %Rounds to whole pixel
logical = sqrt((cxill(count)-cxsa).^2 + (cyill(count)-
cysa).^2) < R_pix + ill_pix;%->
% checks to see if illuminator within a filled subap
if sum(logical) ~= 0 %informs illuminator within an in
valid area
   message = 'Failed!!! Please click farther from a filled
   subaperture';
   beep
else %Places illuminator within array Aperture
   cxi = cxill(count) + 1 - x_min;
   cyi = cyill(count) + 1 - y_min;
   Aperture(cyi - ill_pix:cyi + ill_pix,cxi - ill_pix:cxi
   + ill_pix) ...
= Aperture(cyi - ill_pix:cyi + ill_pix,cxi - ill_pix:cxi
+ ill_pix) + sub_ill*3;
end

if count == 1 && sum(logical) == 0 && cluster == 3 % 3 ill
cluster

   log = 1;
   while log ~= 0
      figure(hex19)
      imagesc(n,m,Aperture);
      colormap(CM)
      title('Please choose direction for second transmitter')
      [cxill(count+n_ill+1),cyill(count+n_ill+1)]=
      ginput(1); %gets pixel value from clicking on figure
      a1 = atan2(cyill(count+n_ill+1) - cyill(count),
      cxill(count+n_ill+1) - cxill(count)); % angle between
      first illuminator and second
      a2 = a1 + 60*pi/180; %angle between first and third
      illuminators
      cxill(count+n_ill+1) = round(cxill(count)
      +R_pix*cos(a1));
%Finds value of 2nd ill
      cyill(count+n_ill+1) = round(cyill(count)
      +R_pix*sin(a1));
      cxill(count+n_ill+2) = round(cxill(count)
      +R_pix*cos(a2));
%Finds val of 3rd ill
```

```
        cyill(count+n_ill+2) = round(cyill(count)+R_pix*-
        sin(a2));
        log = sqrt((cxill(count+n_ill+1)-cxsa).^2 +
        (cyill(count+n_ill+1)-cysa).^2) < R_pix + ill_pix;
        log2 = sqrt((cxill(count+n_ill+2)-cxsa).^2 +
        (cyill(count+n_ill+2)-cysa).^2) < R_pix + ill_pix;
        log = sum(log)+sum(log2); %Checks to see if ill ended
        up in invalid areas
      if log~= 0 %Informs of invalid placement
        message = 'Failed!!! Please   choose   a   direction
        farther from a filled subaperture';
        beep
      else %Places ills in array Aperture
        cxi = cxill(count+n_ill+1) + 1 - x_min;
        cyi = cyill(count+n_ill+1) + 1 - y_min;
        Aperture(cyi - ill_pix:cyi + ill_pix, cxi - ill_pix:
        cxi + ill_pix) ...
    = Aperture(cyi - ill_pix:cyi + ill_pix, cxi - ill_pix:cxi +
    ill_pix) + sub_ill*3;
        cxi = cxill(count+n_ill+2) + 1 - x_min;
        cyi = cyill(count+n_ill+2) + 1 - y_min;
        Aperture(cyi - ill_pix:cyi + ill_pix, cxi - ill_pix:
        cxi + ill_pix) ...
    = Aperture(cyi - ill_pix:cyi + ill_pix, cxi - ill_pix:cxi +
    ill_pix) + sub_ill*3;
          end
        end
      end

    end
    message = 'Please choose next illuminator location';
    colormap(CM)
end
figure(hex19)
imagesc(n,m,Aperture);
title('Aperture and Illuminator Array')
drawnow
%% Effective pupil plane
n_ill = n_ill+cluster; %Number of illuminators
n_ps = n_aps*n_ill; %Number of pupils
%(Effective number of sub_aps based on actual subaps and
number of illuminators)
ap_cen_x = zeros(n_ps,1);
ap_cen_y = zeros(n_ps,1);
```

```
for p = 1 : n_aps %pupil center pixels due to first illuminator
  ap_cen_x(p) = subap_x_cen(sapi(p));
  ap_cen_y(p) = subap_y_cen(sapi(p));
end
for p = 2 : n_ill %cycles through remaining illuminators to
get pupil centers
  difx = cxill(p) - cxill(1); % x coordinate transform from
  changing ill
  dify = cyill(p) - cyill(1); % y coordinate transform from
  changing ill
  for q = 1 : n_aps %cycles through subaps to get pupil center
  pixels
    ap_cen_x(n_aps*(p-1)+q) = subap_x_cen(sapi(q)) -
    difx;
    ap_cen_y(n_aps*(p-1)+q) = subap_y_cen(sapi(q)) -
    dify;
  end
end
x_max = max(ap_cen_x) +p_cs; % Larget x
x_min = min(ap_cen_x) -p_cs; % Smallest x
y_max = max(ap_cen_y) +p_cs; % Larget y
y_min = min(ap_cen_y) -p_cs; % Smallest y
np = x_min:1:x_max; % Pixel numbers in x centered
mp = y_min:1:y_max; % Pixel numbers in y centered
Nxp = length(np); % X pixels
Nyp = length(mp); % Y pixels
pupil = zeros(Nyp,Nxp); %Array for pupil plane

for count = 1 : n_ps % Draws Pupil plane
  cxi = ap_cen_x(count) +1 - x_min;
  cyi = ap_cen_y(count) +1 - y_min;
  pupil(cyi - R_pix:cyi + R_pix,cxi - R_pix:cxi + R_pix) = ...
    pupil(cyi - R_pix:cyi + R_pix,cxi - R_pix:cxi + R_pix) +
    sub_ap;
end
figure
pupil_plane = gcf;
imagesc(np,mp,pupil);
colormap(jet)
title('Synthetic Pupil with Overlap')
%% pupil plane normalization
% pupil plane normalization tells what the theoretical
% resolution will be for a system with multiple illuminators.
mask = pupil > 0; %Finds where pupils overlap
```

```
pupil(mask) = pupil(mask)./pupil(mask);          %normalizes
overlap areas to 1
figure
imagesc(np,mp,pupil);
colormap(gray)
title('Synthetic Pupil')
p_area = sum(sum(pupil))*dx^2; %Gets physical area to com-
pare MTF of
% systhetic pupil to monolithic pupil of same area
%% Freeing up space
clear n m Nx Ny nx ny nxs nys sub_ap ill_pix nxill nyill
sub_ill np mp Aperture mask
%% Array MTF

[Ny,Nx] = size(pupil);
if Ny < Nx %Check to see which dimension is larger to make a
square array
   Ny = Nx;
else
   Nx = Ny;
end
nfy = Ny*2+1; nfx = Nx*2+1;
Ipsf=abs(fftshift(fft2(pupil,nfy,nfx))).^2; % Intensity
point spread function
otf=fftshift(ifft2(Ipsf)); %optical transfer function
m_otf=max(max(abs(otf))); %max of otf
mtf=abs(otf)/m_otf; %modulation transfer function

clear Ipsf otf pupil
% %% Makes shure rounding does nt move the center of the MTF
% from the center pixel
[vals,inds] = max(mtf);
[vals,indx] = max(vals);
[val,indy]=max(mtf(:,indx));
difx = indx - Nx - 1;
dify = indy - Ny - 1;
if difx ~=0 || dify ~=0
mtf = imtranslate(mtf, [difx,dify]);
end
% %%

fx = (-Nx:Nx)/(nfx*dx); %Spatial frequency label
fy = (-Ny:Ny)/(nfy*dx);
figure
imagesc(fx,fy,mtf)
```

```
axis image
colormap jet
title('Three Dimentional MTF')

figure
hor_mtf = mtf(Ny+1,Nx+1:nfx);
plot(fx(Nx+1:nfx),hor_mtf)
axis([0 max(fx) 0 1])
title('Horizontal MTF')
figure
vert_mtf = mtf(Ny+1:nfy,Nx+1);
plot(fy(Ny+1:nfy),vert_mtf)
axis([0 max(fx) 0 1])
title('Vertical MTF')
%% MTF Rotation and optimal directions by area under the MTF
rot_mtf = mtf;
tot_rot = 90;    % degrees of rotation
rot_inc = 1;     % Increment of rotation
max_a = 0;       % value of maximum area under MTF
min_a = 1e12;    % value of minimum area under MTF
countmx = 0;
countmn = 0;
% Long algorithm for coming up with the max and min. It was
% supposed to find degenerate angles as well, but the ares
% are not exactly equal so it does not work. Works to find an
% angle for both, so I didn't worry about fixing it at this
% time
for ang = 0:rot_inc:tot_rot
  rot_mtf = imrotate(mtf,ang,'bicubic','crop');
  sum_h = sum(rot_mtf(Ny+1,Nx+1:nfx));
  sum_v = sum(rot_mtf(Ny+1,Nx+1:nfy,Nx+1));
  if sum_v >= max_a || sum_h >= max_a %Checking for maximum
  area
    if sum_v > sum_h % checks for vertical and horizontal
      s = sum_v;
      m = rot_mtf(Ny+1:nfy,Nx+1);
      angle = ang + 90;
  else
      s = sum_h;
      m = rot_mtf(Ny+1,Nx+1:nfx);
      angle = ang;
  end
  if s == max_a %checks to see if one of the two directions is a
  degenerate
```

```
        countmx = countmx + 1;
        anglemx(countmx) = angle;
    else %if not degenerate sets the new max
        max_a = s;
        mtf_mx = m;
        countmx = 0;
        clear anglemx
        anglemx(1) = angle; %Sets angle of max
    end
end
if sum_v <= min_a || sum_h <= min_a %Checking for minimum
area
    if sum_v < sum_h %Checks to see which direction smaller
        s = sum_v;
        m = rot_mtf(Ny+1:nfy,Nx+1);
        angle = ang + 90;
    else
        s = sum_h;
        m = rot_mtf(Ny+1,Nx+1:nfx);
        angle = ang;
    end
    if s == min_a % Checks for degenerate
        countmn = countmn + 1;
        anglemn(countmn) = angle;
    else %If not degenerate sets new min
        min_a = s;
        mtf_mn = m;
        countmn = 0;
        clear anglemn
        anglemn(1) = angle; %sets new min angle
    end
  end
end
figure
plot(fx(Nx+1:nfx),mtf_mx)
if countmx > 1
  tmx = sprintf('Maximum MTF, Angle = %d with %d Degenera-
  cies',anglemx, countmx-1);
else
  tmx = sprintf('Maximum    MTF,    Angle = %d%c',anglemx,
  char(176));
end
title(tmx)
```

```
axis([0 max(fx) 0 1])
figure
plot(fx(Nx+1:nfx),mtf_mn)
axis([0 max(fx) 0 1])
if countmx > 1
  tmn = sprintf('Minimum MTF, Angle = %d with %d degenera-
  cies',anglemn, countmn-1);
else
  tmn = sprintf('Minimum    MTF,    Angle = %d%c',anglemn,
  char(176));
end
title(tmn)
figure
mtfI = gcf;
imagesc(fx,fy,mtf)
axis image
fxmx = max(fx); fxmn = min(fx);

colormap jet
hold on
plot(fx,fx*tan(anglemx(1)*pi/180),':g',fx,fx*tan(an-
glemn(1)*pi/180),'-.k')
hold off
axis([fxmn fxmx fxmn fxmx])
title('MTF with Maximum and Minimum Slices Marked')

legend(tmx,tmn)
%% Effective Monolithic Aperture MTF
R_eff = sqrt(p_area/pi); %Radius of Effective monolithic
aperture
R_epix = ceil(R_eff / dx); %Radius in pixels
ndx = (-R_epix:R_epix)*dx; %physical vector
[nx, ny] = meshgrid(ndx);
mono = sqrt(nx.^2+ny.^2) < R_eff; %fills  the  array  with
effective monolithic aperture

Ipsf=abs(fftshift(fft2(mono,nfy,nfx))).^2;    %Intensity
point spread function
otf=fftshift(ifft2(Ipsf)); %OTF
m_otf=max(max(abs(otf))); % max of OTF
mtf_e=abs(otf)/m_otf; % MTF

fx = (-Nx:Nx)/(nfx*dx);
fy = (-Ny:Ny)/(nfy*dx);
figure
```

```
plot(fx(Nx+1:nfx),mtf_e(Ny+1,Nx+1:nfx))
title('MTF of Monolithic Aperture with Equal Area to
Synthetic Pupil')
figure
imagesc(fx,fy,mtf_e)
axis image
colormap jet
title('Three dimensional MTF of Monolithic Aperture')
```

Problems and Solutions

4-1. Take a SAL with a real aperture size of 15 cm. Assume that the vehicle carrying the SAL travels horizontally at 150 m/s and that we have a synthetic-aperture time of 4 ms and a squint angle of 60 deg. The bandwidth of the LiDAR is 1 GHz. The wavelength of the LiDAR is 1550 nm, and the range is 50 km.

(a) What resolution does the LiDAR have in all three dimensions?

(b) Change the synthetic-aperture time to 10 ms and provide the 3D resolution.

4-1 Solution:

(a) $L = v\Delta t = (150\,\text{m/s})(4\,\text{ms}) = 0.6\,\text{m}.$

Cross-range resolution is

$$\delta = \frac{R\lambda}{2L + D} = \frac{(50\,\text{km})(1550\,\text{nm})}{2(60\,\text{cm}) + (15\,\text{cm})} = 5.74\,\text{cm}.$$

Range resolution is

$$\Delta R = \frac{c}{2B} = \frac{c}{2(1\,\text{GHz})} = 15\,\text{cm}.$$

(b) Changing the synthetic-aperture time to 10 ms, the range resolution stays the same.

Cross-range is now

$$L = v\Delta t = (150\,\text{m/s})(10\,\text{ms}) = 1.5\,\text{m}.$$

Cross-range resolution is

$$\delta = \frac{R\lambda}{2L + D} = \frac{(50\,\text{km})(1550\,\text{nm})}{2(150\,\text{cm}) + (15\,\text{cm})} = 2.46\,\text{cm}.$$

The angle resolution in elevation is $\vartheta_E \approx \lambda/D$, so the elevation resolution is (50 km)(1500 nm/15 cm) = 51.7 cm. In azimuth we will have higher resolution.

4-2. We want to measure the vibration of an object. We have a LiDAR that is 1 km away from that object and is at a wavelength of 1550 nm. The object is vibrating with an amplitude of 3 μm at a frequency of 10 Hz. What is the maximum Doppler shift frequency we will have to measure? If we can measure down to a Doppler frequency shift of 10 Hz, what percentage of the vibration sine wave will be measured?

4-2 Solution:

$$v = Af_t = (3\,\mu\text{m})(10\,\text{Hz}) = 30\,\mu\text{m/s},$$

$$\Delta f = \frac{2v}{\lambda} = \frac{2(30\,\mu\text{m/s})}{(1550\,\text{nm})} = 38.71\,\text{Hz}.$$

Because

$$\frac{\int \sin x\,dx}{\int \sin x'\,dx'} = \frac{\cos x}{\cos x'},$$

$$\frac{\cos 2\pi\Delta f/f_t}{\cos 2\pi} = \cos 2\pi\Delta f/f_t = \cos 2\pi(38.71)/(10) = 0.6891.$$

Therefore, 68.91% of the vibration sine wave will be measured.

4-3. Plot Doppler frequency versus velocity for illumination by 2000- and 10,000-nm lasers for velocities from 0 to 100 m/s.

4-3 Solution:

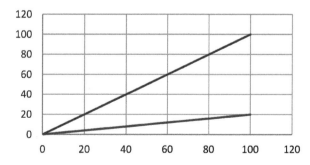

4-4. For a 0.5-μm magnitude of vibration at 1 Hz, calculate the maximum velocity and Doppler shift frequency, assuming 1550-nm wavelength light. What rate of sampling is required for detection?

4-4 Solution:

$$v = Af_t = (0.5\,\mu m)(1\,\text{Hz}) = 0.5\,\mu m/s,$$

$$\Delta f = \frac{2v}{\lambda} = \frac{2(0.5\,\mu m/s)}{(1550\,\text{nm})} = 0.6452\,\text{Hz},$$

$$v = 2\pi f A \cos(2\pi ft).$$

To detect a frequency of 0.6452 Hz, we would need a sample time of 0.775 for Nyquist sampling.

4.5. In Problem 4-4, if the maximum sampling time is 100 ms, what should be altered to make that vibration detectable?

4-5 Solution: The Doppler frequency should be increased to 5 Hz. This can be done by increasing the amplitude, or frequency, of the vibration, or by using a shorter wavelength. The wavelength of the LiDAR is under the control of the user.

4-6. Plot the required sampling time versus vibration amplitude at a vibration frequency of 10 Hz.

4-6 Solution:

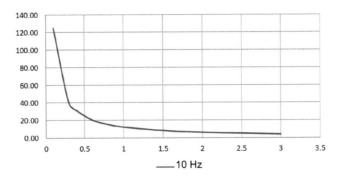

4-7. For a three-illuminator and four-aperture array, repeat the calculation mentioned in Section 4.3 and

    (a) Plot the synthetic pupil-plane aperture, including overlaps.
    (b) Plot the horizontal and vertical MTF plot for the array.

4-7 Solution:

(a)

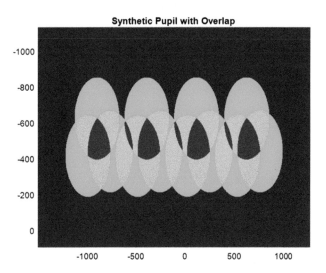

(b)

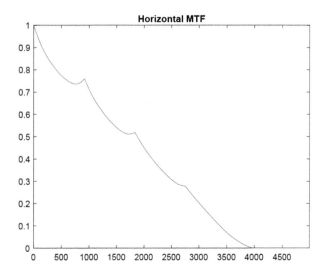

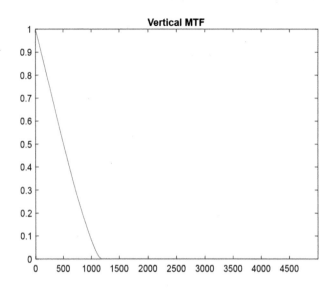

4-8. Find the cross-range resolution for a SAL, with the target moving at Mach 1.8 (relative to sea level) and a squint angle of 40 deg. Use a wavelength of 1550 nm and assume a synthetic-aperture time of 5 ms.

4-8 Solution: Mach 1 is roughly 343 m/s, so Mach 1.8 is 617.4 m/s.

$$L = vt = (5\,\text{ms})(617.4\,\text{m/s}) = 3.087\,\text{m},$$

$$\Delta\vartheta = \frac{\lambda}{2L} = \frac{1550\,\text{nm}}{2(3.087\,\text{m})} = 251.05\,\text{nrad}.$$

4-9. Plot the velocity of a target versus cross-range resolution for an inverse SAL, from 10 m/s to 300 m/s. Assume that the synthetic-aperture time of the LiDAR is 10 ms throughout, and a wavelength of 1550 nm.

4-9 Solution:

```
v=10:0.1:300;
t=0.01;
L=v*t;
w=1.55e-6;
dth = w./(2*L);
plot(v,dth*1e6)
```

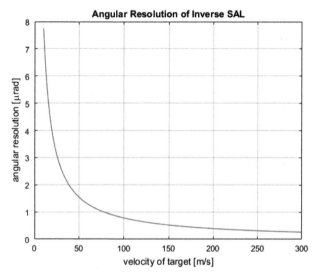

4-10. Plot the angular resolution $\Delta\vartheta$ versus the angular resolution $\Delta\varphi$ for an inverse SAL, with a range of velocities from 1 to 10 m/s, range at 1 km, and assuming 1550-nm wavelength and 10-ms synthetic-aperture time. What is the intersection of these two plots?

4-10 Solution:

```
v=1:0.1:10;
t=0.01;
L=v*t;
w=1.55e-6;
r=1e3;
dth=w./(2*L);
dph=L/r;
plot(v,dth,v,dph)
```

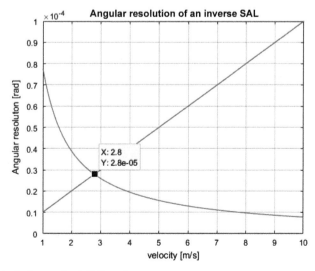

The two plots intersect at 2.8 m/s.

References

1. J. R. Fienup, "Phase retrieval algorithms: a comparison," *Appl. Opt.* **21**(15), 2758–2769 (1982).
2. O. Steinvall, M. Tulldahl, F. Berglund, and L. Allard, "Laser profiling for airborne target classification," *Proc. SPIE* **10636**, 1063602 (2018) [doi: 10.1117/12.2303965].
3. V. W. Abei, "Electron bombarded active pixel sensor," U.S. Patent 6,285,018 (2001).
4. R. A. LaRue, K. A. Costello, and V. W. Aebi, "Hybrid photomultiplier tube with high sensitivity," U.S. Patent 5,374,826 (1994).
5. J. Beck, T. Welch, P. Mitra, K. Reiff, X. Sun, and J. Abshire, "A highly sensitive multi-element HgCdTe e-APD detector for IPDA applications," *J. of Electronic Materials* **43**(8), 2970–2977 (2014).
6. J. Taboada and L. A. Tamburino, "Laser imaging and ranging system using two cameras," U.S. Patent 5,157,451 (1992).
7. American Society for Photogrammetry and Remote Sensing, "ASPRS Positional Accuracy Standards for Digital Geospatial Data," Edition 1, Version 1.0, Nov (2014).
8. National Geodetic Survey Continuously Operating Reference Station (CORS) https://www.ngs.noaa.gov/CORS/.
9. Federal Geographic Data Committee, U.S. Geological Survey, "Geospatial Positioning Accuracy Standards Part 3: National Standard for Spatial Data Accuracy," FGDC STD-007.3-1998.
10. National Geodetic Survey https://www.ngs.noaa.gov/.
11. National Cooperative Highway Research Program (NCHRP), *Guidelines for the Use of Mobile LIDAR in Transportation Applications*, Report 748, Transportation Research Board, National Academies (2013).
12. A. F. Chase and D. Z. Chase, "Detection of Maya Ruins by LiDAR: Applications, Case Study, and Issues," Chapter 22 in *Sensing the Past: From Artifact to Historical Site*, Vol. 16 *Geotechnologies and the Environment*, N. Masini and F. Soldovieri, Eds., Springer International Publishing AG, Cham, Switzerland, pp. 455–468 (2017).
13. U.S. Geological Survey, "National Enhanced Elevation Assessment," Final Report (2012).
14. L. J. Sugarbaker, E. W. Constance, H. K. Heidemann, A. L. Jason, V. Lukas, D. L. Saghy, and J. M. Stoker, "The 3D elevation program initiative: A call for action," U.S. Geological Survey Circular 1399 (2014).
15. W. E. Clifton, B. Steele, G. Nelson, A. Truscott, M. Itzler, and M. Entwistle, "Medium altitude airborne Geiger-mode mapping LIDAR system," *Proc. SPIE* **9465**, 946506 (2015) [doi: 10.1117/12.2193827].
16. National Academy of Sciences, *Laser Radar: Progress and Opportunities in Active Electro-Optical Sensing*, P. F. McManamon, Chair (Committee on Review of Advancements in Active Electro-Optical Systems to Avoid

Technological Surprise Adverse to U.S. National Security), Study under Contract HHM402-10-D-0036-DO#10, National Academies Press, Washington, D.C., p. 63 (2014).

17. National Academy of Sciences, *Laser Radar: Progress and Opportunities in Active Electro-Optical Sensing*, P. F. McManamon, Chair (Committee on Review of Advancements in Active Electro-Optical Systems to Avoid Technological Surprise Adverse to U.S. National Security), Study under Contract HHM402-10-D-0036-DO#10, National Academies Press, Washington, D.C., p. 88 (2014).

18. R. C. Hardie, M. Vaidyanathan, and P. F. McManamon, "Spectral band selection and classifier design for a multispectral imaging laser radar," *Opt. Eng.* **37**(3), 752–762 (1998) [doi: 10.1117/1.601907].

19. M. Vaidyanathan, T. P. Grayson, R. C. Hardie, L. E. Myers, and P. F. McManamon, "Multispectral laser radar development and target characterization," *Proc. SPIE* **3065**, 255–266 (1997) [doi: 10.1117/12.281017].

20. D. Killinger, P. Mamidipudi, J. Potter, J. Daly, E. Thomas, and S. E. Chen, "Laser Doppler vibration lidar sensing of structural defects in bridges," Tenth Biennial Coherent Laser Radar Technology and Applications Conference, Mount Hood, Oregon, June 28–July 2 (1999).

21. D. Killinger, "System and method for multi-beam laser vibrometry triangulation mapping of underground acoustic sources," U.S. Patent 7190635 (2007).

22. B. W. Krause, J. Buck, C. Ryan, D. Hwang, P. Kondratko, A. Malm, A. Gleason, and S. Ashby, "Synthetic aperture ladar flight demonstration," CLEO 2011 Laser Science to Photonic Applications, 1–6 May, Baltimore (2011).

23. P. F. McManamon and W. Thompson, "Phased array of phased arrays (PAPA) laser systems architecture," *IEEE Aerospace Conference Proceedings*, 9–16 March, Big Sky, Montana (2002).

24. R. Sullivan, "Pulse Doppler Radar," Chapter 17 in *Radar Handbook*, Second Edition, M. I. Skolnik, Ed., McGraw-Hill, New York (1990).

25. M. Soumekh, *Synthetic Aperture Radar Signal Processing with MATLAB Algorithms*, John Wiley & Sons, New York, p. 75 (1999).

26. M. A. Richards, *Fundamentals of Radar Signal Processing*, McGraw-Hill, New York, pp. 390–396 (2005).

27. M. I. Skolnik, *Introduction to Radar Systems*, Second Edition, McGraw-Hill, New York, Chapter 14 (1980).

28. S. M. Beck, J. R. Buck, W. F. Buell, R. P. Dickinson, D. A. Kozlowski, N. J. Marechal, and T. J. Wright, "Synthetic-aperture imaging laser radar: laboratory demonstration and signal processing," *Appl. Opt.* **44**(35), 7621–76294 (2005).

29. B. D. Duncan and M. P. Dierking, "Holographic aperture ladar," *Applied Optics* **48**(6), 1168–1177 (2009).

30. J. R. Fienup, "Phase retrieval algorithms: a comparison," *Applied Optics* **21**(15), 2758–2769 (1982).
31. W. G. Carrara, R. S. Goodman, and R. M. Majewski, *Spotlight Synthetic Aperture Radar Signal Processing Algorithms*, Artech House, Norwood, Massachusetts (1995).
32. G. A. Tyler, "Accommodation of speckle in object-based phasing," *J. Opt. Soc. Am. A* **29**(5), 722–733 (2012).
33. D. J. Rabb, J. W. Stafford, and D. F. Jameson, "Non-iterative aberration correction of a multiple transmitter system," *Optics Express* **19**(25), 25048 (2011).
34. J. R. Fienup, "Phase error correction for synthetic-aperture phased-array imaging systems," *Proc. SPIE* **4123**, 47–55 (2000) [doi: 10.1117/12.409285].
35. B. Gunturk, D. Rabb, and D. Jameson, "Multi-transmitter aperture synthesis with Zernike based aberration correction," *Optics Express* **20**(24), 5179–5186 (2012).
36. J. R. Krazcek, P. F. McManamon, and E. A. Watson, "High resolution non-iterative aperture synthesis," *Optics Express* **24**(6), 6229–6239 (2016).
37. D. Rabb, D. F. Jameson, J. W. Stafford, and A. J. Stokes, "Multi-transmitter aperture synthesis," *Optics Express* **18**(24), 24937–24945 (2010).
38. S. Coutts, K. Cuomo, J. McHarg, F. Robey, and D. Weikle, "Distributed coherent aperture measurements for next generation BMD radar," *Fourth IEEE Workshop on Sensor Array and Multichannel Processing*, 12–14 July, Waltham, Massachusetts 2006).
39. P. McManamon, *Field Guide to Lidar*, SPIE Press, Bellingham, Washington (2015) [doi: 10.1117/3.2186106].
40. J. R. Kraczek, "Noniterative Multi-Aperture and Multi-Illuminator Phasing for High-Resolution Coherent Imaging," Ph.D. Thesis, University of Dayton (2017).

Chapter 5
LiDAR Sources and Modulations

5.1 Laser Background Discussion

Active EO sensors employ coherent sources in the wavelength region from the LWIR (around 10 μm) to the atmospheric transmission limit for UV light (around 200 nm). The sources can be based either on lasers or on nonlinear optical systems driven by lasers. Lasers are typically categorized by the type and format of the medium used to generate their output, which at the highest level are gases, liquids, and solids.

Among the solid media used to generate the laser output, materials are further categorized by their electrical characteristics. Solid state lasers employ insulating solids (crystals, ceramics, or glasses) with elements added (dopants) that provide the energy levels needed for laser action. Energy to excite the levels is provided by other sources of light, either conventional sources such as arc lamps, or other lasers, in a process called optical pumping.

Solid state lasers in turn are divided into two broad categories, bulk or fiber, with the latter having recently emerged as an important technology for generation of high average powers with high beam quality, as discussed below.

Even though they are also made from solid state materials, semiconductor lasers are considered as a separate laser category. While the lasers can be made to operate by optical pumping, if the semiconductor material can be fabricated in the form of an appropriate p–n junction, it is possible to pass electrical current through the junction and generate laser output directly. These diode lasers are by far the most widely used form of semiconductor laser and have led to major advances in source technology for active EO sensors.

One of the main benefits touted for lasers is their coherence. Unfortunately, the word coherence is not meaningful without an adjective in front of it. Spatial coherence means that laser beams can be very narrow. Temporal coherence means that the laser beams can have a narrow spectral linewidth. Temporal coherence can also be expressed as the coherence length of a laser. Figure 5.1 shows spatial modes for lasers with various levels of spatial coherence.[1]

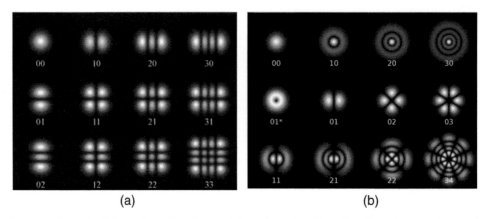

Figure 5.1 Spatial modes of a laser: (a) rectangular Hermite–Gaussian modes and (b) circular Laguerre–Gaussian modes (reprinted from Ref. 1).

The (0,0) modes can be close to what we call diffraction limited, meaning that the laser beam is spatially narrow. If we were talking in microwave jargon, we would say that these lasers have high antenna gain. The (0,0) mode, whether using rectangular or circular modes, will have angular resolution close to the diffraction limit given approximately by Eq. (3.1): $\vartheta \approx \lambda/D$, where λ is the wavelength, and D is the beam diameter at the aperture. Some lasers, however, have higher-order modes and spread out faster in angle, as indicated in Fig. 5.1. For flash-illuminated LiDARs, higher-order modes can be acceptable, as discussed in Chapter 4.

Temporal coherence tells us how monochromatic a laser beam is. A LiDAR that beats the return signal against a local oscillator (LO) is called a heterodyne, or coherent, LiDAR and needs a narrow linewidth for beating between lasers to produce interference. The coherence length of a laser is a measure of how far the laser beam has to travel before a full wavelength of phase shift occurs between the extreme frequencies in the laser linewidth. Interference, or heterodyne mixing, does not occur beyond the coherence length. For a LiDAR, the laser pulse has to travel to the target and return; therefore, without delaying a copy of the emitted signal, the coherence length of the laser in a coherent LiDAR must have a coherence length longer than the round-trip distance to the target.

If you delay a sample of the outgoing signal so that you can use it as the LO, the coherence length of the laser used in a coherent LiDAR has to be more than twice the range depth of the target. You can calculate coherence length C_1 by the speed of light c divided by linewidth B:

$$C_1 = \frac{c}{B}. \tag{5.1}$$

If two laser beams are beat against each other with less than one coherence length separation in the distances traveled, constructive and

Table 5.1 Coherence lengths.

Laser bandwidth	Laser coherence length
1 kHz	300 km
10 kHz	30 km
100 kHz	3 km
1 MHz	300 m
10 MHz	30 m
100 MHz	3 m
1 GHz	30 cm
10 GHz	3 cm

destructive interference will occur between the two beams. As we get closer to the coherence length, the contrast of the fringes gradually decreases as the range gets longer. With more than one coherence length of travel before interfering, the beams there will exhibit random interference, so phase information is lost. The speckle fringes will have zero contrast. Table 5.1 shows the coherence length for lasers of certain bandwidths.

Speckle is a form of interference, so narrow-linewidth lasers exhibit speckle. Lasers with a broad linewidth speckle average over many different wavelengths according to

$$B = \frac{c}{\lambda^2} \Delta\lambda, \tag{5.2}$$

which shows how to convert from laser linewidth in wavelength to laser linewidth in frequency, or bandwidth. Laser filters are usually specified in wavelength, but for heterodyne LiDARs, the bandwidth is usually quoted in frequency. A very narrow filter on receive does not usually result in as narrow a linewidth as when using heterodyne detection. This can be seen from Table 5.2. A 10-GHz linewidth, temporal heterodyne LiDAR would be very broadband but is very narrow when looked at from the width of a wavelength filter. A wavelength-based filter with a 10-GHz bandpass would be <0.1 nm in width. A filter this narrow will probably be a volume holographic filter with a narrow acceptance angle. Most filters resulting from multiple coatings will not be this narrow.

Table 5.2 Frequency and wavelength conversion for certain laser linewidths.

$\Delta\lambda$ (nm)	B (GHz)
0.1	13
1	126
10	1265
100	12650

In this chapter, lasers that might be used in current and future LiDARs are emphasized. Historic lasers for LiDAR are discussed in the history chapter. Lasers of current interest for LiDARs are primarily diode lasers, fiber lasers, and bulk solid state lasers. Quantum cascade lasers may also be used. Gas lasers have faded from LiDAR use, although as late as the first decade of this century, a CO_2-based LiDAR was developed for synthetic-aperture LiDAR. Figure 5.2 shows a diagram of a basic laser.

Pulsed LiDARs often use a solid state material that has a long upper-state lifetime, so it is possible to Q switch the laser to create a nanosecond-class series of pulses. Using a long upper-state lifetime requires fewer laser diodes and lower cost to generate the same energy per pulse. High range resolution in a LiDAR requires short laser pulses or large bandwidth modulation. Solid state lasers might be Nd:YAG at 1.064 µm or various solid state lasers operating at 1.5 or 2 µm to be more eye safe. Very recently, fiber lasers have become popular for use with LiDARs. Fiber lasers can have more than 30% efficiency, even based on wall plug efficiency. They are restricted to lower peak powers than bulk solid state lasers because of the small cross-section of the fibers. Fiber lasers in LiDAR normally operate at >10-kHz pulse repetition rate or higher. This can limit peak power, which can be an issue in laser fibers. Fibers have a small area, so high peak powers can cause Brillion nonlinear interactions, or even more serious issues. Lastly, diode lasers—the most efficient form of laser—can be used with LiDAR. Fiber or bulk solid state lasers are usually pumped using diodes. Diode lasers often do not have narrow linewidths or a narrow spatial beam. Fiber or solid state lasers can be thought of as coherence converters, taking lower–spatial- and temporal-coherence diode light and making a higher-quality laser.

5.2 Laser Waveforms for LiDAR

5.2.1 Introduction

Some fundamental laser issues are encountered when developing a LiDAR waveform. Direct-detection LiDARs cannot measure phase. They measure time of flight for range measurement by using pulses. When using a pulsed

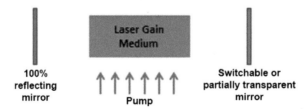

Figure 5.2 A simple diagram of a basic laser, which requires a gain medium, some method of pumping the medium, and a resonator for obtaining stimulated emission (adapted from Ref. 32).

waveform, there is an advantage to using a laser capable of being Q switched if significant LiDAR range is desired. With a Q-switched laser, it is possible to store a large amount of energy in the laser medium over a period of time and then release it all in a short pulse that is capable of accurately measuring range. For example, a Nd:YAG laser has a 220-μs upper-state lifetime and can emit a Q-switched pulse in a few nanoseconds. This is a pulse compression of about a factor of 100,000. It is possible to accumulate diode laser energy for 220 μs and emit it in about 2 ns. A laser not capable of Q switching (when used for a direct-detection LiDAR) will be acceptable for a short-range LiDAR but not for a medium or long-range LiDAR because it will be difficult to get enough laser energy into a short pulse. Some auto-LiDARs with a range of ~100 m or less use direct diode illumination. Longer-range auto-LiDARs with a 200-m range usually use a Q-switched fiber laser. Kilometer-range (or longer-range) pulsed LiDARs will have to take advantage of the energy storage from Q switching. It is difficult to reach high peak power without using Q-switching lasers.

A limitation with pulsed lasers is the required range resolution. Short pulses allow better range resolution. Equation (1.11) gives range resolution as $\Delta R = c/2B$, where B is the bandwidth, and $1/B$ is the pulse width. The better the required range resolution the shorter the pulse into which we have to pack a certain amount of energy.

If we have a coherent LiDAR, we can use continuous-wave (cw) or high-duty-cycle waveforms. This can be done because range can be measured based on the changing frequency (or phase) of the laser, so we do not have to use pulsed lasers. For coherent LiDARs, however, a narrow-linewidth laser is required. If the laser is too broad in wavelength, it is impossible to measure phase by beating the return signal against the LO. Direct-detection LiDARs prefer short pulses, so a high–peak-power laser is required for long range. Coherent LiDARs need narrow-linewidth lasers. These are fundamental laser parameters to consider when developing LiDARs. Most lasers now start with diode lasers as pumps because diode lasers are the cheapest and most efficient lasers. Diode lasers have problems both with high-power short pulses and with narrow linewidth. Neither of these issues is significant when using diode lasers to pump other lasers, but both can be significant when directly using diode lasers for LiDARs, especially long-range LiDARs.

To accurately measure range, we need a high-bandwidth waveform that determines the range resolution [Eq. (1.11)]. We need to be able to determine how long it took for light to travel from the LiDAR transmit aperture to the target and back to the LiDAR receive aperture. To accurately measure velocity using a coherent LiDAR, we need to measure the change in frequency due to Doppler shift. Measuring low frequencies requires a long measurement period. The simplest measurement period we can use for understanding LiDAR waveform is a single pulse. As a rule of thumb, the speed of light is

0.3 m (or 1 ft) per ns. A laser pulse goes to the target and returns, so a 1-ns pulse will provide 0.15-m range resolution, or about 6 in. Range resolution is more precisely limited by the convolution of the pulse shape and the target, so pulse rise time can be a factor as well as pulse length. The range profile of the target gets into the calculation as well. For a coherent LiDAR we can measure velocity on a single pulse by using the Doppler shift:

$$\Delta V = \frac{\lambda * \Delta f}{2},$$ (5.3)

where ΔV is the velocity resolution, and Δf is the change in frequency due to the Doppler shift. The smallest Δf we can measure is one over twice the time of the measurement, so a difficulty with a short pulse that provides high range resolution is that you cannot measure velocity accurately. A 1-ns pulse would allow measurement of 1-GHz Δf, which for a λ of 1.5 μm would measure velocity down to 750 m/s. To measure high range resolution and high velocity resolution, we can use a pulse doublet (Fig. 5.3) or another high time–bandwidth product waveform. We use a single pulse width to accurately measure range, and a pulse train to measure Doppler shift and therefore velocity. The minimum number of pulses in a pulse train is a doublet. This is sometimes called a polypulse waveform if sampling requires more than two pulses.

Frequency modulation (FM) chirp and pseudo-random–code modulation are additional potential waveforms. With FM chirp, you can chirp in one direction or both directions. In pseudo-random–code modulation you randomly change frequency or phase over the extent of the bandwidth. There are advantages to each of these waveforms, which are further discussed next.

5.2.2 High time–bandwidth product waveforms

An ideal coherent LiDAR waveform will allow precise measurement of both range and velocity with a single waveform. This is called a high time–bandwidth product waveform. For example, we could have two 1-ns pulses separated by 10 μs, providing 0.15-m range resolution and 7.5-cm/s velocity resolution for 1.5-μm light. This is referred to as having a time–bandwidth product of 10,000 because we have increased our velocity measurement capability by a factor of 10,000 while keeping the same range resolution.

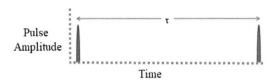

Figure 5.3 Pulse doublet waveform (adapted from Ref. 32).

5.2.2.1 Polypulse waveforms

A simple pulse doublet can be used (see Fig. 5.3) with the width of one pulse being used for range precision; however, the width between the front of the first pulse and the back of the last pulse can be used for velocity measurement resolution. The phase measured in the first pulse must be related to the phase measured in the second pulse. The laser must have a coherence time that equals or exceeds the time separation of the pulses. The coherence time is the period of time over which the phase of the signal is correlated between the phase at the beginning and the end of the coherence time. It is calculated as one over the linewidth B of the laser. The coherence length is the coherence time multiplied by the speed of light. If the pulses are separated by more than the coherence length of the laser, the pulse doublet (or polypulse) measurement technique will not work. If phase has been randomized between pulses, the doublet pulse waveform will not allow velocity measurement. The linewidth of the laser used must be less than one over the pulse separation. More than two pulses can also be used if additional sampling is required to measure Doppler shift frequencies of interest. This is called a polypulse waveform.

5.2.2.2 Linear frequency modulation

Range resolution is inversely proportional to the total signal bandwidth B [see Eq. (1.2)]. For a coherent LiDAR, it is possible to linearly chirp the frequency of the carrier wavelength in order to measure range. It is possible to linearly chirp in only direction, or to chirp both up and down in frequency, as shown in Fig. 5.4.

Normally, to have a certain bandwidth and therefore a matching range resolution, it is necessary to have a detector that can respond to the bandwidth B. However, that can be expensive or impossible, depending on what range resolution is used in the design goal. One way to reduce the need for high-bandwidth detectors is to chirp the LO as well as the emitted signal. This method is shown schematically in Fig. 5.5 and has used been used to obtain 26-μm range resolution with a 5-THz–bandwidth waveform.[2] It is impossible with current detector technology to have a 5-THz–bandwidth

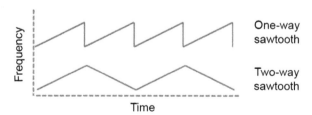

Figure 5.4 Possible one-way and two-way linear frequency chirps.

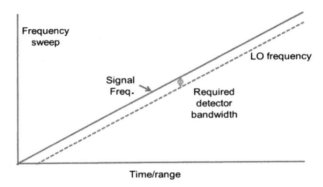

Figure 5.5 Linear FM waveform that chirps both the signal and the LO (reprinted from Ref. 32.

detector, but it is possible to have range resolution associated with 5 THz of bandwidth.

A second way to obtain high bandwidth and excellent range resolution without having detectors at that bandwidth is to use multiple lasers at the same time, as shown in Fig. 5.6, where each laser has a chirped frequency of B, but it is possible to obtain range resolution constant with B_{eff}. Each laser is chirped over a portion of the bandwidth required to obtain the desired range resolution. In this case, we not only do not need a detector with a wide bandwidth, but we do not need a modulator that can generate the full bandwidth. This process is sometimes called stretch processing.

5.2.2.3 Pseudo-random–coded LiDAR

A pseudo-random–coded waveform has the same bandwidth requirement to obtain a certain range resolution as any other waveform. For a

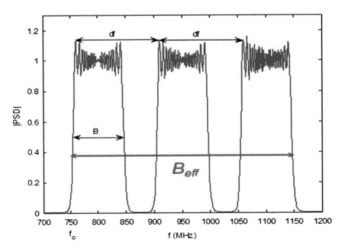

Figure 5.6 Stretch processing (adapted from Ref. 32).

pseudo-random–coded waveform, it is not practical at this time to reduce the required detector or modulator bandwidths like we can using a linearly frequency–modulated (LFM) waveform. In theory, one could do that for a case where the LO and the return signal are matched in time, but to date this type of correlation has not been accomplished. However, a recent Master's thesis did something similar.[3] A pseudo-random–coded waveform can have more flexibility than a LFM waveform and can be generated in a simple way. It is possible to modulate amplitude, frequency, or phase in a pseudo-random fashion. Frequency and phase modulations of course have a relationship. Also, if multiple transmitters are used, each transmitter can be assigned a unique, specific code. Code-division multiple access (CDMA) is one embodiment of a pseudo-random–coded waveform. CDMA can be understood using the analogy of a group of people speaking the same language, and other languages than this being perceived as noise and rejected. Time-division multiple access (TDMA) is another embodiment. This is the equivalent of people taking turns to speak. Frequency-division multiple access (FDMA) is a third embodiment and can be understood with the analogy of people speaking at different pitches. Pulse-position modulation can have a single waveform or can have periodic codes. When periodic codes are used, one code length is called a chip. Figure 5.7 shows a single pseudo-random–coded waveform, and Fig. 5.8 shows a periodically coded waveform.

One issue with pseudo-random–coded waveforms is how to find an "ideal" code. A code could be designed for optimum range resolution, optimum velocity resolution, or some combination of the two. There are various algorithms to address this issue.

5.2.3 Radiofrequency modulation of a direct-detection LiDAR

With the advent of laser communication over fibers, we now have laser detectors with tens of gigahertz of bandwidth. We can therefore modulate a laser carrier at the X band and directly detect it.[4] This is then a simple direct-detection LiDAR that does not use an optical LO but has some of the characteristics of a coherent LiDAR. Once the signal is detected, we can beat

Figure 5.7 Single pseudo-noise (PN) code (reprinted from Ref. 32).

Figure 5.8 Periodic PN code (reprinted from Ref. 32).

the detected signal down to an intermediate frequency (IF) and process it just like we would process a microwave radar signal. With advances in analog-to-digital (A/D) converters, we can digitize the RF return and do the processing, including creating an IF frequency, in the digital domain. Doppler sensitivity is based on the RF modulation frequency, not on the optical carrier. Range resolution is based on the bandwidth of the RF modulation. Imposing RF modulation on the optical carrier has the simplicity of direct detection and can leverage radar IF, detection, and processing techniques. Figure 5.9 shows the imposition of RF modulation on top of an optical carrier frequency.

An optical carrier will be at 200 THz (1.5 μm) or maybe 300 THz (1 μm), whereas a RF modulation could be at the X band (~10 GHz) or maybe the K_u band (~16 GHz). Lower RF frequencies could also be used. With high-bandwidth detectors becoming available, it could even be possible to use K_a band (~35 GHz) modulation. The RF modulation we impose on an optical carrier can be frequency (FM) modulated, amplitude (AM) modulated, or phase modulated.

5.2.4 Femtosecond-pulse modulation LiDAR

While it is not common, someone could make a LiDAR using very short pules, say from a titanium sapphire (Ti:sapphire) laser. As a result, we will briefly include discussion of femtosecond (fs)-pulse Ti:sapphire lasers. Let us say that we have a series of 5-fs pulses. If the LiDAR has pulses approximately this long, it can have a range resolution of 0.75 μm, exceeding even the range resolution achieved with a linear FM chirp and stretch processing. Of course, no detector is capable of measuring such a short pulse. When LiDARs send out these very short pulses, they usually do so using a dispersive element just before pulse emission. This element compresses the pulse by taking a very broadband laser

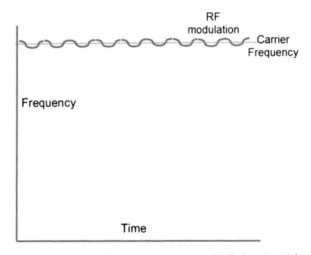

Figure 5.9   RF modulation of a direct-detection LiDAR (reprinted from Ref. 32).

pulse and piling the multiple frequencies of the laser into a single pulse in time.[5] Without this last element, the probability of optical damage in the LiDAR would be very high due to high peak power. A dispersive compressive element is used just before emission and on receive. In order to be able to detect such short pulses, a dispersive element must dramatically expend the pulse width prior to detection. This might allow a sufficiently high detector bandwidth. At the time of this writing, I am not aware of anyone who has built a LiDAR using this technique, but this may be possible.

5.2.5 Laser resonators

A resonator is required to have a light beam pass through a gain medium many times to create the amplified stimulated emission of a laser. If, for each time around the resonator, there is positive amplification, meaning brighter light at the end of a pass than at the beginning, then stimulated emission results in a laser. A simple stable resonator is shown in Fig. 5.10. Stable resonators do not have an output unless additional elements are inserted to Q switch out the laser energy, or they use a partially reflecting mirror at one end. For example, a polarizing element can be inserted, and a Pockels cell can be used for switching the polarization to create an output. In that case, the signal keeps increasing until the output path is created and the energy is dumped from the cavity. This is Q switching; the Q of the cavity is changed suddenly from a high value to a low one to cause light output.

Some high-power lasers use unstable resonators, as shown in Fig. 5.11. In this case, there is some loss on every path because some of the laser energy misses one of the end mirrors.

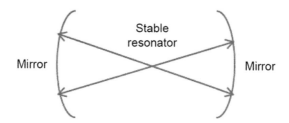

Figure 5.10 A stable resonator (reprinted from Ref. 32).

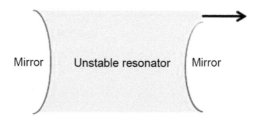

Figure 5.11 An unstable resonator (reprinted from Ref. 32).

Figure 5.12 Unfolded stable resonator (reprinted from Ref. 32).

A resonator can be virtually unfolded as a series of lenses in a row, as shown in Fig. 5.12. This series of lenses can be modeled using matrices; using this method, a stable resonator is modeled using Jones matrices. A thin lens can be modeled in Jones matrices as

$$\begin{bmatrix} 1 & 0 \\ -\frac{1}{f} & 1 \end{bmatrix},$$

where f is focal length. A straight section can be modeled as

$$\begin{bmatrix} 1 & L \\ 0 & 1 \end{bmatrix},$$

where L is the the length of the cavity. A resonator can be thought of as a series of thin lenses separated by straight lines of length L. A full period of the resonator in Jones calculus is a half-lens followed by free space followed by a half-lens and then another free space.

5.2.6 Three-level and four-level lasers

In order to have stimulated emission, population inversion is required. Energy is provided to pump a molecule from a ground state to an upper state. A three-level laser is shown in Fig. 5.13, where a molecule is pumped from the ground state to an upper state. The upper state quickly decays by some

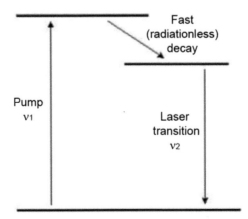

Figure 5.13 Energy diagram of a three-level laser (reprinted from Ref. 32).

nonradiative method to an intermediate state that is significantly above the ground state in energy. Spontaneous and stimulate emission can occur from this state to the ground state. The ideal case would have a small energy difference between the initial upper state and the longer-lifetime upper state to which the molecule decays after nonradiative decay. The quantum defect is the amount of energy lost in radiationless decay. It is a loss of energy so should be minimized.

The main issue with a three-level laser system is that there is a high population in the ground state unless the laser is pumped with high power. Stimulated emission only occurs when the upper laser state has a higher population than the lower state, so a high pump power is required to achieve stimulated emission in a three-level laser system. The ground state has to be depopulated or absorption will occur rather than stimulated emission. Ruby ($Cr^{3+}:Al_2O_3$) is a good example of a three-level laser.

A four-level laser eliminates the requirement to depopulate the ground state by lasing between an upper state and a lower state that is slightly above the ground state. Of course, this increases the quantum defect loss, but if there is fast decay into the ground state from the somewhat higher state, the population of the laser target level will remain low. Figure 5.14 shows the energy diagram of a four-level laser. Some lasers are called quasi-four-level lasers because the two lower levels are very close, and the laser needs to be cooled to prevent thermal agitation from moving molecules from the ground state to this somewhat higher state. The advantage of four-level lasers is that they do not require pumping with as much power as three-level lasers. Nd:YAG is a good example of a four-level laser.

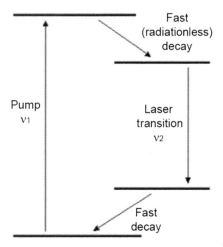

Figure 5.14 Energy diagram of a four-level laser (reprinted from Ref. 32).

5.2.7 Laser-pumping considerations

Laser diodes are pumped electrically, but most other lasers used today for LiDAR are pumped using light. Historically, flashlamps were used to pump lasers using light, but they were inefficient and unreliable. A flashlamp usually generates broadband light similar to a blackbody; however, most laser media only absorb, and therefore capture, narrowband light, making flashlamps an inefficient way to pump a laser, even if the flashlamp is electrically efficient. Now, almost all light-pumped lasers are pumped with laser diodes. Laser diodes can be both efficient and narrowband, so they can hit the right spectral region to efficiently raise a laser medium to an elevated energy state, ready for stimulated emission. If the laser pump source is spectrally matched to the laser medium, then the main issues for pumping efficiency will be the efficiency of the pump source and the spatial overlap of the pump and the laser modes. End pumping can provide very high spatial overlap efficiency between the pump and the laser mode, but often cannot be scaled to high-power lasers. Figure 5.15 is a conceptual diagram of a laser diode end pumping a fiber laser, with cladding around the gain medium.

By converting incoherent laser diode energy into spatially and temporally coherent laser power, side pumping can allow many laser diodes to be used. However, side pumping does not have as much spatial overlap as end pumping, so is not as optically efficient as a way of pumping a laser. Side pumping can, however, pump to much higher power lasers because more diode lasers can be used for pumping a laser medium. Figure 5.16 conceptually shows side pumping of a laser medium using laser diodes as pumps.

5.2.8 Q-switched lasers for LiDAR

A laser with a long upper-state lifetime can build up energy in a cavity with no release. Then, suddenly, the Q of the cavity can be changed, switching out the stored energy in a single pulse. This pulse can be nanoseconds long, but can contain substantial amounts of energy, resulting in high peak power. This

Figure 5.15   End-pumping geometry (reprinted from Ref. 32).

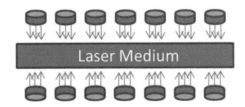

Figure 5.16   Side-pumping geometry (reprinted from Ref. 32).

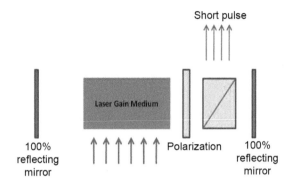

Figure 5.17 Geometry of an actively Q-switched laser using a Pockels cell (adapted from Ref. 32).

method is called Q switching because we are switching the Q of the cavity to release the energy stored in the laser medium. Q is proportional to the ratio of the energy stored in the resonator at any given time to the energy lost from the resonator in a single round-trip cycle. Q switching is a great way to efficiently generate high-energy nanosecond pulses. One method of Q switching is shown in Fig. 5.17. A Pockels cell can be used to switch from almost no output from the cavity during a round trip to almost all of the energy being suddenly dumped out of the cavity. So in Q switching, the laser is switched from a high-Q cavity to a low-Q cavity. A Pockels cell can be switched quickly so is a common method of active Q switching. Some methods of passive Q switching use a saturable absorber, which is a material whose transmission increases when the brightness of light exceeds some threshold.

To make Q switching useful, the laser medium has to store energy in the upper state. Lasers such as diode lasers or dye lasers cannot be usefully Q switched because they do not use a laser medium that stores energy in the upper state. High-power–pulsed lasers are typically only used in LiDARs when the laser material allows significant energy storage in an upper state. Short-range pulsed LiDARs do not need to be Q switched because the required energy per pulse is low, e.g., at a range of 100 m.

5.2.8.1 Pockels cells

A Pockels cell is an EO crystal that changes the index of refraction for light of one polarization when a voltage is applied. Pockels cells will be further discussed in Chapter 7 on beam steering. Depending on the crystal, voltage can be applied in different directions. For the EO effect, birefringence is linearly proportional to the electric field. This is in contrast to the Kerr effect, which is usually smaller and has a square relationship. At higher voltages, sometimes the Kerr effect can cause a larger index of refraction change, such as in KTN, which is an oxide of potassium (K), tantalum (Ta), and niobium (Nb). Pockels cells can switch in a few nanoseconds, but they require high voltage (often thousands of volts) and high current. They also may have a limited angle of acceptance. Some researchers have looked at breaking up Pockels cells into multiple layers to reduce the required voltage per layer; however, none have published their results as of this writing. Many types of crystals can be used as Pockels cells. The most common Pockels cell material is $LiNbO_3$. The required voltage depends on the crystal used. Pockels cells are very useful for Q-switching lasers, and Q switching is very useful for longer-range LiDARs that use a pulsed waveform. A Pockels cell is used for active Q switching. Passive Q switching provides less control of the pulse timing. Also, Pockels cells can be used as a variable waveplate. Variable waveplates can switch the polarization of a laser signal. This can be useful for multiple LiDAR applications. Variable waveplates can be made using uniform voltage applied across the crystal, creating a phase delay.

5.2.9 Mode-locked lasers for LiDAR

Laser temporal modes typically operate with random phase relationships that fluctuate with slight changes in operating conditions. If one places a time-dependent loss element in the laser cavity at exactly the spatial frequency of the mode spacing, after some time, the lowest-loss condition for the laser leads to the modes acquiring a fixed phase relationship among the different frequencies. Figure 5.18 shows spatial modes in a laser cavity. These spatial modes result in mode locking. We see an integral number of spatial modes trapped in the laser cavity. When this is translated from the frequency domain to the time domain, this phasing leads to the generation of a steady train of pulses at a rate that is inverse to the mode spacing. The phasing leads to pulses that time themselves to pass through the loss element at its minimum loss point.

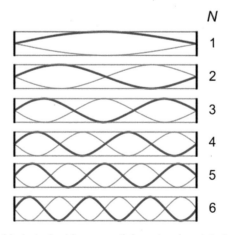

Figure 5.18 Mode-locked laser spatial modes (reprinted from Ref. 32).

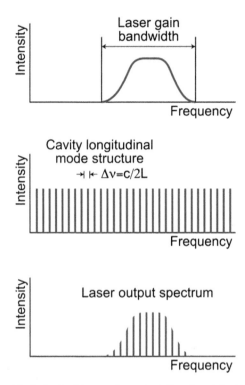

Figure 5.19 Mode-locked laser characteristics (reprinted from Ref. 32).

Figure 5.19 shows the laser gain bandwidth of a given laser type in the top graph, the cavity longitudinal modes in the center graph, and the fact that when you put the modes together, you get many narrow lines in the output spectrum. This general process is called mode locking. The more modes that are locked together, the shorter the pulse. How quickly the loss element changes from a high state to a low state influences the number of modes. The major influence on the pulse duration is from the gain linewidth of the laser. That puts an upper limit on the number of modes that can lase and, therefore, the number of modes that can be locked or, equivalently, a lower limit on the pulse duration. The product of the minimum pulse duration in seconds and the gain linewidth in hertz is a constant that depends on the exact temporal shape of the pulse and falls in the range of 0.3–0.4. For Nd:YAG, the minimum pulse width is about 2 ps, while for Ti:sapphire, the value is 3.5 fs, and for Cr:ZnSe, it's about 7 fs. Pulses that have the minimum value or duration for a given spectral linewidth are called time–bandwidth limited.

Typical mode-locking schemes could employ an electro-optic or acousto-optic loss modulator, but they have sinusoidal variations in loss and generally are not able to generate pulses that fully occupy the available linewidth of the laser. Passive loss elements based on semiconductors (or recently, graphene), which show a rapid reduction in loss with intensity, can produce much shorter pulses. A breakthrough in mode locking of the large-bandwidth Ti:sapphire laser occurred with the accidental discovery that the laser medium itself could provide a fast-switching loss element based on the nonlinear refractive index of sapphire. This led to generation of 100-fs–duration pulses, and subsequent refinement of this technique combined with advances in the optical resonator mirrors have led to Ti:sapphire lasers that can generate pulses approaching the 3.5- to 4.0-fs limit. The pulses contain only a few optical cycles for a pulse with an estimated pulse width of 6.5 fs.[6]

Mode-locked lasers provide periodic short pulses in regular series. Pulse separation is related to cavity length L according to

$$\tau_m \geq \frac{2L}{c}, \tag{5.4}$$

where τ_m is the time taken for the light to make one round trip of the laser cavity. Pulse width is related to laser linewidth according to

$$\Delta t \approx \frac{0.44}{PB}, \tag{5.5}$$

depending on the specific pulse shape, and where P is the number of modes. The Fourier transform of the specific frequency line shape will result in the pulse width. Line broadening can be homogeneous or inhomogeneous. Homogeneous broadening is a type of emission spectrum broadening in which all atoms radiating from a specific level under consideration radiate with equal intensity.[7] If it is homogeneous broadening, the frequncy line shape is Gaussian. If it is inhomogeneous broadening, the frequency line shape is Lorentzian. Collisional or pressure broadening would be homogeneous. The presence of dopants tends to promote inhomogeneus broadening. For N modes locked with a frequency separation of Δv, the overall mode-locked bandwidth is NB.

Even when a solid state laser generates a single spatial or transverse mode, it can generate a large number of longitudinal modes. The resonant frequencies of the longitudinal modes of an optical cavity correspond to a large integer times the speed of light divided by the round-trip optical path length of the cavity. For a 50-cm–long cavity, the frequencies are spaced by about 0.3 GHz. For Nd:YAG, the linewidth of the laser gain is about 180 GHz, so, in principal, that laser could oscillate on hundreds of longitudinal modes. In reality, the modes compete for power in the laser medium, and in principle, if the laser is allowed to reach a true steady state, only one mode would remain operational. More typically, the modes occupy slightly different spatial regions in the material, and Nd:YAG lasers may have multiple modes that span on the order of 10 GHz. Remarkably, this is true even for materials like Ti:sapphire that have a linewidth $500 \times$ larger because of the competition among modes for gain.

Typical mode-locked oscillators operate at a pulse rate around 100 MHz with an average power of 1–10 W, leading to per pulse energies of 10–100 nJ (nanojoules). For a Ti:sapphire laser producing 10-fs pulses, each pulse has a peak power of 1–10 MW (megawatts), comparable to the peak power of high-energy, Q-switched, solid state lasers. If one sends this pulse into a Ti:sapphire amplifier system, amplification to the relatively modest energy of 1 mj (millijoule) leads to a peak power of 100 GW (gigawatts). At typical beam sizes, even if the effluence is comparable to the saturation effluence level, the high electric fields in the pulse will lead to substantial nonlinear effects in the material as well as possible dielectric breakdown.

5.2.10 Laser seeding for LiDAR

High-power lasers that we might want to use for a LiDAR often do not have the proper linewidth or beam quality. Spatial beam quality is a combination of how close the laser beam is to being diffraction limited and whether the beam is spatially smooth or ragged. One way to improve linewidth and beam quality of a laser is to seed it. To seed a laser, the power of the seed laser should be high enough to cause the higher-power laser to lock to the lower-power signal instead of starting with random stimulated emission. In the absence of a seed, random fluctuations will build up in the laser cavity, starting with spontaneous emission. The stimulated emission can occur at any wavelength and any mode, consistent with the laser medium and the laser cavity. Seeding a laser is a way to constrain the laser to oscillate with certain parameters. We start with the seed laser signal at some power level. A good seed laser should have the laser properties required by the design, but at a lower power. For example, the seed laser might have good spatial coherence (beam width and shape) or good temporal coherence (a narrow linewidth and long coherence length). It stimulates emission in the new laser, resulting in a signal that is locked to the incoming seed laser. Because the laser cavity is seeded rather than building up from random fluctuations, the laser signal has additional constraints beyond those of the cavity and the laser medium. Figure 5.20 is a diagram showing a seed laser inserted into a Q-switched laser.

Seeding can also be done with nonlinear processes. It is often done with the single line of an optical parametric oscillator (OPO) to obtain a narrow-linewidth seeded-laser output from the OPO. Seeding can encourage certain spatial modes in a laser. Seeding may be preferred to amplification for some laser media, e.g., when the laser gain in the medium is low. Often seeding or master oscillator power oscillator (MOPO) configurations are used in coherent LiDARs because of the need for specific laser parameters. The primary reason for using either seeding or master oscillator power amplifier (MOPA) configurations is that it is much easier to generate narrow-linewidth,

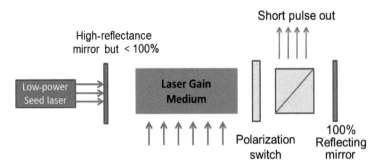

Figure 5.20 A conceptual diagram of a seed laser (adapted from Ref. 32).

phase-stable lasers at these lower power levels. The goal is to transfer the characteristics of the low-power laser to those of the high-power laser. Seeding or amplifying are two approaches to accomplish this goal.

5.2.11 Laser amplifiers for LiDAR

Instead of seeding a laser to obtain the exact, required high-power–laser parameters, it is possible to start with a master oscillator (MO) and amplify it. The properties of a good MO should be like the properties of a good seed laser. Once again, we want a low-power laser that has the beam quality and linewidth required in a higher-power laser. Often this is referred to as a master oscillator power amplifier (MOPA) configuration. Figure 5.21 shows a MO followed by an amplifier. Notice in this case that the amplifier is just the gain medium with no resonator. Depending on how long the gain medium is and what its gain is, the amount of amplification in a given stage will vary. It is likely that low-power amplification stages will have higher gain than high-power amplification stages. For high-power stages, it is more of a power-adding process than linear amplification. One issue is isolation. If you do not isolate the amplifier from the MO, feedback from the amplifier might distort the MO. Depending on how powerful the MO is, it might take multiple amplification stages to reach the desired power level. Gain per stage is usually limited to approximately a factor of ten. This is one reason that it is often necessary to have multiple gain stages, as shown in Fig. 5.22.

With a MOPA, the efficiency of the MO and of the early-stage amplifiers does not count significantly toward the overall laser efficiency. It is the efficiency of the final amplifier stage that drives the overall efficiency. Assume an amplification factor of ten in the final high-power stage. In that case, only

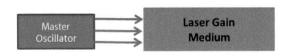

Figure 5.21 A MO followed by a laser amplifier (adapted from Ref. 32).

Figure 5.22 Multistage laser amplification (adapted from Ref. 32).

10% of the power is developed in the MO and early stages. Whether those early stages are 30% efficient or 3% efficient will not significantly influence overall efficiency.

High-gain laser media are useful for amplifiers. The laser gain in a medium is determined by the properties of the gain medium and the power level used for pumping the medium. If you pump with sufficiently high power, it is possible to get high gain out of many materials. For seeding lasers, the gain in laser medium is not a significant issue because the seed laser will have a resonator, which means many passes back and forth through the gain medium. With an amplifier, there is only a single pass through the gain medium. With high-gain media, the amplifier does not need to be long. Significant signal gain can be achieved in a short gain region.

Figure 5.23 shows a multistage MOPA configuration that could be called a laser manifold. This configuration can be used in the multiple-input, multiple-output (MIMO) LiDAR approach discussed in Chapter 4. The transmitters should be coherent with respect to one another, meaning that there is a fixed phase relationship between each transmitter. There should be a single MO with narrow linewidth (long coherence length). Then the LO is split into many different identical signals. A unique tagging modulation can be placed on each signal. Each signal is amplified to the level needed for LiDAR operation. For coherent LiDAR, it will also be necessary to have local oscillators (LOs) at each receive subaperture, requiring further splitting of the original MO signal. LO signals will be significantly lower power than the transmitted signal, so they may not require amplification and may be frequency shifted. As described earlier for temporal heterodyne detection, it is common to frequency shift the LO, whereas in spatial heterodyne detection, the LO is usually not frequency shifted. MIMO techniques can be used with

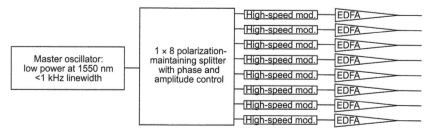

Figure 5.23 Multiple coherent laser transmitters in a multistage MOPA configuration (EDFA is erbium-doped fiber amplifier) (reprinted from Ref. 32).

either spatial or temporal heterodyne LiDAR. If frequency shifting is required, all of the LOs could be shifted at one time. It is possible to start with an MO and split it into the LO signals and the transmitter signals; then the LO signals can be frequency shifted, or not, and further split into the required number of LOs. Additionally, after the laser has be further split into the number of laser signals, tagging modulation can be imposed, and then each signal can be amplified.

5.3 Lasers Used in LiDAR

The use of lasers for active EO sensing started with the use of the first laser to be operated (1960), the solid state ruby laser.[8] With the development of techniques to generate nanosecond-duration pulses, ruby lasers provided the first example of laser rangefinders. Since 1960, almost every type of laser has been employed in demonstrations of active EO sensors. At this point, with a few exceptions, LiDARs now employ either diode or solid state lasers, the latter often being combined with nonlinear optics. This laser distribution is expected to be the case for the foreseeable future due to a favorable combination of output format, operating wavelength, relatively high efficiency, ruggedness, compact size, and reliability.

Nonlinear optics employ crystals with special properties that can convert a laser output to shorter wavelengths by harmonic conversion or to longer wavelengths by parametric processes. Parametric conversion processes have the added advantage that the wavelengths generated can be tuned using a variety of techniques, which is useful for sensors that require specific wavelengths, e.g., sensors that detect specific gases. In some cases, parametric and harmonic processes can be combined for added wavelength tuning.

Historically, CO_2 lasers were used in LiDARs (see Chapter 2 on the history of LiDAR). Nd:YAG lasers were used for designators in the earlier years of LiDAR. At the time when designators were initially used, all Nd:YAG solid state lasers were pumped by flashlamps. Flashlamps were not very reliable and made inefficient lasers because they create a very broad-spectral-range pulse, and only a very small portion of that pulse can be used to pump the laser medium. More recently, diode lasers are available as pumps. Now virtually all LiDARs use diode lasers or a related laser type called quantum cascade lasers. Diode lasers and quantum cascade lasers use electricity to pump them into a higher state. Bulk solid state and fiber lasers use laser diodes or can use quantum cascade lasers in certain bands. Diode lasers can be used directly as the illuminator but can also be used to pump the laser medium that is used for the illuminator. Pump media used may be bulk solid state materials such as crystals, or fibers. Laser diodes are usually >50% efficient, but have the drawback of low pulse energy, poor beam divergence, and broad linewidths. Fiber lasers are currently likely to be 30% efficient and can have a diffraction-limited beam and narrow linewidths. Pulse energy can be much higher than laser diodes but will be limited because the cross-section of a fiber is small, so very high peak powers will develop nonlinear effects in the fiber, or even cause damage. Bulk solid state lasers might be typically 20% efficient. They can have excellent linewidth, beam width, and peak powers, depending on how they are designed.

5.3.1 Diode lasers for LiDAR

Diode lasers are used in almost all LiDARs today. In most cases, diode lasers are used as pumps for solid state lasers. In some cases, the diode lasers are used directly as emitters for the LiDAR. Diodes have the highest wall plug efficiency of any laser. Of course, this is expected since in most cases the other laser types start with a diode laser pump; therefore, any loss makes other laser types less efficient. Most diode lasers used for LiDAR are interband lasers, which use transitions from one electronic band to another. Recently, cascade diode lasers have been developed as an important LiDAR source for the MWIR and LWIR wavelengths. Edge-emitting diode lasers have a thin gain layer in one dimension, i.e., high beam divergence in that dimension, meaning that it takes fast optics to capture the beam in the thin dimension. A fast-axis collimating (FAC) lens is normally used to capture the diode laser energy. If the gain region is allowed to be too large in the horizontal direction, the laser becomes multimode. Because of the small gain area, it is difficult to get high peak power from edge-emitting diode lasers. Vertical-cavity surface-emitting lasers (VCSELs) emit from the top surface rather than the edge, allowing better 2D coupling into a fiber, but also have size limitations due to mode control. Also, none of the diode laser materials exhibit long upper-state lifetime storage, so diode lasers are either cw, or quasi-cw, where quasi-cw means approximately a factor of four higher peak power if the diodes are allowed to cool between pulses. To use diode lasers directly for LiDAR, due to peak power limitations, it is ideal to have high–duty-cycle waveforms. If the diode laser has a narrow linewidth, it is possible to use a diode laser with FM chirp or pseudo-random–coded waveforms, but making diode lasers with a narrow linewidth is difficult. One pair of materials that is commonly used in laser diodes is gallium arsenide (GaAs) with aluminum gallium arsenide ($Al_xGa_{1-x}As$).

5.3.1.1 Interband, edge-emitting diode lasers

Interband diode lasers employ electronic transitions between energy states in the conduction band and the valence band of the semiconductor crystal. The semiconductor material used for the diode laser must have a direct optical bandgap; i.e., the electronic transitions giving rise to laser operation must occur without the assistance of mechanical vibrations (phonons) in the semiconductor. This eliminates the most common semiconductor material, silicon, as well as a related material, germanium.

Laser operation in direct-bandgap semiconductors occurs by optical transitions between the lowest-lying electronic states in the conduction band and the highest-lying states in the valence band, with the requirement that there be a higher density of states occupied in the conduction band compared to the valence band. This nonequilibrium condition, called a population inversion, occurs in diode lasers through the injection of sufficient electrical current into the lasing region.

Fabrication of a diode laser requires that the semiconductor be available in both n-type materials (where electrons are the majority current carrier) and p-type materials (where holes are the majority current carrier), which is achieved by the addition of certain impurities to the materials. While there are many semiconductors with direct bandgaps, for a variety of reasons, many of these cannot be doped to make both n- and p-type materials to form diodes; notably, this includes all of the so-called II-VI binary semiconductors such as CdS, CdTe, ZnS, ZnO, and ZnSe. (The Roman numerals refer to the relative positions of the elements in the periodic table). Infrared lasers have been operated based on IV-VI materials such as PbS and PbSe but require cryogenic cooling for efficient operation. To date, III-V binary, direct-bandgap semiconductors such as GaAs, GaSb, InAs, InSb, and most recently GaN (as well as alloys of these crystals with other III-V elements), are by far the materials most widely used for diode lasers. Noncryogenic operation in the 350- to 2000-nm region is possible with devices based on III-V materials.

Figure 5.24 shows the key features in a very simplified diagram of an edge-emitting, interband diode laser. Electrical current passes through a wire bond to a top stripe contact that confines the current in two dimensions. In the device shown in the diagram, the current passes (is injected) through p-type

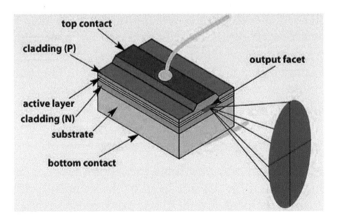

Figure 5.24 Diagram of the structure of an edge-emitting, interband diode laser (reprinted from Photonics.com: http://www.photonics.com/Article.aspx?AID=25099).

material into the active layer, fundamentally a junction between p- and n-type material, then through n-type material to an electrical contact to complete the circuit. Laser gain occurs in a thin, 3D region, comprising the active layer, with the third dimension set primarily by the width of the stripe contact. The injected current and the structure of the device act to form an optical waveguide that confines the laser light to the same region as the gain, both vertically (perpendicular to the junction plane) and horizontally (parallel to the junction plane.) Most edge-emitting diode lasers operate with the laser cavity formed by the cleaved ends of the waveguide region. One end is coated with a deposited stack of dielectrics to form a highly reflecting mirror, while the other end is coated to provide only a small amount of reflectivity. Typical lengths between the two faces are in the sub- to several-millimeter range. In some cases, one diode end has an antireflection coating to enable use of an external mirror and/or tuning element to obtain more control of the diode wavelength.

The details of the junction are much more complex than the simple p–n junction implied by Fig. 5.24. Major developments in diode laser performance since the first demonstrations of operation in the early 1960s have been the result of developing the fabrication technology to make multilayer semiconductor structures (heterostructures) that better confine the lasing region and reduce the current needed for laser operation, as well as increase the efficiency in converting electrical power to laser power. As the thickness of the layers has been reduced with improved processes to the tens-of-nanometer level, quantum effects that fundamentally change the nature of the semiconductor-material energy levels have been utilized to make further improvements in performance. In addition to structuring in the plane of the junction, additional material structuring in the horizontal direction to confine the current and laser power have led to improved laser properties.

A critical and fundamental characteristic of a diode is the nature of the region where laser action occurs. In the vertical direction, the region of laser emission is limited to a dimension of about 0.5 μm, set by the fundamental nature of the junction. This is comparable to or smaller than the wavelength of the light emitted, which assures that the light (fast axis) emitted in the vertical direction is diffraction limited, consisting of only a single transverse mode. In Fig. 5.24, the beam dimension is larger in the vertical direction because the fast-axis light rapidly diverges from the small emitting region. To effectively capture all of the power from the diode laser requires fast optics, at least for the fast-axis light, and this is accomplished with specialized aspheric optics that are often fastened directly to the diode-laser package. As mentioned, these optics are called fast-axis collimating (FAC) lenses.[9]

The limiting feature of the small dimension is that even relatively low–absolute-power levels lead to very high power densities at the surface of the edge-emitting diode laser (on the order of 10 MW/cm^2) and can lead to catastrophic optical destruction (COD) of the diode laser beyond certain power levels. COD results from the inevitable defects at the surfaces of semiconductors, which absorb the laser light, heat up, and with enough power, melt the material at the surface.

Regardless of the cause of the limited power, one can increase the power output of an edge-emitting diode laser by increasing the width of the lasing region along the horizontal direction, which reduces the intensity at the surface for a given absolute power level and also reduces the generated heat density. Unfortunately, beyond a dimension of several microns, the light emitted in the direction along the plane of the junction becomes multimode.

An upper limit to the power from a single emitting region is found when the laser action, rather than occurring through the cavity formed by the cleave ends, starts in a perpendicular direction along the width of the stripe. Typical junction widths for so-called broad-stripe lasers are in the 100- to 400-μm range. State-of-the-art diode lasers operating in the 900-nm–wavelength region can generate 10–15 W of cw power with a 100-μm stripe width.

In order to produce diode lasers with higher average power, starting in the 1980s, the semiconductor laser industry took advantage of improvements in materials quality and lithographic techniques to manufacture multiple diode lasers on one piece (chip) of semiconductor material. Figure 5.25 is a schematic of a multi-emitter diode bar, for which the most common overall horizontal dimension is 1 cm. Bar fabrication requires use of a means to suppress laser action along the length of the bar; an additional manufacturing challenge is in mounting the bar on a heat sink to efficiently remove the heat, while also keeping the entire bar in one plane so that all of the emitters line up exactly. If they do not line up, the net divergence of all of the beams after collection by external optics increases. Key bar parameters are the number of emitters per bar and the stripe width of each emitter, which together determine the fill factor for the bar, with typical stripe widths in the 100- to 200-μm

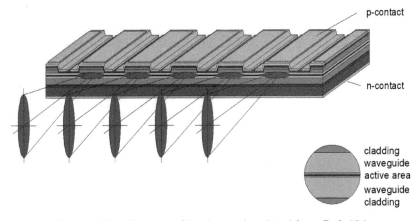

Figure 5.25 Diagram of bar lasers (reprinted from Ref. 10.)

range. Unlike single emitters, the power output of bars running in the cw mode is thermally limited, although recent improved thermal engineering of bar mounts and cooling may lead to COD becoming a limit as well. The thermal limits to power can be overcome if the diodes are run in pulsed mode, which is sometimes referred to as a quasi–continuous-wave (QCW) mode. Typically, this mode is used for pumping solid state lasers, with pulse widths ranging from 0.1- to 1 ms and on-time (duty cycle) ranging from 1 to 10%.

5.3.1.2 Interband, vertical-cavity diode lasers

Another class of interband diode lasers has been developed more recently than the edge-emitting devices just discussed and is based on improved semiconductor-processing technology. Because of their lower power output, vertical-cavity, surface-emitting diode lasers (VCSELs) are not as desirable for LiDAR use. The narrower linewidths could possibly make them interesting for coherent LiDAR at some point in the future. Several designs of these new VCSEL devices are shown in Fig. 5.26. The key structure in the device is constructed by building up a series of semiconductor layers. The direction of optical gain in the junction is perpendicular to the plane of the junction. Since the path length is very short, VCSELs require much higher-reflectivity mirrors in the laser cavity to permit oscillation, with the output coupling mirror having less than 1% transmission. The bottom and top mirrors [distributed Bragg reflectors (DBRs) in Fig. 5.26] consist of alternating layers of high– and low–refractive-index semiconductor materials; the active region, employing quantum-well–structure semiconductor heterojunctions, is sandwiched between the DBRs. Electrical contacts on the top and bottom allow electrical pumping with current passing through the structure, and a circular hole etched into one of the contacts allows the output beam to emerge. The entire

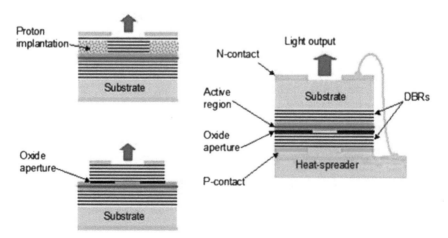

Figure 5.26 Several different VCSEL structures [from Princeton Optronics (now ams AG) web page[11]].

length of the VCSEL is on the order of 10 μm for devices operating in the 900-nm region. To generate a diffraction-limited output in that region, the emitting-region diameter has to be kept below 4 μm, with power levels on the order of 5-mW; however, higher multimode powers are possible as a result of recent work, with Princeton Optronics claiming operation of a 5-W, 976-nm, cw laser with a 300-μm aperture.[11]

Because of the extremely short optical cavity, the devices operate on a single frequency and can be turned on and off in the 10- to 100-ps range, allowing direct current modulation for data rates in the tens of gigahertz. Since the fabrication process is similar to that of semiconductor ICs, it is possible to construct many devices at one time, providing a major cost advantage over edge-emitting diode lasers. Finisar Corp.[12] claims to have shipped more than 150,000,000 VCSEL devices for use as the source in an optical mouse. Future large-scale applications are likely in optical inter-connects for high-speed electronics and in fiber-to-the-home systems.

VCSELs had been limited in wavelength coverage to the 650- to 1300-nm region, primarily because the GaAs-based technology that is used to manufacture the electrically pumped devices cannot extend beyond that region. Recently, devices based on InP semiconductor that operate in the 1500-nm region have been reported. These devices will find wide use for higher-speed fiber applications where the low dispersion of silica fibers at 1500 nm allows long-distance links of 10 GHz and higher.

5.3.1.3 Quantum cascade lasers

If it is desired to have a LiDAR operating in the MWIR band, then quantum cascade lasers (QCLs) are an interesting possibility. The power levels of QCLs are low but improving; over time, QCLs will likely become one of the main infrared laser sources for infrared LiDAR applications. Because QCLs do not have any energy storage and cannot be Q switched to obtain high-energy pulses, high–duty-cycle waveforms will be required to make efficient LiDARs. QCLs rely on an artificial gain medium made possible by quantum-well structures, which are often referred to as being band-structure engineered. For LiDAR applications, we expect the infrared region to be of interest. It is possible that shorter infrared wavelengths will be available with new materials. The name quantum cascade laser is due to the cascade process of one electron interacting with multiple quantum wells as it crosses the structure. The one electron creates multiple photons, corresponding to the number of quantum wells in the device, thereby greatly improving QCL efficiency. QCLs are able to operate in the MWIR and LWIR regions without the need for cryogenic cooling. QCLs can act as very good seed lasers for nonlinear frequency conversion devices (OPAs, etc.)

Typical devices are fabricated on InP and employ GaInAs/AlInAs quantum wells, which are formed from nanometer-thickness layers, typically

10–15 in number. Devices based on other combinations of materials have been demonstrated to have inferior performance. Since the energy separation of the states is a function of the structure, it can be adjusted to generate a wide range of wavelengths, covering approximately 3.5–30 μm and 60–200 μm, with the latter falling in the terahertz region. The gap in long-wavelength coverage is due to the strong phonon absorption region of the semiconductor that overcomes the gain in the quantum wells. The short-wavelength limit is set by several factors, including fabrication difficulties with the required quantum-well alloys and losses due to scattering among different energy states in the conduction band. Alternative systems based on different materials are now under development and will allow for operation at shorter wavelengths.

Laser operation comes through injection of a current of electrons through the quantum-well structure, which results from a voltage applied across the structure. Both the efficiency and gain would be low if the device employed just one quantum well since the energy of the emitted photon is small compared to the energy required to put an electron in the conduction band. This drawback is overcome by fabricating a structure consisting of multiple (typically 25–75) quantum wells in series with another heterojunction structure, the injector through which the electron tunnels from one quantum well to another.

Figure 5.27 is a simplified, partial schematic of a QCL structure with distance as the horizontal axis and energy as the vertical axis. It shows the

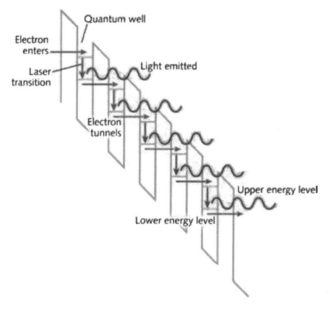

Figure 5.27 Quantum cascade laser conceptual diagram (from Laser Focus World: http://www.laserfocusworld.com/articles/2010/10/photonic-frontiers-quantum-cascade-lasers-prepare-to-compete-for-terahertz-applications.html).

energy levels for the electron with voltage applied across the structure length that creates the gradient in energy. As with edge-emitting interband lasers, the laser power is emitted perpendicular to the current flow through the layers, so a given quantum well interacts with only a thin slice of the laser mode.

One the remarkable features of QCLs is that they are able to operate in the MWIR and LWIR regions without the need for cryogenic cooling, even in the cw mode. This is not possible with interband diode lasers at the same wavelengths, since the high thermal population of the conduction band turns such narrowbandgap diodes into, essentially, short circuits at ambient temperatures.

5.3.1.4 Interband cascade lasers

A related semiconductor laser that has features of both interband and cascade diode lasers is the aptly named interband cascade laser (ICL). Operation of this device relies on multiple quantum-well structures, but now both electrons and holes are involved, and the quantum-well transitions are between an upper state in the conduction band and a lower state in the valence band. The material system commonly used involves either GaSb or InAs substrates, with alloys of InAs, GaSb, and AlSb employed for the quantum-well and injector structures. Two advantages of the ICL over the QCL are (1) shorter-wavelength operation at room temperature from 2.9 μm to 5.7 μm, with cryogenic operation at 2.7–10.4 μm and (2) the ability to lase with much lower electrical powers (about 30$\times$) than QCLs since the transition lifetimes are much longer. As with conventional interband edge-emitting diode lasers, one would expect difficulty with long-wavelength operation at room temperature, so QCLs have an advantage in this case.

5.4 Bulk Solid State Lasers for LiDAR

Nd:YAG designators were some of the first LiDARs, albeit bistatic LiDARs. Nd:YAG lasers usually operate at 1.064 μm. The eye cannot see 1.064-μm light but can focus it on the retina. The damage threshold is about 10,000$\times$ lower than light at 1.5 μm or longer, which cannot penetrate the eye to focus on the retina. A number of bulk solid state lasers are used for LiDAR in the so-called eye-safe region of 1.5 μm or longer. Er:YAG is a good example of an eye-safe laser around 1.5 μm. Tm:YAG or HoTm:YAG are special cases of longer-wavelength lasers because a pump diode near 0.8 μm can generate two laser photons, decreasing the quantum deficit. Both Tm:YAG and HoTm:YAG lase near 2 μm.

The laser transitions for solid state lasers occur (with a few exceptions) between energy levels of ionized atoms from the 3D transition-metal group (Sc to Zn) or from the ionized rare-earth, or lanthanide, series of elements (La to Yb). The atoms are hosted in solids (crystals, glasses, or ceramics) and are added as dopants to the mix of elements used to make the solids. Those ions that give rise to laser operation are referred to as active ions.

A unique class of solid state media has recently emerged and is based on divalent transition metals doped in semiconductors, the most well-developed of which is the dopant Cr^{2+} in ZnSe. As with Ti:sapphire, the vibronic transitions lead to very broad linewidths, but the transitions are centered in the MWIR region, around 2500 nm for the case of Cr:ZnSe, which has been tuned from about 2000–3500 nm.

The remainder of the discussion on bulk solid state lasers considers fiber lasers.

5.4.1 Fiber lasers for LiDAR

A typical fiber laser is cylindrical with a gain region surrounded by cladding (Fig. 5.28). A pre-form is used and then stretched in a drawing tower to make it longer but smaller in the cross directions. Typically, a diode is placed at the end of the fiber, with a lens to focus light into the fiber and the cladding, although other pumping schemes may be used. Often light ricochets around both the cladding and gain medium, only pumping when it passes through the gain medium. A single-mode fiber has a small cross-section, which limits the peak power available in fiber lasers. Recent photonic crystal fibers have been developed that allow higher peak power while maintaining a single mode. Average power is usually not a large problem because the length allows good heat dissipation. Due to the peak power limitation, fiber lasers are good primarily for higher–repetition-rate LiDARs and for cw LiDARs.

The most prevalent peak-power–damage issue is damage to the fiber tip, but once light is in a fiber, nonlinear effects such as Brillouin and Raman scattering can be significant. These are not actual damage, but rather a distortion of the signal. Fiber lasers can have 30% efficiency and are becoming more and more prevalent in LiDAR not only because of this efficiency but also because it is easier to wire a sensor together with fibers and couplers than it is to align free-space optics. Also, the telecommunications industry uses fiber lasers, so many inexpensive and reliable fiber lasers and couplers are available, especially near 1.5 μm. Geiger-mode APD–based LiDARs use fiber lasers with a 10 kHz or greater repetition rate. Materials with energy storage can be used in fiber format, so you can Q switch fiber lasers. Fiber lasers usually have very high beam quality because of their long and thin geometry.

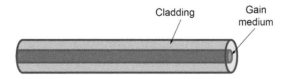

Figure 5.28 Conceptual diagram of a fiber laser (reprinted from Ref. 32).

5.4.1.1 Higher–peak-power waveguide lasers for LiDAR

The most stressing case for LiDAR peak power is a low–repetition-rate pulsed LiDAR for long-range flash LiDAR that illuminates a large area at one time. If you have 60- or 100-Hz operation and need high range resolution, your LiDAR will have hundreds of millijoules per pulse in a nanosecond pulse. For long-range operation, this could be joules per pulse in a nanosecond—very high peak power—and the LiDAR usually needs at least moderate beam quality. However, for flash illumination, the laser beam may be spread out, mitigating its required spatial quality.

Two approaches allow higher–peak-power waveguide lasers to be used for LiDAR. One approach uses photonic crystal fiber lasers; the second uses 1D waveguide lasers. Both laser types can have a single mode while increasing the cross-section of the laser gain medium. The photonic crystal fibers have a pattern of holes that can control the laser modes. The 1D waveguide is only constrained in a single dimension but needs a resonator in the other dimension unless it is only used as a power amplifier. Both of these approaches allow the cross-sectional area of a waveguide to become larger while still constraining the laser modes. If you are not concerned about having a well-controlled laser mode, you can make the waveguide as large as you want, but you lose the spatial quality of the laser beam. LiDARs usually require controlled laser modes, although with flash illumination it may be possible to reduce spatial coherence requirements on the laser source since a larger area is being illuminated. Fiber tip damage may still be an issue even with photonic crystal fibers, but you should not have to focus the light to as small an area. Multiple narrow-linewidth fiber lasers can be coherently combined as another method to obtain higher peak power, but once they are combined, the same power limitations discussed earlier apply.

5.4.2 Nonlinear devices to change the LiDAR wavelength

Often we would like to have a LiDAR wavelength that is not the same as the wavelength of existing lasers, or that can be generated more efficiently by using a high-efficiency laser at one wavelength and then changing the wavelength. The most common method of changing wavelength is to use an optical parametric oscillator (OPO) or an optical parametric amplifier (OPA). An OPO is a laser resonator, and an OPA is an amplifier. Both take one high-energy photon and divide the energy between two new photons. The wavelength equation for this energy division is

$$\frac{1}{\lambda_p} = \frac{1}{\lambda_s} + \frac{1}{\lambda_i}, \tag{5.6}$$

where λ_p, λ_s, and λ_i are the pump, signal, and idler wavelengths, respectively. The signal and idler labels depend on user interest. To have the nonlinear effects necessary, we need high peak power or long gain length in a material that has the right properties. The nonlinear conversion gain has to be higher in an OPA than in an OPO because OPAs are single-pass devices. All three of the waves need to stay in phase so that interaction can occur. A decade or more ago, people started to make engineered materials to keep the three waves in phase. The first of these engineered materials was periodically poled lithium niobate (PPLN). In PPLN, when the waves begin to walk out of phase, the domains in the crystal are flipped such that the domains move back into phase. Figure 5.29 shows a diagram of a PPLN sample with alternating domains flipped. One issue with PPLN is that layers tend to be thin, limiting the peak power out of an OPA. PPLN also absorbs in the upper region of the MWIR, so periodic structures with broader transmission bands have been pursued. In addition to OPOs and OPAs, other nonlinear frequency conversion processes such as sum and difference frequency generation can convert the wavelength of light.

The development of lasers enabled the high optical intensities required to observe nonlinear optical effects in materials. Generally speaking, light passing into a material interacts with the electrons and (at longer wavelengths) the vibrational states of the atoms in the material. For transparent materials, the main interaction causes the electrons to move (be polarized) at the same frequency as the incident light, in linear proportion to the electric field of the light. The interaction between the light and the polarization leads to the speed of light in a material slowing down by an amount characterized by the (linear) refractive index of the material. The intensities of light generated by lasers allows for observation of nonlinear effects in the electronic polarization, where the polarization has elements proportional to the square of the electric field, the cube, and so on.

One can use symmetry arguments to show that squared (second-order) terms in the electronic polarization in crystals are possible only in materials that lack inversion symmetry. This implies that if one moves from an atomic location in one direction in the crystal and compares the surrounding environment to the environment one would experience by moving in exactly the opposite direction, there is at least one direction where the environment would be different. Many materials lack inversion symmetry, including all of the III-V and II-VI compound semiconductors, as well as crystalline quartz (SiO_2). Not included, however, is silica glass, which, like all truly amorphous

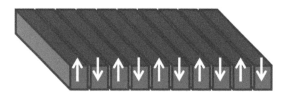

Figure 5.29 Periodically poled material for OPOs or OPAs (reprinted from Ref. 32).

materials, exhibits second-order nonlinearities only over very short lengths of material.

In the following two subsections, two important nonlinear processes are discussed, both of which depend on second-order nonlinear effects in crystals: harmonic generation and optical parametric generation.

5.4.2.1 Harmonic generation and related processes

The immediate effect of the presence of second-order terms in the polarization is the generation of an optical frequency that is twice the incident frequency. The first observation of this was in a quartz crystal with a ruby laser beam incident,[13] where a faint amount of light was detected at the second harmonic of the 694.3-nm ruby laser: at 347.15 nm.

Analysis of the second-order nonlinear process shows that the power in the second harmonic can grow along the path of light in the crystal if the input, fundamental wavelength light and the second-harmonic light can stay in phase. If they do not stay in phase, the power initially grows along the path in the crystal and then starts to decrease as the two beams get out of phase. The power falls to zero in a distance called the coherence length and then continues to grow and fall in an oscillatory manner along the path of the light. In all materials, de-phasing generally occurs due to the change in linear refractive index (dispersion) with wavelength, which causes the fundamental and second-harmonic light to get out of phase.

In birefringent crystals, the refractive index for light in the crystal depends on the direction of the light in the crystal as well as its polarization. If one arranges the direction of light and the polarizations of the fundamental and the second-harmonic light to have the same refractive index in the crystal, the second-harmonic power will stay in phase with the fundamental power and grow to become a substantial fraction of the fundamental power. This technique is called birefringent phase matching and was developed shortly after the first demonstration of second-harmonic generation in crystal quartz.

In general, light that propagates in a birefringent crystal exhibits a unique walk-off effect, in which the flow of power in the light beam deviates from the propagation direction. In phase-matching schemes, either the fundamental beam or the harmonic beam may exhibit walk off, which serves to limit the length of crystal over which the harmonic-generation process occurs. In rare crystals, phase matching can take place where the two beams propagate along one of the so-called principal axes of the crystal, in which case there is no walk off. This is called noncritical phase matching (NCPM), while the more general case is called critical phase matching (CPM). Often, NCPM is achieved by adjusting the crystal temperature to tune the refractive indices to the correct value. One advantage of NCPM is that the sensitivity of phase matching to beam angle is much less than for CPM, making crystal alignment less of an issue; however, more importantly, NCPM can provide efficient operation with non–diffraction-limited beams that have a relatively high spread of angles.

Not all materials that lack inversion symmetry have the necessary birefringent properties to achieve phase matching. Even fewer materials can achieve NCPM. In addition to the necessary symmetry and birefringent properties, crystals used for efficient harmonic generation must be highly transparent at the fundamental and harmonic wavelengths to avoid heating, which not only creates optical distortion, but can disturb phase matching by changing the material refractive properties. While the ability to temperature tune the phase matching can be an advantage, materials that have a strong temperature dependence on phase matching may not be able to generate high average powers since they will de-tune even if the crystal absorbs a small amount of power. Given that high peak powers are often used, the material must also be resistant to optical damage. Unfortunately, materials that exhibit high nonlinearity are also susceptible to optical damage from unwanted processes created by higher-order nonlinear effects that generate free electrons in the material. Some materials are also susceptible to photorefractive effects that are created by the interaction of high-intensity light with impurities in the material, leading to the creation of trapped electrons, an internal electric field, and a change in refractive index from this field that destroys the phase-matching condition.

The second-order term that allows generation of a harmonic can be viewed as the combination of two wavelengths (of the fundamental light) to produce a third wavelength at the second harmonic. A more general interaction is possible where the two wavelengths are different; with appropriate phase matching, it is possible for a crystal to generate a third wavelength that is the sum (sum frequency generation) or the difference of the two wavelengths (difference frequency generation). The most common interaction of this type is in the so-called third-harmonic generation, where one crystal generates some fraction of light in the second harmonic, and another crystal combines the second harmonic with the fundamental light to produce the third harmonic. (Although it might be possible to utilize the third-order nonlinearity of a material to do this in one crystal, this nonlinearity is much weaker in materials, and a sequential process involving the second-order nonlinearity is much more efficient.) Fourth-harmonic generation involves the use of two second-harmonic crystals in series, with the second arranged to double the harmonic generated by the first.

Given the many requirements for effective harmonic generation, it should not be surprising that the number of crystals that have been used with success in harmonic frequency, sum frequency, and difference frequency generation applications is limited. Table 5.3 lists the most widely used materials, along with some key properties and comments. The materials are listed in order of their first reported development. The quantity used to characterize the degree

Table 5.3 Common birefringent crystals used for harmonic generation and related generation processes.

Crystal (acronym)	Figure of merit d^2/n^3	Transparency (nm)	Comments
KH_2PO_4 (KDP) potassium dihydrogen phosphate	0.045	200–1500	Several variants available; can be grown in very large sizes. Major use in fusion driver systems (such as the National Ignition Facility) for third-harmonic generation.
$LiNbO_3$ (LN) lithium niobate	30	330–5500	Large nonlinearity; major limitation due to photo-refraction: not suited for generation of visible light except when doped with magnesium.
$K(TiO)PO_4$ (KTP) potassium titanyl phosphate	2.4	350–4500	Major application in high-power second-harmonic generation of Nd-doped light where high beam quality is not required, or in low-power devices such as laser pointers; some long-term degradation due to formation of light-induced defect centers.
β-BaB_2O_4 (BBO) beta barium borate	1.3	196–2600	Robust material with capability for CPM over a wide range of wavelengths; requires high beam quality for efficient operation; used primarily for fourth-harmonic generation of nanosecond-duration lasers or general harmonic generation with femtosecond-to-picosecond pulsed lasers.
LiB_3O_5 (LBO) lithium borate	0.18	160–2600	Very robust material; can perform NCPM for Nd-doped lasers with temperature phase matching; widely used in commercial cw and pulsed laser products for green- and UV-light generation.

of second-order nonlinearity is a tensor called the d coefficient and relates the nonlinear polarization in the material to the product of the electric fields of the incident beam or beams. In general, this quantity depends on the direction and polarization of the beams in the crystal. Analysis of the conversion efficiency of materials to the second harmonic shows that it depends on the square of the d coefficient for the particular interaction divided by the cube of the material refractive index; this quantity is often used as a simple figure of merit for a nonlinear material, although the efficiency of the process depends on a number of other properties set by phase matching in the material.

The list of materials in the table is limited to those used with lasers in the 800- to 2000-nm range. The semiconductor-based materials can be used for harmonic generation of CO_2 and similar gas lasers but are more widely employed for parametric generation, which is covered in the next subsection.

Given all of the listed constraints for nonlinear materials to work well, it may not be surprising that new nonlinear-crystal development features major advances on a ten-year timescale at best and seems to have slowed down in the last two decades. The materials listed in Table 5.3 emerged between 1960 and the late 1980s, and no major new materials have emerged since then. Advances in improving the size and quality of the materials listed do continue.

Although LBO does not have the largest figure of merit among the materials shown, it is not subject to photorefractive and other optical damage issues, and also exhibits an unusually low level of absorption at both the fundamental and harmonic wavelengths. As a result, LBO has been used for generation of harmonic powers of several hundreds of watts at visible wavelengths, and tens of watts at the Nd-laser third harmonic, being limited more by the properties of the drive laser than the material itself. Conversion efficiencies are in the 50–65% range for second-harmonic systems and in the 30–50% range for third-harmonic devices. LBO is also widely employed in commercial, cw, diode-pumped Nd-doped lasers with intracavity doubling to generate cw diffraction-limited powers at green wavelengths as high as 20 W.

One significant advance in nonlinear devices has been successful in exploiting the properties of existing crystals. Early in the initial work in nonlinear optics (during the 1960s), theoretical studies showed that another way to phase match nonlinear generation involved making an artificial crystal that consisted of alternating layers of material arranged such that the sign of the nonlinear coefficient for the process of interest alternated in a periodic manner. As noted, without normal phase matching, power along the beam path alternates in a distance given by the coherence length. If a crystal is properly structured, instead of power converting back to the fundamental due to dephasing, the change in sign of the nonlinearity results in power continuing to convert to the harmonic. While the buildup of harmonic power per length of material is not as rapid as with normal phase matching, this quasi–phase matching (QPM) allows efficient harmonic generation in nonbirefringent materials and has features in terms of angular acceptance that are common with NCPM.

The first attempts at QPM employed materials that were polished into thin slices and bonded together in periodically alternating orientations. However, losses at the interfaces were too high for effective generation. Another possible approach for certain materials that exhibit the ferroelectric effect employs high-voltage electric fields to orient the crystal structure. Through a combination of (1) lithography patterning of electrodes, (2) a setup involving electrodes that alternate at a period of a few to tens of microns, and (3) application of an appropriate voltage, one can permanently pole a ferroelectric crystal with an alternating pattern of the signs of the nonlinear coefficient.[14] Such crystals are said to be periodically poled (PP) and include those listed in Table 5.3, LN, and KTP, as well as variants such as $LiTaO_3$ and Mg:LN, which offer better resistance to photorefractive degradation. While these ferroelectrics can also employ birefringent phase matching, QPM provides phase matching over a much greater range of wavelengths (limited by the transparency of the material) and allows for the use of crystal orientations that maximize the nonlinear coefficient. One drawback of ferroelectric QPM is that the thickness of the material is limited by the length of material over which poling is possible, typically on the order of 1 mm, although recent advances have shown that more than 5-mm thicknesses are possible in LN doped with MgO.[15]

Another approach to fabricating QPM materials employs semiconductors, most prominently GaAs,[16] where lithography and semiconductor processing technology is employed to form a patterned, thin substrate of material. One can then employ thick-film growth techniques on this substrate to fabricate QPM crystals with (to date) 1- to 2-mm thicknesses.

As with the development of nonlinear crystals, progress in new QPM materials has been slow and requires dedicated, long-term efforts. Practical ferroelectric and semiconductor devices each emerged from 10-year-long, combined academic–industry programs. While QPM nonlinear materials can be used for harmonic generation, they have seen more widespread use in parametric generation and will be discussed in more detail in the next section.

5.4.2.2 Optical parametric generation

When there is no phase matching in a crystal with a second-order nonlinearity, the harmonic wavelength converts back into the fundamental wavelength as it propagates in the crystal. In fact, this process whereby the harmonic generates longer-wavelength light is a special case of optical parametric generation, which is a descendant of a process first observed and employed at radio/microwave frequencies. In general, the same second-order nonlinearity that can produce shorter-wavelength light also allows a process in which a beam of light at one wavelength converts into two beams, where, for energy-conservation reasons, the sum of the frequencies of the two beams has to sum to the frequency of the original beam. The optical parametric process thus allows for generation of longer-wavelength light. Of the two beams created, the shorter-wavelength beam is called the signal, while the longer-wavelength beam is the idler; the original beam is called the pump. These three beam designations are taken from microwave terminology.

The interaction of the pump, signal, and idler beams shows that, with phase matching of the three waves, application of the pump light in the crystal leads to optical gain for both the signal and idler waves along the path of the pump beam. The signal and idler waves build up together, so if one uses the parametric process to amplify an input at the signal wavelength, creating an OPA, power builds up in the idler wave, and *vice versa*. (In terms of photons, every generated signal photon is accompanied by an idler photon.) If one employs an optical cavity around the crystal for either the signal or the idler, or for both, the device is called an OPO. An OPO whose cavity is resonant for either the signal or the idler is said to be singly resonant, while a device that resonates at both frequencies is said to be doubly resonant. Doubly resonant devices are infrequently employed because of the instabilities that arise in the output due to interactions between the signal and idler frequencies. When the pump intensity is high enough and the interaction length is long enough, the gain in the OPA may be high enough to allow significant signal/idler power to build up from noise, and the device is referred to as an optical parametric generator (OPG). Parametric generation for which the signal and idler beams have the same wavelength is called degenerate amplification, and gain is maximum in this case.

In contrast to lasers in general, optical parametric devices have no energy storage, as gain is present only when the pump power is present. The ability of these devices to generate peak power comes about purely from the peak power of the pump. Energy conservation requires, at best, that the sum of the energy in the signal and idler outputs equals that of the pump, with the energy apportioned according to the wavelengths/frequencies of the signal and idler. In the amplification process, as the signal and idler beams build up in power, the power in the pump beam drops or is depleted. In theory, it is possible for all of the energy in the pump to convert to signal and idler power; however, in practice, due to nonuniform power distribution in the pump beam, parametric devices perform on the order of 50% of full conversion, although some devices have pushed this to 90%. Barring any optical absorption in the crystal, the parametric process is completely lossless, and the sum of the power in all three beams remains constant. As a result, in theory, no heat is deposited in a parametric device.

Given that the gain is only on during pumping, constructing OPOs needs to account for the time needed for beams to pass back and forth in the optical resonator. If the resonator is too long, the gain may be depleted before enough passes back and forth occur to build up power from the noise. For many Q-switched solid state pump lasers, the gain is effective for about 10–20 ns; therefore, the cavity length typically must be 10 cm or less. OPO efficiency is reduced because the pulse takes longer to build up due to no power being extracted from the pump light.

For a given crystal and pump-beam orientation, the phase-matching process leads to specific signal and idler wavelengths. As with harmonic generation, CPM and NCPM processes exist, the latter being much more desirable. The ability to tune the generated signal and idler wavelengths by adjusting the angle or temperature of the crystal (or by adjusting the pump wavelength using tunable pump lasers) is one of the most important features of optical parametric devices.

The first optical parametric devices were demonstrated in the 1960s and looked to be a major breakthrough in coherent sources. However, the limitations of materials like LN due to photorefraction, along with issues of optical damage from the high powers required for operation, prevented widespread use of OPOs and gave them a reputation for being unreliable.

The development of more robust materials (e.g., BBO and LBO) have made OPOs more practical. Commercial devices for scientific applications are currently based on tripled, Q-switched, Nd:YAG pump lasers (driving BBO-based OPOs) and provide nanosecond-duration tunable light over the 410- to 2500-nm range.

The material KTP was initially developed for frequency doubling of Nd:YAG lasers; however, examination of its phase-matching characteristics showed that it could be used as a NCPM OPO, also pumped by Nd-doped lasers, to generate a signal wavelength around 1550 nm. The device thus acts to convert a very unsafe (in terms of eye damage) laser for applications like active sensing into a much eye-safer pulsed source. Originally, KTP crystals could only be grown in very small (millimeter-dimension) sizes through hydrothermal processes, but development of flux-growth technology allowed scaleup to centimeter-sized materials. At that point, construction of high-energy OPOs was possible. To date, the highest-energy OPO (450 mJ of signal at 1570 nm, pumped by 1.1 J from a Q-switched Nd:YAG laser[17]) is based on a 1-cm–aperture KTP crystal.

One drawback with KTP is that the idler wavelength is absorbed in the material, limiting average powers to <10 W due to thermal effects. Motivated by this issue, crystal growers developed flux growth of an isomorph of KTP, $K(TiO)AsO_4$ (KTA), which has similar phase-matching properties but is highly transparent at the idler wavelength, allowing signal average-power scaling to 33 W (330 mJ at 100 Hz),[18] and which held the record for some period as the highest–average-power OPO. At present, KTP OPOs have replaced Er:glass lasers for eye-safe rangefinder applications due to their higher performance and higher pulse-rate capabilities.

As noted, semiconductor nonlinear crystals provide transparency much farther into the infrared than the oxide materials listed in Table 5.3 and are suited for optical parametric generation at long wavelengths, particularly beyond 4 μm. With a few exceptions (CdSe being one), the III-V and II-VI materials lacking inversion symmetry also lack birefringence, and one has to look at more-complex compounds such as the ternary (three-element) chalcopyrite structure (I-III-VI$_2$) materials to obtain birefringent phase-matching crystals. Semiconductor materials have the advantage of (generally) much higher nonlinear coefficients, but at the expense of small bandgaps compared to oxide materials. Some materials are completely opaque to common Nd-doped pump lasers, while others suffer from multiphoton absorption with those pumps. In general, due to their structure, chalcopyrites have complex thermal expansion properties and are a challenge to grow.

Because of the strong interest in the development of tunable infrared sources for countersensor applications, considerable investment has been made to develop chalcopyrite materials. At present, the most widely used material is $ZnGeP_2$ (ZGP), which combines a very high nonlinearity (figure of merit of 200) with good thermal conductivity. ZGP is not transparent to Nd-doped lasers but can be pumped by 2000-nm–region lasers (as well as OPOs) and can be phase matched to generate a wide variety of wavelengths from 2.5 to 10 μm, although the material has some absorption starting for wavelengths beyond 8 μm. Recently reported ZGP-based OPOs pumped by a Q-switched, 35-kHz pulse rate, Ho:YAG hybrid laser have generated average powers of 27 W in the 4-μm–wavelength region.[19]

For longer-wavelength OPO operation, covering the LWIR 8- to 12-μm atmospheric window, there are other semiconductor materials with transparency extending beyond the range of ZGP, including CdSe and GaSe. Systems that generate these wavelengths employ tandem OPO schemes in which a solid state laser pumps one OPO and the signal or idler wavelength is then used to pump the OPO covering the LWIR range.

The development of QPM materials has no doubt had the highest impact on recent development of OPOs, OPAs, and OPGs. For PPLN, the ability to obtain NCPM-like phase matching over a wide range of wavelengths through the use of different poling periods, along with the relatively high nonlinearity, have made it possible to operate Nd-laser–pumped OPOs with very low thresholds—low enough for cw operation as opposed to pulsed operation. For the 1000-nm and longer pump wavelengths, there appears to be minimal (if any) degradation due to photorefractive effects, especially if the material is held above room temperature. PPLN-based optical parametric devices provide efficient wavelength generation out to wavelengths around 4 μm, beyond which the absorption in LN becomes a limiting issue. Variants on PPLN such as periodically poled KTP (PPKTP) and periodically poled stoichiometric $LiTaO_3$ (PPSLT) provide some variation in OPO performance but offer no extension to longer wavelengths.

The more recent development of GaAs-based QPM structures, called orientation-patterned GaAs (OP-GaAs) provides a material that is equivalent to ZGP in terms of a very high figure of merit (also around 200), longer-wavelength transparency to 17 μm, and all the advantages of QPM. As with ZGP, OP-GaAs cannot be pumped with Nd-doped lasers, and 2000-nm pumps provide the best option. OP-GaAs–based OPOs are currently under active investigation and have shown thresholds low enough for cw operation[20] with 5.3 W of combined signal (3.8 μm) and idler (4.7 μm) power when pumped by a 24.7-W Ho:YAG laser. In addition, the materials can be readily pumped by the relatively low peak powers available from pulsed fiber lasers.[21] Given the current limited aperture size (1- to 2-mm height) of this material compared to ZGP, it remains to be seen if OP-GaAs can provide the high-energy pulses needed for some applications.

5.5 Fiber Format

Optical fibers employ highly transparent optical materials that are structured to form a waveguide for light. The simplest fiber designs employ a circular cross-section with a central circular region (core) that has a higher refractive index than the surrounding material (cladding). Through the process of total internal reflection at the interface between the high- and low-index regions, light incident at a sufficiently low angle to the interface is reflected with no loss of power, and the only loss of power for the light contained in the core is due to the inherent absorption and scattering losses in the core material. Typical core diameters for fibers supporting only the lowest-order (diffraction-limited) mode are on the order of 5–10 μm. Through the development of glass-processing techniques, fibers based on silica (SiO_2) glass capable of transmitting light with minimal loss over kilometer-length distances became available in the 1970s and have developed to the point that fiber-optic–based media are the basis for data transport in the telecommunications industry.

Although it was recognized in the 1960s that the glass in fibers could be doped with rare earths and provide optical gain for communications systems, there was no practical pump source for the fibers, and the early materials were lossy. Improvements in fiber-related glass technology in the 1980s and development of reliable diode pump lasers made it possible to deploy Er-doped fibers as amplifiers for telecommunications applications, where they eliminated the need to employ electronic repeaters after a certain distance in a fiber link.

Erbium doping was chosen because the 1550-nm wavelength for the first excited-state–ground-state transition provides good overlap with the lowest-dispersion–wavelength region for silica-based fibers, thereby maximizing the data rate that can be transmitted. [The same dispersion that provides useful stretching of pulses for chirped-pulse amplification (CPA) systems is a major limitation when the goal is to keep the pulse widths short in high–data-rate-transmission systems.) The problem of up-conversion that limits performance of Er-doped bulk lasers is eliminated in fibers by keeping the Er doping level low, and by using the very long absorption paths for pump light that is coupled into the core of the fiber. The other issue with Er-doped systems—the partial occupation of the lower laser level—is not a problem for fibers because large intensities of pump light can be established at moderate pump-power levels in the small cores; therefore, population inversions well over 50% are easily obtained.

When pumping the core of a fiber laser using single-mode fibers to efficiently couple in pump light, the pump must be single mode as well. For pumping a diode laser, the powers required for telecommunications fiber amplifiers are well within the limits of single-mode devices. When power extraction is the goal, the amount of power that can be extracted from the core-pumped fiber for diode-pumped systems is limited. A major breakthrough in fiber-laser technology that was first demonstrated in the late 1980s was the recognition that pump light could be coupled into a much larger cladding region surrounding the core if the cladding, in turn, had an outer cladding of a low-index material, typically a specially formulated polymer (plastic) material, or, in some cases, silica glass with fluorine added.

Figure 5.30 is a schematic of a so-called double-clad fiber laser. Pump light is launched into the cladding region, and as it propagates, it passes through the core region, where it gets absorbed and pumps the active ions. The characteristic length over which a certain fraction of the pump is absorbed is increased over that for straight core pumping (in the best case) by the ratio of the cladding area to the core areas. With a purely cylindrical outer surface for the pump cladding and a core located in the center, some of the pump light will never pass through the core, so actual double-clad fibers deviate from cylindrical surfaces to ensure that all of the pump light does pass through the core.

With the development of double-clad fibers, fiber lasers were able to provide the same sort of brightness enhancement of pump light that bulk solid state lasers routinely accomplish, albeit with pumping limited to diode lasers. As high-power diodes improved in performance and a means to couple the power of many diodes into a multimode fiber was realized, the power available from fiber lasers increased, crossing the 1-kW level for Yb-doped fibers in 2003.[22] Single fibers are now commercially available with multimode powers of 50 kW and single-mode powers exceeding 10 kW.

Since fibers are made of glass, it may seem puzzling that they can run at such high powers, given the limits of bulk glass lasers. Four key requirements to high-power generation in fibers are as follows: (1) Heat produced in the core from laser operation has a comparatively short (submillimeter) path to travel to the outer surface of the fiber where it can be removed by active cooling. (2) Heat generation is typically spread over a meters-long distance, so the thermal loading per centimeter is modest. (3) The thermal gradient in the core itself is generally small, and the largest gradients are in the cladding region where laser operation occurs. (4) The fibers consist primarily of fused silica. The thermal lensing that does occur from the gradients is generally a small perturbation of the fiber waveguide profile, which dominates when determining the spatial-mode properties of the laser beam.

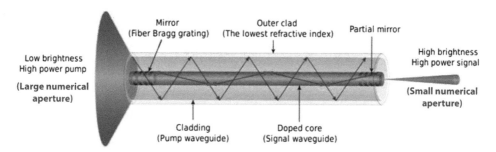

Figure 5.30 Schematic of a pumped double-clad fiber laser (from Wikipedia: https://en. wikipedia.org/wiki/File:Fl.svg under license CC0 1.0 Universal Public Domain).

There are ultimate thermal limits to fiber power; for operation near 1000 nm, the use of Yb-doped fibers is favored over Nd-doped fibers because of the lower heat generated in laser operation. As with Er-doped fibers, the partial occupation of the lower Yb laser is not an issue in fiber laser operation.

The small core size and long operating lengths that allow high cw power to be output from fiber-format lasers also provide the biggest challenge for certain modes of operation. The long lengths and high intensities in the core promote nonlinear effects that are generally not seen with bulk lasers. For single-frequency generation, stimulated Brillouin scattering (SBS) of light, which involves an interaction with low-frequency (acoustic) phonons in the glass, acts to amplify light in the fibers in a direction opposite to the laser propagation. This can lead to destruction of optics at the beginning of a fiber-laser amplifier system as well as limits to the power output. For typical core sizes and operating lengths in fiber lasers, the SBS process may start with only tens of watts of power. SBS has a very limited gain linewidth, on the order of tens of megahertz. SBS can therefore be circumvented by operating the fiber laser with a large spread of operating frequencies, or by adding frequency modulation onto a single-frequency input. Power levels on the order of 1 kW from Yb:fiber lasers have been produced via the frequency-modulation technique.[23]

Beyond SBS, for broadband fiber lasers, stimulated Raman scattering (SRS) provides an ultimate limit to power. Traditional, small-core (5- to 100-μm diameter) fibers can be used in the 100-W level for typical lengths. High-power fiber lasers are operated with fibers having core sizes in the 20- to 30-μm range, while SRS limits these lasers to on the order of 10 kW.

For the pulsed mode of operation, SRS is clearly a limit, but fiber lasers are also much more subject to amplified spontaneous emission (ASE) problems when fibers are long due to the long lengths of the gain region and the small volume of active material. Although rare earths have long storage times, conventional fibers are challenged to generate amplified pulse energies beyond 0.1–1 mJ because of ASE.

The glass host material for fibers produces large gain linewidths for the rare-earth dopants and supports generation and amplification of pulses with durations on the order of 100 fs. At high pulse energies beyond the limits from SRS, the fiber materials' weak dependence of the refractive index on the electric field (over typical fiber lengths) leads to a distortion of the pulse spectrum (and eventually of the shape) through a process called self-phase modulation (SPM). As with bulk lasers, CPA techniques have been used with fibers to increase the energy outputs with 100-fs–duration pulses, but the limits are considerably below those for bulk lasers.

All of these limits to operational modes can be mitigated by increasing the core size of the fiber. For single-mode operation, to use a larger core requires that the difference in refractive index between the core and cladding be reduced, which also lowers the range of angles that the fiber can guide (the

NA). At some point, the refractive index difference becomes so small that fabrication of glass with the required uniformity and control of the refractive index become impractical, and the amount of bending the fiber can endure becomes limited. One approach to increasing the single-mode capabilities of fibers is to fabricate a large-core fiber with a large NA that would typically support multiple transverse modes, and then purposely wrap the fiber around a spool (typically about 10 cm in diameter), which creates high losses for all but the lowest-order, diffraction-limited mode. Fiber core diameters in the 20- to 30-μm–diameter range for operation around 1000 nm can be used to generate single-mode output even when the fiber should be propagating in multiple modes.

Conventional fibers are fabricated from solid glass, but developments in fiber technology have allowed development of so-called photonic crystal (PC) fibers, which have a microstructure on a scale of the wavelength of light that consists of regular patterns of air holes in the glass cladding around the core. This is done by making the glass used to pull the fiber from capillaries rather than from solid glass. The microstructure has unique properties compared to solid materials: it allows for fine control over the mode properties of the fiber and, ultimately, enables the use of larger core sizes for single-mode fibers than is possible with solid glass. (PC fibers can also be made with very small cores when enhancement of nonlinear effects is desired.) The larger core sizes do not prevent the problem of excessive bend loss. The largest-core PC fibers are typically made with very thick glass claddings that prevent bending, so the active material is in the form of a long (typically 50–100 cm), 1-mm–diameter rod that must be handled and mounted with care.

The use of PC fiber configurations has enabled single-mode generation with effective diameters of the optical mode in the 50- to 100-μm–diameter range, as well as a large increase in the peak power available from fiber lasers. Recent results with Yb-doped fibers include generation (at 1040 nm) of near-single-mode, 27-mJ-energy, sub–60-ns-duration pulses at a 5-kHz rate for 130 W of average power. The final amplifier stage had a 1.3-m–long rod stage.[24] For short-pulse operation, the same sort of large-mode PC fiber was employed with a CPA configuration to produce 500-fs–duration, 2.2-mJ pulses at a 5-kHz rate.[25]

While these results show that reasonable peak powers can be generated by fibers, the use of rigid, rod-type structures raises the question of whether the devices are more closely related to bulk lasers than fiber lasers. Additionally, in terms of engineering, the use of gain media with meter-long lengths presents engineering challenges for ruggedizing the technology. The appeal of conventional fiber lasers lies in the fact that the gain material can be easily bent to allow compact packaging, and that optical components in the laser can be fused together like telecommunications systems to eliminate alignment problems, an arrangement not possible with rod fibers. Finally, limits to average powers from rod-type gain media have been found that are related to thermal effects such as lensing, making these structures closer to bulk lasers in operational issues.

To date, the majority of work on high-power fiber lasers has involved Yb-doped silica glass, which exhibits optical–optical efficiencies in the 70–90% range. Er-doped silica works well for low-power telecommunications applications; however, the Er ion transitions that one could use to pump the laser are relatively weak, and too high a doping level leads to up-conversion processes that reduce efficiency and gain. For typical double-clad designs at appropriate doping levels, the absorption lengths for the pump light are at least an order of magnitude larger than for typical Yb-doped fibers, leading to long fibers that create more severe power limits from nonlinear processes. The most powerful Er-doped fiber lasers employ the same Yb-sensitization process used with bulk lasers to reduce the length of fiber needed to absorb pump light. To date, this has allowed generation of around 300 W of output in the 1550-nm–wavelength region, with the limit set by the laser action from the Yb ion sensitizers, which starts at high pump powers and steals excitation.[26]

Thulium (Tm)-doped silica fiber lasers in the 2000-nm–wavelength region have recently advanced to cw power levels exceeding 1 kW.[27] What allowed this development was the recognition that, as with bulk-crystal lasers, if the Tm ion doping level is made high enough, pumping the ion at 790 nm generates two excited Tm states for one pump photon, leading to optical–optical efficiencies exceeding 60%, as well as the ability to use the high-power diode lasers available at 790 nm. To achieve the necessary high Tm-doping levels, one needs to add a high level of Al ion impurities to make more sites available for the laser ions. The nonlinear processes that limit powers for Yb-doped fibers are reduced at the longer wavelength; additionally, for the same fiber NA, one can operate a single-mode fiber with twice the core diameter of a Yb:fiber laser. The net effect is that the power limits from SBS and SRS are considerably higher. To date, the highest, cw, true single-frequency output from any fiber laser is the 600-W level obtained from a Tm:fiber laser.[28] The technology for Tm-doped silica fibers is not advanced to same level as for Yb-doped devices, particularly regarding PC fibers, and the prospect for Tm-doped lasers to provide high peak powers and high-energy, ultrafast pulses looks excellent.

Silica glass has thermo-mechanical properties that far exceed other glasses capable of being pulled into low-loss optical fibers. However, there is interest in so-called soft glasses for certain applications that need fibers doped to high concentrations of active ions, or fibers with reduced multiphonon relaxation, or for operation at longer wavelengths than is possible from silica. Powers from soft-glass fibers have tended to have limits around 100 W or less.

One of the most sought-after goals is the development of single-crystal fibers made from materials like YAG; however, major technical challenges need to be overcome to fabricate practical, single-mode waveguiding structures and, beyond that, double-clad structures for high-power pumping. The advantages of YAG and other crystals as a fiber material compared to silica include a much higher thermal conductivity, which would permit the use of shorter-length fiber before thermal limits set in. Crystal fibers would have narrower absorption and emission linewidths, and, according to a recent study, a much higher threshold for SBS, at least in YAG.[29] A YAG crystal fiber in shorter lengths would allow for generation of much higher, single-frequency powers than is possible with silica glass material.

The materials challenges associated with fiber-based lasers are major, and advances in high-power fiber lasers have been able to leverage the substantial interest in development of low-loss silica fibers for telecommunications. The primary challenge is in making any fiber material sufficiently low loss that light is barely attenuated by scatter and other imperfections over a path of 10 m. This is much more difficult than making laser crystals with a typical path on a 1- to 10-cm scale.

Advances in diode-laser technology will contribute to further advances in fiber lasers. The cladding diameter for double-clad designs must be large enough to guide the desired pump power. An interest in employing diode lasers for direct cutting and welding of materials has motivated the development of schemes such as wavelength beam combining of multiple diode arrays[30] or techniques to combine high-brightness, single-emitter diodes[31] to increase the brightness of diode-based sources. These schemes will allow double-clad fiber lasers to employ smaller-diameter claddings and reduce the length of fiber required to absorb the pump power, which will act to increase any nonlinear limits to power. Ultimately, thermal limits will set in, but for many fiber laser systems, those limits have not been reached.

Problems and Solutions

5-1. Suppose that a resonator for a HeNe laser has two concave mirrors ($R_1 = 10$ m, $R_2 = 5$ m). The distance between the two mirrors is 50 cm.

(a) What is the spot size of beam?
(b) Where is the beam waist located?

5-1 Solution:
We have $d = z_2 - z_1 = 0.5$ m > 0, $R_1 = 10$ m, $R_2 = 5$ m, $\lambda = 633$ nm.

$$z_0^2 = \frac{d(-R_1 - d)(R_2 - d)(R_2 - R_1 - d)}{(R_2 - R_1 - 2d)^2} = 1.58\,\text{m}^2 \rightarrow z_0 = 1.26\,\text{m},$$

$$z_1 = \frac{d(R_2 - d)}{(R_1 - R_2 + 2d)} = -0.16,$$

$$z_2 = \frac{d(R_1 + d)}{(R_1 - R_2 + 2d)} = 0.34.$$

The beam waist is at $z = 0$, which is 0.16 m away from the left mirror.

$$w_0 = \left(\frac{\lambda}{\pi}\right)^{0.5} \left[\frac{d(-R_1 - d)(R_2 - d)(R_2 - R_1 - d)}{(R_2 - R_1 - 2d)^2}\right]^{\frac{1}{4}} = 5.034 \times 10^{-4}\,\text{m}.$$

The spot size at the left mirror is

$$w_1 = \left(\frac{\lambda}{\pi}R_1\right)^{0.5} \left[\frac{d(R_2 - d)}{(-R_1 - d)(R_2 - R_1 - d)}\right]^{\frac{1}{4}} = 5.072 \times 10^{-4}\,\text{m}.$$

The spot size at the right mirror is

$$w_2 = \left(\frac{\lambda}{\pi}R_2\right)^{0.5} \left[\frac{d(-R_1 - d)}{(R_2 - d)(R_2 - R_1 - d)}\right]^{\frac{1}{4}} = 5.213 \times 10^{-4}\,\text{m}.$$

(a) As seen, the spot size of the beam at the left mirror is 5.072×10^{-4} m and at the right mirror is 5.213×10^{-4} m. The minimum spot size is 5.034×10^{-4} m.

(b) The location of the beam waist is $z = 0$, which is 0.16 m away from the left mirror.

5-2.

(a) Calculate the coherence length of a low-pressure mercury lamp with a spectral filter at 546 nm having a spectral width of 4 GHz.

(b) Calculate the coherence length of a single-mode Ar laser at 488 nm with 1-MHz bandwidth.

5-2 Solution:

(a) $L_c = c/\Delta v = 0.075$ m,
$\tau_c = 1/\Delta v = 2.5 \times 10^{-10}$ s.

(b) $L_c = c/\Delta v = 300$ m,
$\tau_c = 1/\Delta v = 10^{-6}$ s.

5-3. A resonator contains a homogeneously broadening gaseous gain medium with parameters of $A_{21} = 108$ s^{-1}, $\tau_1 = 1$ ns, $\tau_2 = 5$ ns, and $\lambda_0 = 1.06$ μm.

(a) What is the shortest pulse width obtainable from this laser when mode locked?

(b) Develop the equations to describe the transition between all levels.

5-3 Solution:

(a)

$$\Delta v = \frac{1}{2\pi}\left(\frac{1}{\tau_1}+\frac{1}{\tau_2}\right) = 1.91\times 10^8\,\text{Hz},$$

$$\Delta T = \frac{1}{2\pi\Delta v} = 0.834\,\text{ns}.$$

(b) We have N_0 to N_3. N_2 is the metastable state. Therefore,

$$\frac{dN_3}{dt} = R_{03} - \frac{N_3}{\tau_{32}} - \frac{N_3}{\tau_{31}} - \frac{N_3}{\tau_{30}} = R_{03} - N_3\left(\frac{1}{\tau_{32}}+\frac{1}{\tau_{31}}+\frac{1}{\tau_{30}}\right) = R_{03} - \frac{N_3}{\tau_3},$$

$$\frac{dN_2}{dt} = \frac{N_3}{\tau_{32}} - \frac{N_2}{\tau_2} - N_2 B_{21}\rho(v) + N_1 B_{12}\rho(v),$$

$$\frac{dN_1}{dt} = \frac{N_2}{\tau_{21}} + \frac{N_3}{\tau_{31}} - \frac{N_1}{\tau_{10}} + N_2 B_{21}\rho(v) - N_1 B_{12}\rho(v),$$

$$\frac{dN_0}{dt} = \frac{N_3}{\tau_{30}} + \frac{N_2}{\tau_{20}} + \frac{N_1}{\tau_{10}} - R_{03},$$

where
$N_T = N_0 + N_1 + N_2 + N_3,$
$\frac{1}{\tau_3} = \frac{1}{\tau_{32}}+\frac{1}{\tau_{31}}+\frac{1}{\tau_{30}},$
$\frac{1}{\tau_2} = \frac{1}{\tau_{21}}+\frac{1}{\tau_{20}}.$

All transitions between all levels (including parasitic transitions) are considered in the aforementioned equations.

5-4. Plot the beam diameter as it expands over a 100-m range, given $\lambda_0 = 1.064$ µm, $\lambda = 1.550$ µm, 0.970 µm, and a 10-mm beam-waist diameter.

5-4 Solution:
Using the equation

$$w(z) = w_0 \left[1 + \left(\frac{\lambda z}{\pi w_0^2} \right)^2 \right]^{1/2},$$

the following plot was created.

```
w0 = 0.010;
l1 = 1.064e-6;
l2 = 1.55e-6;
l3 = 0.97e-6;
z = 1:0.1:100;
w1 = w0*sqrt(1+(l1*z/(pi*w0^2)).^2);
w2 = w0*sqrt(1+(l2*z/(pi*w0^2)).^2);
w3 = w0*sqrt(1+(l3*z/(pi*w0^2)).^2);
plot(z,w1*1e3,z,w2*1e3,z,w3*1e3)
```

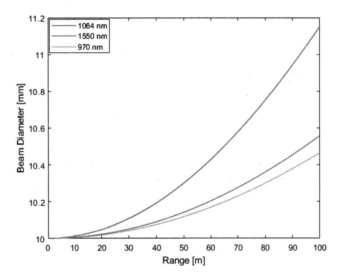

5-5. Plot the radius of curvature as it evolves over 100 m, using the same parameters as in Problem 5-4.

5-5 Solution:

$$R(z) = z \left[1 + \left(\frac{\pi w_0^2}{\lambda z} \right)^2 \right].$$

```
w0 = 0.010;
l1 = 1.064e-6;
l2 = 1.55e-6;
l3 = 0.97e-6;
z = 10:0.1:100;
r1 = w0*(1+(pi*w0^2./(l1*z)).^2);
```

```
r2 = w0*(1+(pi*w0^2./(12*z)).^2);
r3 = w0*(1+(pi*w0^2./(13*z)).^2);
plot(z,r1,z,r2,z,r3)
```

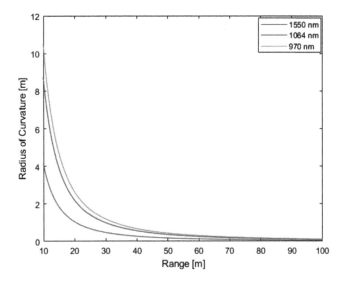

5-6. Plot the evolving Guoy phase over 100 m, using the same parameters as in Problems 5-3 and 5-4. A Gaussian beam acquires a phase shift that differs from that for a plane wave with the same optical frequency. This difference is called the Gouy phase shift, where z_R is the Rayleigh length, and $z = 0$ corresponds to the position of the beam waist.

5-6 Solution:

$$\phi(z) = \tan^{-1}\left[\frac{\lambda z}{\pi w_0^2}\right].$$

```
w0 = 0.010;
l1 = 1.55e-6;
l2 = 1.064e-6;
l3 = 0.97e-6;
z = 10:0.1:100;
p1 = atan(l1*z/(pi*w0^2));
p2 = atan(l2*z/(pi*w0^2));
p3 = atan(l3*z/(pi*w0^2));
plot(z,p1*180/pi,z,p2*180/pi,z,p3*180/pi)
xlabel('Range [m]')
ylabel('Guoy Phase [degrees]')
```

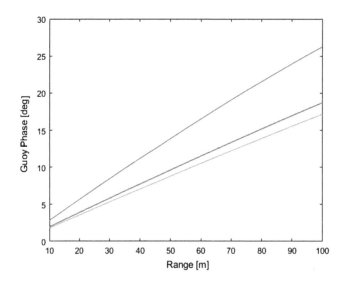

References

1. OPI Online Course: Laser and Non-linear Optics: Optical Resonators and Gaussian beams: http://www.optique-ingenieur.org/en/courses/OPI_ang_M01_C03/co/OPI_ang_M01_C03_web.html.
2. Z. W. Barber, F. R. Giorgetta, P. A. Roos, I. Coddington, J. R. Dahl, R. R. Reibel, N. Greenfield, and N. R. Newbury, "Characterization of an actively linearized ultrabroadband chirped laser with a fiber-laser optical frequency comb," *Optics Letters* **36**, 1152–1154 (2011).
3. J. Zmnicki, "Spatial Heterodyne Imaging Using a Broadband Source," M.S. thesis, University of Dayton (2018).
4. B. L. Stann, W. C. Ruff, and Z. G. Sztankay, "Intensity-modulated diode laser radar using frequency-modulation/continuous-wave ranging techniques," *Optical Engineering* **35**(11), 3270–3278 (1996) [doi: 10.1117/1.601067].
5. Y. Liu, Y. Wu, C.-W. Chen, J. Zhou, T.-H. Lin, and I. C. Khoo, "Ultrafast pulse compression, stretching-and-recompression using cholesteric liquid crystals," *Optics Express* **10**(24), 10458 (2016).
6. I. D. Jung, F. X. Kärtner, N. Matuschek, D. H. Sutter, F. Morier-Genoud, G. Zhang, U. Keller, V. Scheuer, M. Tilsch, and T. Tschudi, "Self-starting 6.5-fs pulses from a Ti:sapphire laser," *Optics Letters* **22**(13), 1009–1011 (1997).
7. M. Bass, Ed.-in-Chief, *Handbook of Optics: Design, Fabrication, and Testing; Sources and Detectors; Radiometry and Photometry*, Third Edition, McGraw Hill Education, New York, p. 16.5 (2009).
8. T. H. Maiman, "Stimulated optical radiation in ruby," *Nature* **187**, 4736, 493–494 (1960).

9. T. Possner, B. Messerschmidt, A. Kraeplin, V. Bluemel, B. Hoefer, and P. Schreiber, "Assembly of fast-axis collimating lenses with high power laser diode bars," *Proc. SPIE* **3952**, 392–399 (2000) [doi: 10.1117/12. 384424].

10. http://www.jenoptik.com/products.

11. https://ams.com/vcsel.

12. http://www.finisar.com.

13. P. A. Franken, A. E. Hill, C. W. Peters, and G. Weinreich, "Generation of optical harmonics," *Physical Review Letters* **7**, 118 (1961).

14. M. M. Fejer, G. A. Magel, D. H. Jundt, and R. L. Byer, "Quasi-phase-matched second harmonic generation: tuning and tolerances," *IEEE J. Quantum Electron.* **28**, 2631 (1992).

15. H. Ishizuki and T. Taira, "High-energy quasi-phase-matched optical parametric oscillation in a periodically poled $MgO:LiNbO_3$ device with a 5 mm $\times$ 5 mm aperture," *Optics Letters* **30**, 2918 (2005).

16. L. A. Eyres, P. J. Tourreau, T. J. Pinguet, C. B. Erbert, J. S. Harris, M. M. Fejer, L. Becouarn, B. Gerard, and E. Lallier, "All-epitaxial fabrication of thick, orientation-patterned GaAs films for nonlinear frequency conversion," *Applied Physics Letters* **79**, 904 (2001).

17. G. A. Rines, D. M. Rines, and P. F. Moulton, "Efficient, high-energy, KTP optical parametric oscillators pumped with 1 micron Nd-Lasers," in *Advanced Solid State Lasers*, T. Fan and B. Chai, Eds., Vol. **20**, *OSA Proceedings Series*, paper PO9 (1994).

18. M. S. Webb, P. F. Moulton, J. J. Kasinski, R. L. Burnham, G. Loiacono, and R. Stolzenberger, "High-average-power $KTiOAsO_4$ optical parametric oscillator," *Optics Letters* **23**, 1161 (1998).

19. A. Hemming, J. Richards, A. Davidson, N. Carmody, S. Bennetts, N. Simakov, and J. Haub, "99 W mid-IR operation of a ZGP OPO at 25% duty cycle," *Optics Express* **21**, 10062 (2013).

20. L. A. Pomeranz, P. G. Schunemann, S. D. Setzler, C. Jones, and P. A. Budni, "Continuous-wave optical parametric oscillator based on orientation patterned gallium arsenide (OP-GaAs)," in *Conference on Lasers and Electro-Optics 2012*, OSA Technical Digest, paper JTh1I.4 (2012).

21. C. Kieleck, M. Eichhorn, A. Hirth, D. Faye, E. Lallier, and S. D. Jackson, "OP-GaAs OPO pumped by a Q-switched Tm, Ho:silica fiber laser," in *Conference on Lasers and Electro-Optics/International Quantum Electronics*, OSA Technical Digest (CD), paper CWJ2 (2009).

22. Y. Jeong, J. Sahu, D. Payne, and J. Nilsson, "Ytterbium-doped large-core fiber laser with 1.36 kW continuous-wave output power," *Optics Express* **12**, 6088 (2004).

23. C. Robin, I. Dajani, C. Zernigue, A. Flores, B. Pulford, A. Lanari, and S. Naderi, "Pseudo-random binary sequence phase modulation in high

power Yb-doped fiber amplifiers," *Proc. SPIE* **8601**, 86010Z (2013) [doi: 10.1117/12.2004486].

24. F. Stutzki, F. Jansen, A. Liem, C. Jauregui, J. Limpert, and A. Tünnermann, "26 mJ, 130 W Q-switched fiber-laser system with near-diffraction-limited beam quality," *Optics Letters* **37**, 1073 (2012).

25. T. Eidam, J. Rothhardt, F. Stutzki, F. Jansen, S. Hädrich, H. Carstens, C. Jauregui, J. Limpert, and A. Tünnermann, "Fiber chirped-pulse amplification system emitting 3.8 GW peak power," *Optics Express* **19**, 255 (2011).

26. Y. Jeong, S. Yoo, C. A. Codemard, J. Nilsson, J. K. Sahu, D. N. Payne, R. Horley, P. W. Turner, L. M. B. Hickey, A. Harker, M. Lovelady, and A. Piper, "Erbium:ytterbium codoped large-core fiber laser with 297 W continuous-wave output power," *IEEE J. Sel. Top. Quantum Electron.* **13**, 573 (2007).

27. T. Ehrenreich, R. Leveille, I. Majid, K. Tankala, G. Rines, and P. F. Moulton, "1-kW, all-glass Tm:fiber laser," SPIE LASE Conference 7580, Session 16: Late-Breaking News (presentation only) (2010).

28. G. D. Goodno, L. D. Book, and J. E. Rothenberg, "Low phase-noise, single-frequency, single-mode 608 W thulium fiber amplifier," *Optics Letters* **34**, 1204–1206 (2009).

29. B. T. D and A. V. Smith, "Bulk optical damage thresholds for doped and undoped, crystalline and ceramic yttrium aluminum garnet," *Appl. Opt.* **48**, 3509 (2009).

30. B. Chann, R. K. Huang, L. J. Missaggia, C. T. Harris, Z. L. Liau, A. K. Goyal, J. P. Donnelly, T. Y. Fan, A. Sanchez-Rubio, and G. W. Turner, "Near-diffraction-limited diode laser arrays by wavelength beam combining," *Optics Letters* **30**, 2104 (2005).

31. http://www.directphotonics.com.

32. P. McManamon, *Field Guide to Lidar*, SPIE Press, Bellingham, Washington (2015) [doi: 10.1117/3.2186106].

Chapter 6
LiDAR Receivers

6.1 Introduction to LiDAR Receivers

A good way to calculate LiDAR range is to use the equations in Chapter 3 to calculate how many photons are received in each detector given a certain emitted laser power, a certain aperture diameter, certain atmospheric conditions, etc. In this chapter, we calculate how many photons are required to detect a target in that detector, or pixel. Various detector technologies and approaches can change the required number of photons for detection in a pixel. The level of change can be orders of magnitude. While there are limits to how few photons are needed for detection, it is instructive to compare various technology approaches with respect to the required number of photons for detection. There is a push to do photon counting, using one (or, on average, less than one) photon for detection. The required energy to 3D map a given area may also be calculated to compare various technology approaches. A number of different detection approaches are currently in use and are compared in this chapter.

The purpose of a LiDAR receiver is to convert photons returned from the target into information. The type of LiDAR and its associated processing determine the exact information sought by the receiver. All LiDAR receiver types need to compete against noise. Much of this chapter will therefore cover approaches to extract signal in the presence of noise. By amplifying the returned signal, we can increase the signal-to-noise ratio (SNR). Signal-related noises such as background noise will be amplified, so amplification does not increase SNR with respect to those noises; however, it does help with certain noises, such as surface dark current noise because most of such noise is not amplified. Amplification can be done before detection using an optical preamplifier, although that is not very common in LiDAR. A very common approach to discriminate against this class of noises is to have gain in the receiver that amplifies the photocurrent generated by the signal. Avalanche photodiodes (APDs) multiply the number of electrons generated when a photon hits a detector. If we amplify the returned signal, additional noise, called excess noise, is introduced during the amplification process, but overall

SNR is usually improved. We have two common forms of amplified, direct-detection LiDAR receivers: linear-mode avalanche photodiodes (LMAPDs) and Geiger-mode avalanche photodiodes (GMAPDs). LMAPDs linearly amplify the number of electrons generated, or the current, as the name indicates. GMAPDs always amplify the number of electrons to some large, maximum signal level. GMAPDs and LMAPDs have become very important in LiDAR systems. Often GMAPDs are used with very low probability of detection, needing many pulses to achieve the required detection probability by using coincidence processing.

To accurately measure range, we need a high-bandwidth detector or some equivalent high-bandwidth timing device because we measure range by measuring the time of flight of light to the target and return. This is one of the major differences between a LiDAR detector and a passive detector. Measuring range is important for 3D LiDAR. The higher the bandwidth of the detector the more precise the range measurements if there is a matching laser waveform unless something is done to make the bandwidth of the LiDAR effectively higher. This type of higher LiDAR bandwidth was discussed in Section 5.2 on laser waveforms. The need for high bandwidth in most LiDAR detectors is a significant factor in LiDAR detector development. There are also LiDARs that do not accurately measure range. We have gated 2D LiDARs that only measure angle/angle information. These LiDAR detectors do not need the high-bandwidth capability. We can have a small number of detectors that sequentially point to various angles in angle/angle space, or we can have a large number of detectors that cover a large angle/angle area at a time with many detectors. The first approach is called scanning LiDAR, and the second approach is called flash LiDAR. As arrays of high-bandwidth detectors have become more available, flash LiDAR has become more popular. Scanning LiDARs require faster steering of the illumination beam and the associated receiver FOV.

For some LiDAR applications, we will need a high frame rate to provide sufficient sampling in range. Also, some 3D imaging LiDARs can store only one range value, but others can store many range values. If a long string of range values is measured and stored, the LiDAR may be called a full-waveform LiDAR.

For coherent LiDAR, the linewidth of the laser used and the spatial shape of the local oscillator (LO), which matches the shape of the return signal, can be important characteristics. Also for temporal heterodyne detection, we will need to sample often enough to measure the beat frequency between the returned signal and the LO. This beat frequency is called the intermediate frequency (IF). For spatial heterodyne systems, we will need detector spacing that is close enough to sample the spatial beat frequency between the LO and the return signal.

6.2 LiDAR Signal-to-Noise Ratio

We have to address how much energy per waveform is required for detection. To have detection, we need a certain SNR:

$$SNR = \frac{\langle i_s^2 \rangle}{\langle i_n^2 \rangle}, \tag{6.1}$$

where i_s is the signal current, and i_n is the mean noise current. We need to have a ratio of the mean squared signal current to the mean squared noise current that equals some threshold value. The mean squared signal current is

$$\langle i_s^2 \rangle = G^2 \mathfrak{R}^2 P_s^2 \tag{6.2}$$

for detect detection, and

$$\langle i_s^2 \rangle = 2\eta_h G^2 \mathfrak{R}^2 P_{LO} P_s \tag{6.3}$$

for coherent, or heterodyne, detection, where G is the preamplifier gain, $\mathfrak{R}$ is the detector responsivity, P_s is the signal power, P_{LO} is the LO power, and η_h is the heterodyne mixing efficiency. We can further define the detector responsivity as

$$\mathfrak{R} = \frac{\eta_q e}{h\nu}, \tag{6.4}$$

where η_q is the detector quantum efficiency, e is the charge on an electron, h is Plank's constant, and ν is the carrier frequency of the light. Noise is denoted as

$$\langle i_n^2 \rangle = \langle i_{shot,sig}^2 \rangle + \langle i_{shotLO}^2 \rangle + \langle i_{bk}^2 \rangle + \langle i_{dk}^2 \rangle + \langle i_{th}^2 \rangle, \tag{6.5}$$

where i_{bk} is the shot noise on the background current, i_{dk} is the shot noise on the dark current, and i_{th} is thermal noise current. We separate signal and LO shot noise terms because direct detection will not have LO shot noise.

Received optical power converts to current, which has to be squared for direct detection to calculate the power term used in the SNR equation. The received optical power is related to the rate of arrival of the received photons by

$$P_s = \frac{Nhc}{\lambda T_m}, \tag{6.6}$$

where T_m is the period of time over which the measurement is made, and N is the number of photons per pixel received during that measurement time. For pulsed LiDAR systems, T_m will be the receiver gate time. The laser pulse width may be narrower than the gate time, as shown in Fig. 6.1. When optical

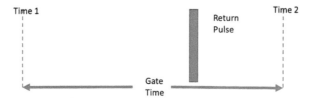

Figure 6.1 Return pulse width and gate time.

signals hit the detector, they create currents in the receiver electronics that have associated mean levels and noise fluctuations. When the signal and noise sources are considered, the resulting electrical SNR is given by

$$SNR = \frac{\langle i_s^2 \rangle}{\langle i_n^2 \rangle} = \frac{G^2 \mathfrak{R}^2 P_s^2}{\langle i_{shot,sig}^2 \rangle + \langle i_{shotLO}^2 \rangle + \langle i_{bk}^2 \rangle + \langle i_{dk}^2 \rangle + \langle i_{th}^2 \rangle}. \qquad (6.7)$$

Each of these noise sources is discussed later in Section 6.2.

In each detector there is a gate time during which the detector can detect a return laser pulse. This gate time will be wider than the return pulse (see Fig. 6.1). A gate is opened at Time 1 and closed at Time 2. To be detected, the return laser pulse must occur any time within the gate time. Signal and background noise can be received whenever the detector gate is open, so a wider gate time will receive more noise. Dark current and thermal noise can continue even when the gate is closed. A gate is also very useful in discriminating against undesired returns, such as from foliage, or backscatter from aerosols or fog.

6.2.1 Noise probability density functions

When considering LiDAR detectors, noise will be a significant issue, so we now discuss noise probability density functions. Common noise density functions that we might encounter include Gaussian, Poisson, and negative binomial distributions. A Poisson distribution gives the probability of exactly x events occurring during a period of time if the events take place independently at a constant rate. This is a common type of occurrence. For example, dark current would meet this criterion. A discrete Poisson probability density function for the distribution of k photons is given by

$$q(k, M) = \frac{M^k e^{-M}}{k!}, \qquad (6.8)$$

where, in this case, M is the mean number of detections. For large M, the Poisson distribution approaches a Gaussian distribution, as shown in Fig. 6.2. This is called the central limit theorem. The probability density then becomes

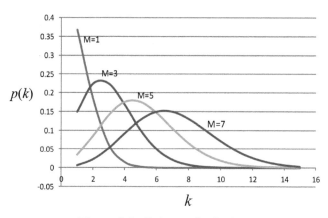

Figure 6.2 Poisson distribution.

$$p(k) \approx \frac{1}{M\sqrt{2\pi}} e^{-(k-m)^2/2M^2}. \tag{6.9}$$

As M gets bigger and bigger, the probability density function moves more into the middle and becomes symmetric and a Gaussian. Figure 6.2 shows that, even for $M = 7$, the shape of the Poisson distribution approaches a Gaussian shape, and Eq. (6.9) shows that the distribution becomes Gaussian. A negative binominal distribution occurs if events occur as a random variable that is represented by a gamma distribution. A diffuse, or rough, target results in a gamma distribution for the integrated intensity at the receiver.[1]

An approach for achieving active grayscale using the illuminator photons is as follows. We can assume that the object of interest can be characterized by a mean reflectivity value with a distribution about the mean. We assume that the reflectivity distribution can be characterized by a width, with some distribution, such as uniform or Gaussian. We divide the distribution into a defined number of reflectivity levels (gray levels). For example, we could assume a mean object reflectivity of 0.1, or 10%, with a uniform distribution between 0.05 and 0.15, or between 5% and 15%. In each case, the reflectivity width would be divided by the number of grayscales desired. Then the LiDAR system is required to discern a variation in reflectivity of (0.15 − 0.05)/ 64 = 0.00156 for the case using 6 bits of grayscale. For each detection modality we must be able to distinguish one gray level from another, even in the presence of noise in the LiDAR receiver. A Gaussian distribution over the same region would have some gray levels closer together than in a uniform distribution. This concept is shown in Fig. 6.3 in terms of the least significant bit (LSB) for grayscale. The LSB is the lowest bit in a series of numbers in binary. It is located at the far right of a string.

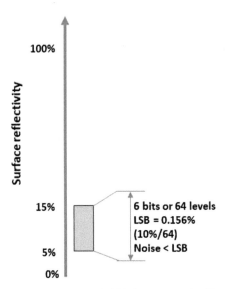

Figure 6.3 Required least-significant bit (LSB) for grayscale. To achieve 6-bit grayscale, the system must have image noise less than the LSB (adapted from Ref. 50).

6.2.2 Thermal noise

Thermal noise, also known as Johnson or Nyquist noise, is electronic noise generated by the thermal agitation of charge carriers (usually the electrons) inside an electrical conductor. This noise occurs regardless of applied voltage. The energy from thermal noise is given by

$$E_{th} = kTB, \tag{6.10}$$

where k is Boltzmann's constant $(1.3806488 \times 10^{-23}\text{-m}^2\text{-kg-s}^{-2}\text{-K}^{-1})$, T is temperature in Kelvin and is often assumed to be 300 K, and B is the bandwidth of the detector. Higher temperature creates more thermal noise. The current resulting from thermal noise is given by

$$\langle i_{th}^2 \rangle = \frac{4kTB}{R_L}, \tag{6.11}$$

where R_L is the load resistance. Thermal noise is not amplified by preamplifier gain, so it is one of the noises that is mitigated by having gain. Thermal fluctuations occur at all temperatures above absolute zero. Thermal noise is very important in radar because the energy of each photon in this case is very small; however, thermal noise is less important in LiDAR because the photons here contain more energy compared to radar, promoting the importance of a competing noise called shot noise. For 263 K shot noise is about 1/30th of a photon in energy, with a wavelength of 1550 nm, as shown in

Table 6.1 The value of kT in photons.

k	T (K)	kT	λ (nm)	$h\nu$
1.38×10^{-23}	273	3.77×10^{-21}	1550	1.28×10^{-19}
1.38×10^{-23}	300	4.14×10^{-21}	1550	1.28×10^{-19}

Table 6.1. We see that for 263 K, thermal noise is about $1/30^{\text{th}}$ of the energy of a single photon.

Thermal noise is distinct from shot noise, which consists of additional current fluctuations that occur when a voltage is applied and a macroscopic current starts to flow. The radiofrequency (RF) community tends to measure thermal noise in decibels above one milliwatt (dbm), or

$$P_{\text{thdbm}} = 10 \log_{10}(1000 \, KTB), \tag{6.12}$$

where 1000 is included because power is referenced to milliwatts rather than to watts.

6.2.3 Shot noise

Shot noise is due to the quantization of an electromagnetic wave. Short wavelengths (high frequencies) have more energy per photon because

$$E_{\text{p}} = h\nu \tag{6.13}$$

is the energy in a photon. In optics we use wavelengths more often than frequencies, so we rephrase Eq. (6.13) as

$$E_{\text{p}} = \frac{hc}{\lambda}. \tag{6.14}$$

The quantization phenomenon occurs for microwave frequencies, but a joule of energy has many more photons at microwave frequencies. At a wavelength of 1.5 μm, a single photon has 1.3×10^{-19} J versus 6.6×10^{-24} J at 10 GHz (3-cm wavelength). There is a small variation in the arrival rate of photons. This variation provides a noise we call shot noise. Shot noise is a Poisson-distributed noise. For large numbers, the Poisson distribution approaches a Gaussian, or normal, distribution, as shown in Fig. 6.2. Shot noise in actual observations becomes indistinguishable from Gaussian noise except when the elementary events (photons, electrons, etc.) are so few that they are individually observed. Since the standard deviation of shot noise equals the square root of the average number of events M, the SNR is given by

$$SNR = \frac{M}{\sqrt{M}} = \sqrt{M}. \tag{6.15}$$

Thus, when M is very large, the SNR is very large as well, and any relative fluctuations in M due to other sources are more likely to dominate over shot noise. The shot noise due to the signal return is

$$\langle i^2_{shot,sig} \rangle = 2eG^2 F \mathfrak{R} P_s B, \tag{6.16}$$

where F is the excess noise factor associated with the preamplifier gain. Excess noise is the noise added during the signal amplification process.

For coherent LiDAR with a high-power LO, the shot noise associated with the LO, which is identical in form to the signal shot noise equation [Eq. (6.16)], is often a dominant noise source. The shot noise is proportional to the square root of the signal, or the square root of the LO power.

6.2.4 Background noise

Background noise is just like signal, but it is signal we do not want. Since it is a form of signal, it will be amplified in the same manner as signal when we use amplification to discriminate against some noise sources. The sun is often a strong source of background noise, so we will calculate its impact. In the visible range, the sun is a strong background noise. Often we can reduce background noise by using a narrowband filter and a narrowband laser.

The sun is approximately a 6050 K blackbody. The diameter of the sun is 1.39×20^9 m, or 9.6×10^{-3} radians (rad) as viewed from the earth, and it is 1.496×10^{11} m from the earth. The sun provides about 1.36 K-W/m^2 above the atmosphere. A narrower wavelength filter rejects more sun light.

The radiance per wavelength L_λ of a blackbody radiator of a certain temperature is calculated as

$$L_\lambda = \frac{2hc^2}{\lambda^5} \left[\exp\left(\frac{hc}{\lambda K T} \right) - 1 \right]^{-1}, \tag{6.17}$$

where we can see a very strong dependence on wavelength, and, again, c is the speed of light (3×10^8 m/s), and k is Boltzmann's constant (1.38×10^{-23} J/K). Equation (6.17) is plotted in Fig. 6.4. We can see the spectral trends of sun radiance, assuming a 6050 K blackbody for the sun. We also can see that the peak radiance of the sun is at 0.5 μm, and 1.55 μm has about 0.15 times the peak sun radiance, so a typical LiDAR wavelength of 1.5 μm is not influenced as much by the sun as a visible LiDAR would be. To calculate the background radiation from the sun, the sun needs to reflect off of the area being viewed by a pixel, usually using Lambertian scattering. It is the reflected radiation of the sun that provides the major background radiation source. Solar glints can provide higher spectral background radiation. The background noise is

$$\langle i^2_{bk} \rangle = 2eG^2 F \mathfrak{R} P_{bk} B, \tag{6.18}$$

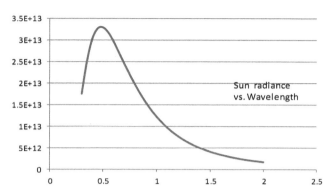

Figure 6.4 Distribution of irradiance from an object at the temperature of the suns' surface.

which is identical in form to the signal shot-noise term. Background noise is a form of signal.

6.2.4.1 Spectral filter technology

Commercially available narrowband filters can be placed in the receiver optical path to block unwanted background light. We assume that the narrowest achievable bandwidth at a reasonable cost is approximately $\sigma_{\min} = 1$ nm for collimated light at normal incidence (e.g., Alluxa offers a filter bandwidth of 0.6 nm for collimated light at 1064 nm[2]). This bandwidth minimum is based on a survey of available filters at the time of this writing but may change with time. As the sensor FOV is increased and rays at larger angles from the optical axis need to be accommodated, the range of effective wave vectors that must be accommodated widens; the wider filter bandwidth passes more scene luminance, introducing more noise. The shift of the resonance wavelength with angle can be modeled as a Fabry–Pérot resonator, as given in the following equation and in Fig. 6.4:

$$\Delta\lambda = \lambda_0 \left[\sqrt{1 - \frac{\sin^2 \vartheta_c}{n_{\mathrm{eff}}^2}} - 1 \right]. \tag{6.19}$$

Figure 6.5 indicates that the required filter bandwidth for a material effective index is $n_{\mathrm{eff}} = 2$. The widest sensor FOV occurs when a single array images the entire area; the angular distance to the corner of the array is

$$\theta_c = DAS \sqrt{\frac{N_x^2 + N_y^2}{4}}, \tag{6.20}$$

where DAS is the detector angular subtense of a pixel, and N_x and N_y are the number of pixels in each direction in the focal plane array (FPA). A larger flash-imaging FPA makes it harder to have a very narrow-bandwidth filter.

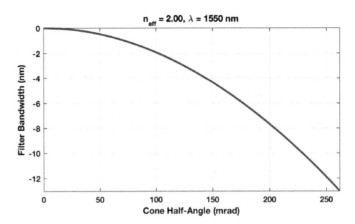

Figure 6.5 Wavelength shift of a resonant filter with incident angle (reprinted from Ref. 50).

As an example, take $N_x = N_y = 128$, and $DAS = 2.5$ mrad; we then have $\theta_c = 0.226$ rad, resulting in a filter bandwidth of 9.3 nm. Recent research suggests the development of new filters that are much more tolerant to angle. Holographic grating filters can also be used to obtain narrower wavelength filters, but only with very narrow acceptance angles.

6.2.4.2 Calculation of background from solar flux

The sun is approximately a 6050 K blackbody as viewed from the earth. The diameter of the sun is 1.39×10^9 m, or 9.6×10^{-3} rad as viewed from the earth at a distance of 1.496×10^{11} m from the sun. For blackbody radiation, the flux is

$$L_\lambda = \frac{2hc^2}{\lambda^5} \left[\exp\left(\frac{hc}{\lambda kT}\right) - 1 \right]^{-1}, \tag{6.21}$$

where, again h is Planck's constant, c is the speed of light, and k is Boltzmann's constant. [Equation (6.21) is the same as Eq. (6.17) and is repeated here for convenience.]

Figure 6.6 shows a very strong dependence on wavelength for blackbody radiation. In this plot, we can see the spectral trends of sun radiance, assuming a 6050 K blackbody for the sun. The wavelength of 1.55 μm has about 15% of the peak sun radiance. Shorter wavelengths are closer to peak sun radiance. Light from the sun hits the earth and is the main source of heat for the earth. The average temperature of the earth surface is close to 300 K. For passive sensors, it is interesting to look at the blackbody curve for a 300 K blackbody because that is close to the average temperature of the ground on the earth.

A 300 K blackbody curve is shown in Fig. 6.7. We can see from the figure that the peak radiation is in the 8- to 12-μm region for a 300 K blackbody. Longwave thermal imagers, or forward-looking infrared (FLIR) imagers, are in

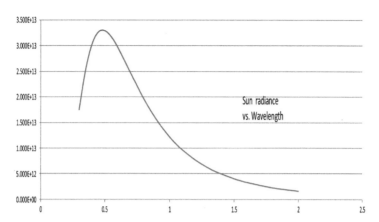

Figure 6.6 Spectral radiance of a 6050 K blackbody (the effective temperature of the sun as viewed from the earth).

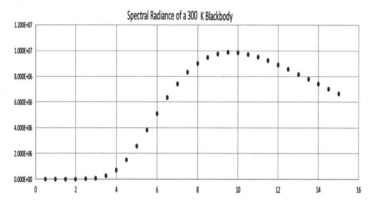

Figure 6.7 Spectral radiance from a 300 K blackbody, which is close to the temperature of the earth surface.

this band. They can see at night because, instead of using direct sun radiation, they use background LWIR radiation. MWIR radiation in the 3- to 5-μm region can also be used but is not as bright. MWIR thermal imagers have recently become more popular due to increased diffraction-limited resolution, and are available as large-format detector arrays. As we can see, MWIR radiation is less than LWIR radiation, but more than near-IR radiation.

We would like to determine how much of a background noise source sun radiation is for a LiDAR. The units of I_λ are W/(sr-m^2-m). We can multiple by the area of the sun to determine the full irradiance per unit wavelength emitted by the sun:

$$I_{\text{sun}} = \pi R_{\text{s}}^2 L, \tag{6.22}$$

where R_{s} is the radius of the sun. We can then take that full radiance-per-unit wavelength from the sun and spread it over the solid angle into which it radiates. The solid angle of a sphere is 4π sr (steradians), but we are taking the solid angle of the sun seen from one location, the earth. We can estimate this solid angle as being π times the radius of the earth squared divided by the distance to the sun squared. Then we can calculate the irradiance per unit wavelength in one square meter on earth as

$$B_{\text{sunsqm}} = \frac{\pi L R_s^2}{R_{\text{e-s}}^2},$$

(6.23)

where $R_{\text{e-s}}$ is the distance from the earth to the sun. We can now convert to photons using

$$E_p = \frac{hc}{\lambda}$$

(6.24)

as the energy per photon. The number of photons received in one square meter is

$$N = \frac{B_{\text{sunsqm}}}{E_p} = \frac{L R_s^2 \lambda}{R_{\text{e-s}}^2 hc}.$$

(6.25)

Table 6.2 shows the number of photons in 1 nm that will come from the sun in 3 ns.

The number of photons hitting one square meter of earth will still need to multiplied by the open gate time for a given camera, which is why in Table 6.2 we state radiance in photons per nanosecond. Figure 6.8 plots the captured number of photons per nanosecond, per meter squared, in 3 nm versus wavelength, using Eq. (6.25). The sun's radiation bounces off of an object and is reflected toward the sensor. We will calculate the number of sun background photons reflected from the ground or another object into the

Table 6.2 Number of photons per square meter in a wavelength band of 3 nm and in a time period of 1 ns.

Wave-length	Radiance per sq m of the sun surface	Radiance from total sun area	Radiance over 1 sq m on earth	Radiance per nm on earth (W)	# photons per nm per sq m on earth	# photons per s per nm per sq m on earth per ns	Captured # of photons per ns in 3 nm
0.4	3.03E+13	1.84E+32	1.448E+09	1.448	1.13E+19	1.13E+10	1.185
0.5	3.28E+13	1.99E+32	1.571E+09	1.571	1.22E+19	1.22E+10	1.286
0.7	2.44E+13	1.48E+32	1.169E+09	1.169	9.11E+18	9.11E+09	0.957
0.9	1.54E+13	9.36E+31	7.375E+08	0.737	5.75E+18	5.75E+09	0.603
1	1.21E+13	7.37E+31	5.809E+08	0.581	4.53E+18	4.53E+09	0.475
1.064	1.04E+13	6.34E+31	4.994E+08	0.499	3.89E+18	3.89E+09	0.409
1.3	6.12E+12	3.72E+31	2.929E+08	0.293	2.28E+18	2.28E+09	0.240
1.55	3.65E+12	2.22E+31	1.748E+08	0.175	1.36E+18	1.36E+09	0.143
1.8	2.29E+12	1.39E+31	1.096E+08	0.110	8.54E+17	8.54E+08	0.090

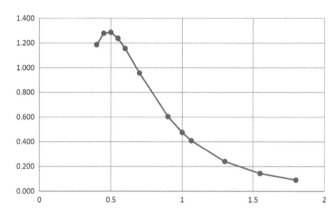

Figure 6.8   Captured number of photons per nanosecond, per millimeter squared, in 3 nm versus wavelength.

receive aperture. Sun radiation reflected off of ground objects is assumed to be Lambertian scattered. Lambertian scattering has a peak scattering straight off of the reflecting surface and then has a cosine fall off in scattered energy, consistent with the reduction in projected area as viewed from that angle. On average, it is as if the scattering is into π steradians.

The amount of background sun radiation scattered into a detector is given by

$$B_{\mathrm{sun}} = \eta_{\mathrm{sun}} B_{\mathrm{sunsqm}} A_{\mathrm{pixel}} \frac{A_{\mathrm{rec}}}{\pi R^2}, \qquad (6.26)$$

where B_{sun} is the solar flux incident on a single pixel in the receiver, η_{sun} is the reflectivity of the surface, B_{sunsqm} is the radiance per square meter on the ground, not counting atmospheric attenuation, A_{pixel} is the projected area of a single pixel, R is range, and A_{rec} is the area of the receive aperture. We can assume 10% reflectivity for the ground, or background, as is sometimes assumed for other hard targets. The area seen by the pixel on the ground is

$$A_{\mathrm{pixel}} = DAS_{\mathrm{az}} DAS_{\mathrm{el}} R^2, \qquad (6.27)$$

where DAS_{az} is the azimuth DAS, and DAS_{el} is the elevation DAS. So we have

$$B_{\mathrm{sun}} = \eta_{\mathrm{sun}} B_{\mathrm{sunsqm}} DAS_{\mathrm{az}} DAS_{\mathrm{el}} \frac{A_{\mathrm{rec}}}{\pi}. \qquad (6.28)$$

For this case then we have

$$B_{\mathrm{sun}} = \eta_{\mathrm{sun}} B_{\mathrm{sunsqm}} DAS_{\mathrm{az}} DAS_{\mathrm{el}} \frac{A_{\mathrm{rec}}}{\pi} = \eta_{\mathrm{sun}} B_{\mathrm{sunsqm}} DAS_{\mathrm{az}} DAS_{\mathrm{el}} R_{\mathrm{rec}}^2. \qquad (6.29)$$

Another issue that will influence the impact of background radiation on a LiDAR is how narrow a filter we can use to filter out the background. Spectral filter technology was discussed in Section 6.2.4.1. Filters can be used that prevent light from hitting the detector. In a heterodyne LiDAR we can also filter electronically because the detector will only respond up to a certain bandwidth.

6.2.5 Dark current, 1/*f* noise, and excess noise

Dark current noise is the noise in a detector when that detector is not illuminated by any light. The charge generation rate in a detector can be related to specific crystal defects within the depletion region of a detector (bulk defects). Bulk dark current participates in preamplifier gain. Bulk defects are the most likely sources of dark current. Surface defects can also contribute to dark current but are much less likely to be a source compared to bulk defects. Surface dark current does not participate in preamplifier gain. Both bulk- and surface-generated dark current primary carriers are Poisson distributed. The pattern of different dark currents can result in a fixed-pattern noise. Dark frame subtraction can remove an estimate of the mean fixed pattern, but a temporal noise still remains because the dark current itself has a shot noise. Dark current is not noise. It is unwanted signal. Like any signal, it contains noise (fluctuations). When you bias a detector, even before you illuminate it with any light, some noise "signal" occurs. In addition, if you mitigate the effects of dark current and other noise by amplification, there can be excess noise associated with the amplification. The noise from dark current is calculated as

$$\langle i_{\mathrm{dk}}^2 \rangle = 2eB(I_{\mathrm{dks}} + I_{\mathrm{dkb}}G^2 F), \tag{6.30}$$

where I_{dks} is surface dark current that is not multiplied in an APD, and I_{dkb} is bulk dark current that is multiplied by the gain of an APD. When you amplify a signal, you add additional noise, which is denoted by the excess noise factor F.

1/*f* noise (also called flicker noise) is a signal with a frequency spectrum such that the power spectral density (energy or power per hertz) is inversely proportional to the frequency. 1/*f* noise is not well understood and is sometimes called pink noise. Figure 6.9 shows 1/*f* noise decreasing with frequency.

6.3 Avalanche Photodiodes and Direct Detection

One way to reduce the effects of noise on a LiDAR is to employ gain either before the detector, or in the detector. We could use a fiber preamplifier,[2] which would amplify the signal prior to it hitting a detector. Historically, people have used image intensifier tubes,[3] which amplify after detection. The

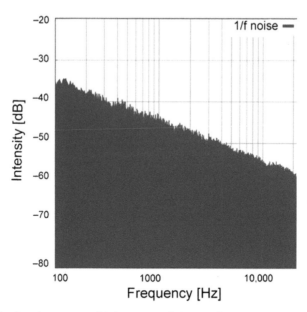

Figure 6.9 1/*f* noise decreases with frequency (adapted from Wikipedia web page on pink noise; created by Ktims and released under license CC BY-SA 3.0).

avalanche photodiode (APD) is the solid state equivalent of the photo-multiplier tube in that both create many electrons from each incident photon. The number of electrons created per absorbed photon is the internal detector gain. The noise associated with the gain process is called excess noise. The gain directly improves the SNR in situations where system noise, such as dark current, limits the performance. Charge gain processes inside detectors exploit the ability to accelerate charged particles in an applied electric field to amplify the number of charge carriers through energetic collisions. One example is photoemissive detectors, in which a primary electron generated by the incident absorbed photon is liberated from the detector photocathode, accelerated through an evacuated space by an applied electric field, and then impacted on a target material, generating additional secondary charge carriers from the primary carrier's kinetic energy. A second type of detector charge gain process takes place inside an APD; the process is impact ionization, in which the primary photoelectrons do not leave the detector material, but undergo ionizing collisions within the semiconductor crystal in a high–electric-field region of a reverse-biased diode junction.

We discuss two classes of APDs as LiDAR detectors: LMAPDs GMAPDs. LMAPDs are operated below their breakdown voltage, generating current pulses that are on average proportional to the strength of the optical signal pulse. LMAPDs normally operate with high-gain current, or charge amplifiers, which develop an output voltage waveform that is proportional to the LMAPD's photocurrent waveform. In contrast, GMAPDs are armed by biasing them above their breakdown voltage, rendering them sensitive to single primary charge carriers. Absorption of one or several photons triggers avalanche breakdown of the GMAPD junction, generating a strong current pulse that is easily sensed, the amplitude of which is limited by a quenching circuit. Immediately following breakdown, the GMAPD's quenching circuit momentarily reduces the applied reverse bias below the GMAPD's breakdown voltage, terminating the avalanche process and allowing trapped carriers to clear the junction before re-arming the GMAPD. If the GMAPD is armed too soon, after-pulsing will occur, resulting in false signals. This creates a dead time after a detection using a GMAPD. Generally speaking, GMAPDs are more sensitive to weaker signals than LMAPDs, but LMAPDs can directly measure signal return amplitude and can resolve optical pulses separated by as little as a nanosecond, depending on the laser pulse width and the APD's linear gain. Certain high-gain LMAPDs, chiefly electron-avalanche HgCdTe APDs, provide enough linear gain to detect single photons without entering avalanche breakdown.

As part of the discussion on APDs, we use a single scenario to judge the amount of energy required to make a 3D map of a given area. The scenario chosen uses a large DAS because that creates more issues that must be addressed due to the large background radiation from the sun.

As stated earlier, one way to reduce the effect of noise in direct detection is to use gain. Expanding the noise terms, we have

$$SNR = \frac{G^2 \mathfrak{R}^2 P_s^2}{2eBG^2 F\mathfrak{R}[P_s + P_{bk} + P_{dkb}] + 2eBi_{dks} + 4ktB/R_L}. \qquad (6.31)$$

Gain after detection will reduce the influence of surface dark current and thermal noise. Background radiation will be amplified by the same amount as the signal. Bulk dark current will also be amplified in an APD; amplification will add noise. For background-limited direct detection, we have

$$SNR = \frac{\eta_D P_s^2}{2h\nu BF P_{dkb}}, \qquad (6.32)$$

where η_D is the detector quantum efficiency, and we have made use of the fact that

$$\mathfrak{R} = \frac{\eta_D e}{h\nu}. \qquad (6.33)$$

In the limit where the signal shot noise dominates the other noise terms, the SNR is given by

$$SNR_{shotlim} = \frac{\eta_D P_s}{2h\nu BF}. \qquad (6.34)$$

Within the limit where the signal shot noise dominates other noise terms, the SNR is directly proportional to the number of photons received. This is the best we can do and is the goal of using gain.

6.3.1 Linear-mode APD arrays for LiDAR

A LMAPD can measure grayscale (see Chapter 1) on a single pulse. The output is proportional to the reflected light. A LMAPD can provide a range profile from a single pulse if range profile storage is built into the ROIC of the device. Some LMAPDs only store the first pulse, some store the last pulse, and some store a limited number of return pulses. Building in this storage capability can make the ROIC physically larger and more complex. Responses from each range bin of interest must be captured and stored to read out a full return waveform. Adding more range bins requires a more complex ROIC. One of the challenges for LMAPDs will be the ROIC development. Another issue for LMAPDs is that the gain may be relatively low, so sensitivity will only be improved by a limited amount. Single-photon sensitivity for LMAPDs is coming but is developing slower than for GMAPDs. Typical LMAPD gain might be a factor of ten, but some state-of-the-art LMAPDs will have gain of 1000. High gain will reduce the influence of noise on the detector. Figure 6.10 shows a linear gain of a factor of four; i.e., the input signal amplitude (shown in blue) is four times as large as the output signal after amplification (shown in green).

LMAPDs are designed with the goal of operating below the breakdown voltage with high gain and low noise. The operation of these APDs produces an average photocurrent that is linearly proportional to the incident optical flux (hence, the name linear mode). Higher gain can produce more sensitivity

4x linear amplification shown

Figure 6.10 A diagram showing the linear gain of a LMAPD.

because it more efficiently masks noise generated downstream in the detection process. LMAPDs with a factor of ten gain in a 128×128 detector camera format are commercially available. These are very useful for shorter-range imaging but less useful for longer ranges when extreme sensitivity is very useful. Some LMAPDs can have up to a factor of 1000 linear gain, thereby approaching single-photon–counting sensitivity. Some of the LMAPDs are made using HgCdTe and require cooling, even when operating in the near-IR region, imposing an additional systems issue. Unlike Geiger-mode systems, which must geo-register and aggregate multiple high–frame-rate returns, linear-mode systems have the advantage of collecting range and intensity information in a single frame, assuming that the ROICs are made to support this (making them true flash-imaging systems). These systems usually operate at lower frame rates with a high probability of detection on each frame (~30 Hz), requiring higher laser energy per pulse. This makes most LMAPDs less friendly to the use of efficient fiber lasers, which are peak power limited so often operate at high repetition rate but lower energy per pulse. Historically, LMAPDs also have not been as sensitive as GMAPDs, requiring higher total laser energy per sample, but a LMAPD LiDAR and a GMAPD LiDAR are designed very differently, so this benefit is not always present. While LMAPDs have the advantage of being able to make a 3D point cloud image on a single laser pulse return, this requires a more sophisticated ROIC to capture and store this imaging data from each detector. Because of the sophisticated ROIC requirement, it is likely that a larger pixel size will be required than is needed for a GMAPD array camera.

6.3.1.1 InGaAs LMAPD arrays

We consider InGaAs LMAPDs with various levels of gain. The 30-μm–square InGaAs LMAPD pixels that are analyzed typically operate at a linear gain of $M = 20$ with 0.2-nA dark current at 263 K, quantum efficiency (QE) of 80%, and an excess noise factor F parameterized by the ionization coefficient ratio $k = 0.2$, resulting in $F = 5.56$ at $M = 20$. Multistage InGaAs LMAPDs have been reported that operate at gains approaching $M = 1000$, with excess noise parameterized by $k = 0.04$, but are not a mature technology.[4] Low–excess-noise LMAPDs made from AlInAsSb[5] and InAs[6] have also been reported, but among the high-gain LMAPDs, electron-avalanche HgCdTe LMAPDs are the most mature. HgCdTe LMAPDs can be manufactured to respond efficiently from the ultraviolet (UV) to the MWIR and can have high linear gains of up to 1000 or more, maintaining an excess noise factor F near 1. The 64-μm HgCdTe LMAPD pixels for which calculations are made can operate at linear gains over $M = 1000$ but are analyzed at $M = 200$, for which the dark current at 100 K is 0.64 pA, $QE = 65\%$, and $F = 1.3$. Two disadvantages of HgCdTe LMAPDs are the need to cool HgCdTe to near 100 K and the high cost.

InGaAs LMAPDs are manufactured from thin films of $In_{0.53}Ga_{0.46}As$ and either $In_{0.52}Al_{0.48}As$ or InP epitaxially grown on InP substrates. The principal functional layers include the relatively narrow-bandgap (0.65 eV) InGaAs absorption layer and the relatively wide-bandgap multiplication layer made from either InAlAs (1.46 eV) or InP (1.35 eV), separated by a space charge layer. The space charge layer ensures that the electric field strength in the absorber remains weak enough to avoid excessive tunnel leakage when the field in the multiplier is strong enough to drive a useful rate of impact ionization.

Figure 6.11 shows a conceptual diagram of a typical LMAPD. This particular configuration is called the SACM (separate absorption, charge, and multiplication) design. The layer ordering of absorber and multiplier relative to the anode and cathode—and the polarity of doping in the charge layer— depend on whether InAlAs or InP is selected as the multiplier material. Holes avalanche more readily in InP compared to electrons, so in an InP-multiplier APD, the absorber is placed next to the cathode, and the charge layer is n type. The layering is opposite for an InAlAs-multiplier APD. APD pixels may be formed either by patterned diffusion of the anode into the epitaxial material, or by patterned etching of mesas from the thin film (in which case the anode layer was doped during epitaxial growth rather than diffused). Metal contact pads are deposited on individual pixel anodes, whereas a common cathode connection through the substrate is commonly used. In etched-mesa designs, the pixel mesa sidewalls are chemically passivated and encapsulated to protect them from environmental degradation. Figure 6.11

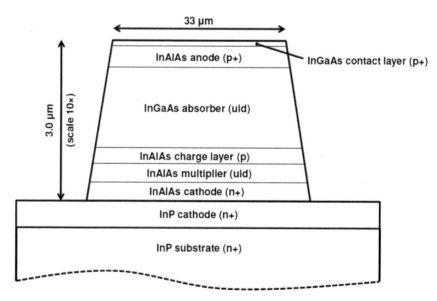

Figure 6.11 A typical InGaAs LMAPD structure in the SACM design (reprinted from Ref. 50).

depicts the structure of an InAlAs-multiplier, etched-mesa InGaAs LMAPD pixel of the type used in the detector array for which calculations are made.

The discussion on InGaAs LMAPDs (Section 6.3.1.1) is based on detector characteristics for Voxtel's commercial InGaAs APD product,[7] whereas the detector characteristics for the HgCdTe LMAPD (Section 6.3.1.2) are those published by DRS Technologies (now Leonardo DRS). In both sections, ROIC characteristics typical of two different design nodes—higher bandwidth, higher circuit noise, and smaller pixel format; and *vice versa*—are used to analyze LMAPD FPA performance.

In general, flash LiDAR ROICs designed to use linear-mode detectors employ a circuit in each pixel that includes: a front-end transimpedance amplifier to convert current or charge from the detector into a voltage signal; various filtering or pulse-shaping stages; and voltage sampling, storage, and readout circuitry. Two main sampling architectures are used: (1) synchronous schemes in which the reflected waveform received by each pixel is regularly sampled with a period on the order of nanoseconds, or (2) asynchronous schemes in which a comparator is used to trigger sampling of reflected pulse amplitude and time-of-arrival when the signal exceeds an adjustable detection threshold. Provided that the signal chain bandwidth is high enough, both the synchronous waveform-recorder sampling scheme and the asynchronous event-driven sampling scheme can support multi-hit LiDAR in which multiple reflections from a single transmitted laser pulse arriving within nanoseconds of each other are separately resolved and timed to penetrate obscurants like foliage. In both cases, sampling is active during a range gate in which target returns are expected, samples are stored locally in each pixel during the range gate, and the accumulated waveform or pulse return data are read out from the array between laser pulses. Higher sample capacity drives the ROIC pixel footprint because of the area required for storage capacitors. In general, the event-driven sampling architecture requires less space to implement because fewer samples must be stored to observe a given number of pulse returns per laser shot. The regularly sampled measurement approach has been called a full-waveform LiDAR for cases where a large number of samples are stored. The sampling architecture analyzed here is the event-driven, asynchronous type, with an in-pixel storage capacity of up to three range-and-amplitude–sample pairs. This matches the foliage "poke through" case analyzed here. Generic characteristics typical of this architecture are applied.

High-bandwidth operation of the signal chain in a flash LiDAR ROIC pixel generally requires high current draw during the range gate. Also, the sourcing and distribution of the supply current becomes more challenging as the array format grows. For these reasons, we analyze two different configurations: (1) high–range-precision (higher bandwidth) operation in which the pixel current draw limits the active format to about 32×32 pixels and (2) operation of a larger-format (128×128 pixels) array with reduced range precision (lower bandwidth). Typical camera frame rates are in the 1- to 10,000-Hz range, but they depend on the array format, the number of samples stored and readout per pixel, and the number of output data channels operated in parallel. The 128×128 pixel format will have a frame rate of about 600 Hz, whereas smaller formats will have a higher frame rate. Aside from differing supply requirements, range precision, and format and frame rates, it should also be noted that operation of the pixel signal chain at different bandwidths will affect absolute sensitivity. Most of the relevant noise sources are wideband; therefore, with everything else being equal, operation of the signal chain with higher bandwidth means more in-band noise and lower sensitivity. However, the signal chain's bandwidth also affects sensitivity to laser pulses of different shapes and duration since the overlap of an input pulse's frequency spectrum with the ROIC's transfer function will determine how efficiently the signal is amplified. Here we assume that the sensor is responding to 4-ns full-width, half-maximum (FWHM) pulses in the calculations for the low-bandwidth configuration and to 1-ns FWHM pulses for the high-bandwidth configuration.

We look at two range precision cases. One requires 25-cm range precision, and the other requires 5-cm range precision. For LMAPDs, the 5-cm range precision case is somewhat challenging. The 25-cm range precision require- ment can be met in a single laser shot using the larger-format, low-bandwidth configuration. However, the smaller, active-format, high–range-precision configuration may be of use for 5-cm range precision, or else multiple samples can be averaged with the larger format to obtain the 5-cm range precision. If the low-bandwidth configuration is used, range measurements from multiple laser shots must be averaged to reduce the standard error of the mean in range below 5 cm. Multiple range measurements will reduce the standard error of the mean by the square root of the number of range measurements such that the minimum number of range measurements N_{Rmin} of the timing standard deviation $\sigma_{\mathrm{t\ ROIC}}$, which must be averaged to achieve a particular timing precision requirement $\sigma_{\mathrm{t\ required}}$, is

$$N_{\mathrm{R\ min}} = \left\lceil \left(\frac{\sigma_{\mathrm{t\,ROIC}}}{\sigma_{\mathrm{t\,required}}} \right)^2 \right\rceil. \tag{6.35}$$

The various types of InGaAs LMAPD FPAs analyzed here make pulse return time estimates by sampling an analog voltage ramp that is distributed to all pixels in the array. Sampling of the ramp is triggered when the rising edge of a signal pulse from a detector pixel passes through an adjustable detection threshold. The threshold level must be optimized to extinguish false alarms arising from circuit noise in the ROIC, convolved with the multiplied shot noise on the APD pixel's dark current and background photocurrent. The ROIC's fundamental timing uncertainty relates both to the voltage noise on the signal that triggers sampling of the ramp (jitter) and to the noise associated with reading the sampled voltage itself. The accuracy of this measurement determines the range resolution:

$$\sigma_{t\,\mathrm{ROIC}} = \sqrt{jitter^2 + resolution^2} = \sqrt{\frac{n_{\mathrm{reference}}}{n_{\mathrm{signal}}}\sigma_{t\,\mathrm{ref}}^2 + \left(\frac{V_{\mathrm{noise}}}{V_{\mathrm{DR}}}\Delta t_{\mathrm{gate}}\right)^2}, \quad (6.36)$$

where $n_{\mathrm{reference}}$ is a signal level at the ROIC pixel input in units of electrons for which the jitter $\sigma_{t\,\mathrm{ref}}$ is known, n_{signal} is the mean signal level for which the jitter is to be estimated, V_{noise} is the read noise on the analog time stamp, V_{DR} is the dynamic range of the time stamp, and Δt_{gate} is the range gate duration. Stronger signals transition more quickly through the comparator threshold, reducing jitter, and a faster ramp rate maps a given magnitude of read noise to a finer temporal resolution, giving rise to the timing precision characteristics calculated in Fig. 6.12.

Figure 6.12 suggests that the low-bandwidth camera configuration will require multiple pulses to range with 5-cm precision. The resolution of a single measurement is not at the level we need, so we will have to average multiple pulses to obtain the required resolution. With a 500-ns armed period and a very weak signal return [100 electrons (e⁻) after avalanche multiplication], the range precision is only about 24 cm. To obtain a standard error of the mean range measurement equal to 5 cm, we would need 25 pulses at this weak signal level. However, only 4 shots are required for 800 e⁻ signal returns to obtain a 5-cm range precision. Even under low-bandwidth operation, for the large-DAS case, we would only need 1 pulse of 100 e⁻ to obtain 25-cm range precision. Note that the 500-ns range gate is similar to the armed times used

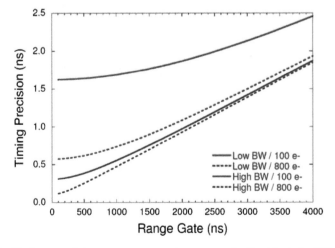

Figure 6.12 Timing precision versus armed range gate for Voxtel's VX 806 camera (reprinted from Ref. 50).

for the GMAPDs (333–666 ns for 50–100 m), which is discussed in Section 6.3.2.

Each pulse return at a given optical signal level has some probability P_{D1} of exceeding the detection threshold. In the 25-cm range precision case, the ROIC's native timing precision is adequate to achieve the range precision requirement. P_{D1} is both the probability of detecting a target surface within a pixel's instantaneous field of view (IFOV) and the probability of ranging to that surface with the required precision. However, if multiple range measurements must be averaged to achieve the range precision requirement of 5 cm, then S total laser shots are transmitted, and the probability of detecting enough pulse returns to achieve a standard error of the mean range that is less than 5 cm is

$$P(N_{\text{success}} \geq N_{\text{R min}}) = \sum_{j=N_{\text{R min}}}^{S} \frac{S!}{j!(S-j)!} P_{D1}^{j} (1 - P_{D1})^{S-j}, \qquad (6.37)$$

where N_{success} is the number of laser shots successfully detected, S is the total number of laser pulses transmitted, and j is the index of the summation. It should be pointed out that Eq. (6.37) gives the probability of detecting enough pulse returns to achieve a particular standard error of the mean range; however, in general, seeing the target surface with less range precision is an easier problem requiring fewer laser shots and/or weaker signal.

Approximating the amplitude distribution of the signal into the pixel comparator as Gaussian, P_{D1} can be approximated as

$$P_{D1} \approx P_{\text{ready}} \left[0.5 - 0.5 \, \text{erf} \left(\frac{n_{\text{th}} - n_{\text{signal}}}{\sqrt{2} \, noise_{\text{total}}} \right) \right], \qquad (6.38)$$

where P_{ready} is the probability that the sensor pixel is able to record a pulse return at the time it arrives, n_{th} is the comparator threshold in units of electrons, and $noise_{\text{total}}$ is the standard deviation of the signal into the comparator, also given in units of electrons; like n_{signal}, n_{th} and $noise_{\text{total}}$ are quantities that correspond to the ROIC pixel input. The total noise is

$$noise_{\text{total}} = \sqrt{noise_{\text{ROIC+dark+background}}^2 + M \, F \, n_{\text{signal}}}, \qquad (6.39)$$

where $noise_{\text{ROIC+dark+background}}$ is the standard deviation of the signal into the comparator in the absence of an optical signal return, M is the mean gain of the APD pixel, and F is the APD pixel's excess noise factor. The excess noise factor for this type of APD [but not for the HgCdTe LMAPDs ($F \sim 1.3$) described in the next section] obeys McIntyre's formula:[8]

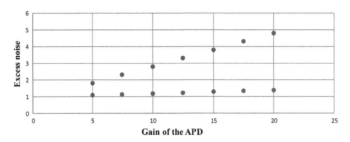

Figure 6.13   Excess noise versus gain.

$$F = M\left[1 - (1-k)\left(\frac{M-1}{M}\right)^2\right].$$

(6.40)

Figure 6.13 plots excess noise versus APD gain. The top set of points is for $k = 0.2$, which is the case for the LMAPD we are considering. In the future, there is a possibility of having $k = 0.2$ in a multistage LMAPD made from InGaAs. If we were able to do that, the excess noise would be as shown in the lower set of points in the plot.

Conceptually, $noise_{\text{ROIC+dark+background}}$ is three separate noise terms added in quadrature—a purely circuit-related noise term and the multiplied shot noise of the APD's dark current and CW background photocurrent.

The arming probability P_{ready} appearing in Eq. (6.41) depends on: (1) when during the range gate the target surface is located (the time value given as t_{target}), (2) the pixel's false-alarm rate (FAR) at that detection threshold setting, and (3) the sample capacity C of the pixel. Since false alarms in an LMAPD receiver circuit are independent stochastic events whose average rate of occurrence is given by the FAR, Poisson statistics apply, and the probability that at least one unused sample storage location is available at the time the return from the target surface is received is

$$P_{\text{ready}} = \sum_{i=0}^{C-1} \frac{(t_{\text{target}} \times FAR)^i \exp(-t_{\text{target}} \times FAR)}{i!},$$

(6.41)

where $C = 3$ is typical of what can fit into a small-pitch ROIC design. To reduce the number of model parameters, one can set t_{target} equal to the range gate duration Δt_{gate}, which corresponds to the conservative case of a target surface at the very end of the range gate.

Like P_{D1}, the FAR depends on the detection threshold n_{th}, but the standard Gaussian approximation for the noise distribution does not accurately model APD noise. Instead, the McIntyre-distributed[9] noise of the APD must be explicitly convolved with the Gaussian-distributed noise of the ROIC to find the amplitude distribution of noise pulses into the pixel comparator:

$$P_{RX}(n) = (P_{ROIC} * P_{APD})[n] \equiv \sum_i P_{ROIC}(i) \, P_{APD}(n - i), \qquad (6.42)$$

where P_{RX} is the amplitude distribution of the noise into the pixel comparator, P_{ROIC} is a Gaussian-like discrete distribution that characterizes the pixel circuit noise, and P_{APD} is the average of McIntyre distributions for the multiplied output of an APD given a certain number of primary input electrons, weighted by the probability that each quantity of primary electrons will result from dark current and background photocurrent generation processes as calculated by Poisson statistics. The convolution is best performed numerically.

Figure 6.14 shows the convolutions calculated for mean APD gains of $M = 5$, 10, 15, and 20 for a $k = 0.2$ InGaAs/InAlAs APD pixel in the large-DAS case and compares the convolutions to Gaussian approximations having the same mean and variance. Whereas correspondence is near the mean, tail divergence is a significant factor for FAR calculations owing to the need to set a detection threshold that extinguishes the great majority of false alarms. Following Rice,[10] the FAR is found from a pre-factor that depends on the pixel signal chain's bandwidth (BW), $noise_{ROIC+dark+background}$, and the value of the convolution at the comparator threshold n_{th}:

$$FAR = \sqrt{\frac{2\pi}{3}} \, noise_{ROIC+dark+background} \, BW \, P_{RX}(n_{th}) \, [\text{Hz}]. \qquad (6.43)$$

In addition to influencing the arming probability P_{ready}, the FAR also determines the probability of a false positive (FP). In the large-DAS case for

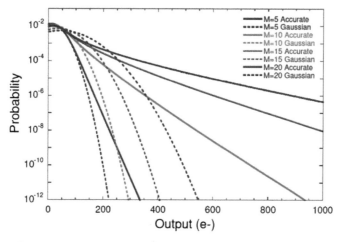

Figure 6.14 Example convolutions of multiplied APD dark current and background photocurrent shot noise with circuit noise (compared to Gaussian approximations having the same means and variances) for the 25-mm aperture, large-DAS case (reprinted from Ref. 50).

which a single range measurement is required to achieve the specified range precision, Poisson statistics give the probability of at least one false positive occurring within the range gate of a given pixel as

$$P_{\mathrm{FP}\,S=1} = 1 - \exp(-\Delta t_{\mathrm{gate}} \times FAR). \tag{6.44}$$

In the case where multiple pulse returns must be averaged to reduce the standard error of the mean range measurement, the coincidence of returns from the same range can be used to reject false positives. If a validation rule of the form 'N_{valid} returns within $\pm t_{\mathrm{error}}$ of a given time-of-arrival' is applied, the probability of at least one false positive consisting of at least N_{valid} time-coincident false alarms occurring anywhere within the range gate (over S total laser shots) is

$$P_{\mathrm{FP}S>1} = 1 - \exp\left[-\Delta t_{\mathrm{gate}} \frac{P(N_{\mathrm{FA}} \geq N_{\mathrm{valid}})}{2\,t_{\mathrm{error}}}\right], \tag{6.45}$$

where the probability of at least N_{valid} false positives out of S laser shots occurring within any given time span $2t_{\mathrm{error}}$ is

$$P(N_{\mathrm{FA}} \geq N_{\mathrm{valid}}) = \sum_{k=N_{\mathrm{valid}}}^{S} \frac{S!}{k!(S-k)!} \times P_{\mathrm{FA}}^{k} \times (1 - P_{\mathrm{FA}})^{S-k}, \tag{6.46}$$

and the probability of at least one false positive occurring within any given $2t_{\mathrm{error}}$ time span per shot is

$$P_{\mathrm{FA}} = 1 - \exp(-2\,t_{\mathrm{error}} \times FAR). \tag{6.47}$$

Equations (6.45) through (6.47) are similar to the calculations we used for GMAPDs with a low probability of detection on a single pulse. To summarize, the probability of successfully measuring range to the required precision depends on the number S of laser shots transmitted, the number N_{Rmin} of range measurements required to achieve that precision, and the per-shot detection probability P_{D1}. The number of range measurements required depends on the signal strength n_{signal}, as does the per-shot detection probability. P_{D1} also depends on (1) the probability that the ROIC pixel's sample capacity has not filled with false alarms by the time a valid target return arrives, (2) the detection threshold n_{th}, and (3) the total noise $noise_{\mathrm{total}}$. The total noise includes a component that depends on the signal strength and a component that is present in the absence of the signal, designated by $noise_{\mathrm{ROIC+dark+background}}$. The analysis is completed by calculation of the FAR, which depends on $noise_{\mathrm{ROIC+dark+background}}$ and n_{th}. For given values of n_{signal} and S, n_{th} can be varied to maximize P_{D1}. The maximum value of P_{D1} is then compared to the critical value of P_{D1} required to achieve a particular probability of measuring range to the required precision (e.g., 90%), and n_{signal} is adjusted until the critical value is just barely reached. This determines the required signal strength at the ROIC pixel input. To translate n_{signal} into photons per pixel at the FPA (i.e., after collection by the camera aperture and after any losses in the optical train), one divides by the product of the mean APD gain (e.g., $M = 20$), the APD's quantum efficiency (80%), and the fill factor of the detector pixel (e.g., 60%).

Figure 6.15 is a plot of the probability of achieving 5-cm range precision in $S = 6$ laser shots using an $M = 20$, $k = 0.2$, $QE = 80\%$ LMAPD detector array operated at 0 °C with 60% optical coupling efficiency in combination with the low-bandwidth–configuration ROIC. Curves corresponding to $C = 3$ (single hit; blue), $C = 2$ (two-hit, single-shot foliage poke through; green), and $C = 1$ (three-hit, single-shot foliage poke through; red) are plotted. The steps in the curves occur at signal levels where the minimum number of range measurements N_{Rmin} that must be averaged to achieve the specified range precision changes by an integer. For example, the probability of detecting 6 out of 6 laser shots at a signal level of 39 photons is much lower than the probability of detecting 6 out of 6 laser shots at a signal level of 40 photons mainly because the number of required detections drops by one (as opposed to the marginally stronger signal return). That is why all three curves drop discontinuously between 39 and 40 photons.

Figure 6.16 is a plot of the probability of achieving 5-cm range precision in a single laser shot using an $M = 20$, $k = 0.2$, $QE = 80\%$ LMAPD detector array with 60% optical coupling efficiency in combination with the high-bandwidth–configuration ROIC. Curves corresponding to $C = 3$ (single hit),

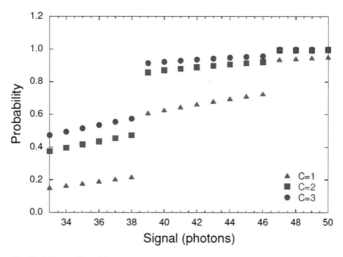

Figure 6.15 Probability of achieving 5-cm range precision in 6 laser shots against bare earth (blue) and with 2-return (green) or 3-return (red) foliage penetration, using an $M = 20$ LMAPD and a full-format mode (reprinted from Ref. 50).

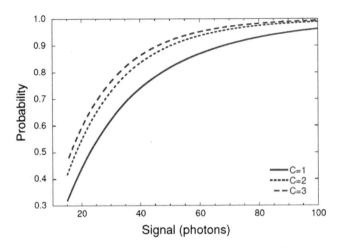

Figure 6.16 Probability of achieving 5-cm range precision in 1 laser shot against bare earth (blue) and with 2-return (green) or 3-return (red) foliage poke through, using an $M = 20$ LMAPD and a high–range-precision mode (reprinted from Ref. 50).

$C = 2$ (two-hit, single-shot foliage poke through) and $C = 1$ (three-hit, single-shot foliage poke through) are plotted. The 16× difference in coverage in angle between the high- and low-bandwidth ROIC configurations should be considered when comparing this result to the low-bandwidth calculation of Fig. 6.15. The low-bandwidth mode frames the entire 128×128 pixel detector array, while the high-bandwidth mode only frames a 32×32 pixel window, one 16th of the number of pixels.

ROICs of this architecture are also capable of grayscale range imaging if they are set up to sample and store the pulse return amplitude at the same time they sample the analog time stamp. In passive imaging systems, the LSB of a sensor's dynamic range is normally mapped to its noise floor such that 6 bits of grayscale imaging would span the range from 1× to 64× the noise-equivalent input level. Passive imaging also assumes natural scene illumination. However, because the flash LiDAR architecture considered here uses an event-driven amplitude sampling scheme, pulse return amplitudes weaker than the comparator threshold will not be sampled. Further, the ROIC's amplifier chain is usually alternating-current (AC) coupled, so natural cw scene illumination will not trigger sampling except through its contribution to the FAR. Grayscale imaging with such a ROIC is active imaging of the reflected laser pulse intensity. As such, the granularity of the grayscale image is still the noise-equivalent input level of the sensor, but the spanned dynamic range is offset from zero by the detection threshold. The dynamic range available for grayscale imaging is smaller than the dynamic range of the signal chain into the threshold comparator.

The grayscale resolution of a conventional passive imager is often expressed as a dynamic range in bits and is calculated from the camera's analog dynamic range by equating the LSB to the camera's noise floor. However, optical signal shot noise increases as the square root of signal level, so a LSB that represents the noise at zero signal (i.e., in the dark) does not quantify the accuracy with which nonzero signal amplitude can be measured; nor is it possible to define an LSB of a fixed size that exactly expresses signal amplitude measurement accuracy for all signal levels within an imager's dynamic range. In contrast, here we quantify grayscale resolution based on there being a 90% probability that any given signal return amplitude measurement lies within a set interval centered on the average return level corresponding to the true target reflectance. The signal interval for which the calculation is made is that spanned by a reflectance bin of specified width.

The mean signal return level in photons per pixel can be rewritten as $N = C(R) \times \rho$ where $C(R)$ is a range-dependent function containing the radiometric aspects of the problem, and ρ is the target reflectance, which runs between $\rho_{low} = 5\%$ and $\rho_{high} = 15\%$. If the range spanned by the target reflectance is divided into N_{bits} range bins, the reflectance bin width is $\Delta\rho = (\rho_{high} - \rho_{low})/2^{N_{bits}}$. The mean signal range spanned by a reflectance bin is therefore

$$\Delta N = C(R) \times \Delta\rho = C(R) \times \frac{\rho_{high} - \rho_{low}}{2^{N_{bits}}}, \tag{6.48}$$

which shows that a reflectance bin of a fixed size results in signal bins of variable size, dependent on $C(R)$. For instance, if $N_{bits} = 6$, then $\Delta N = 6.81$ photons for $C(R) = 5,000$ photons, and $\Delta N = 23.44$ photons for $C(R) = 15,000$ photons, etc.

The bin width in photons, or the width of a grayscale level, increases linearly with mean signal strength, but the amplitude noise [given by Eq. (6.48)] includes a factor of the mean signal strength under the radical. For the scenarios analyzed here, signal shot noise dominates shot noise due to background photocurrent and dark current. Signal shot noise dominates ROIC noise as well. Therefore, Eq. (6.48) can be accurately approximated as

$$noise_{total} \approx \sqrt{M\,F\,n_{signal}} = \sqrt{M\,F\,(N \times QE \times M)} = M\,\sqrt{F \times QE \times N}. \tag{6.49}$$

For calculating the interval over which 90% of signal amplitude measurements will occur, the amplitude distribution in units of electrons that is sampled by the ROIC can be approximated as Gaussian with the standard deviation ($noise_{total}$) given by Eq. (6.49). With this approximation, 90% of signal return measurements will occur within $\pm 1.645 \times noise_{total}$ of n_{signal}, or

$$\Delta n_{90} = 3.290 \times noise_{total}, \tag{6.50}$$

where Δn_{90} is defined as the width in electrons of the interval centered on n_{signal} over which 90% of measurements will occur.

Equation (6.50) allows one to solve for the interval Δn_{90} in electrons given the mean signal per pixel in photons N, which is an input to Eq. (6.49) for the total noise. The interval in units of electrons represented by Δn_{90} can be expressed in units of photons by dividing by a factor of $QE \times M$, which is necessary to compare it to the signal span of a reflectance bin given by Eq. (6.48). The signal level required to achieve a specified grayscale resolution can be found by equating $\Delta N_{90} = \Delta n_{90}/(QE \times M)$ to ΔN from Eq. (6.48) and solving for the value of $C(R)$ that satisfies the equality. However, the signal shot noise from Eq. (6.49) depends on N rather than $C(R)$ and varies from pixel to pixel because N can vary from pixel to pixel, depending on the average reflectance of the target scene within each pixel's IFOV.

To simplify the analysis, we make calculations using the average reflectance of $\rho_{avg} = 10\%$. In this case,

$$C(R)_{req} = \frac{2^{2\,Nbits}}{(\rho_{high} - \rho_{low})^2} \times 3.290^2 \times \frac{F}{QE} \times \rho_{avg}, \qquad (6.51)$$

where F is the function of APD gain for InGaAs LMAPDs and is a fixed value of about $F \approx 1.3$ for HgCdTe LMAPDs. For 6 bits of grayscale resolution (the specified reflectance span) and an effective QE of 56% (corresponding to 60% optical coupling efficiency and 80% detector QE), Eq. (6.51) gives $C(R) = F \times 6.916 \times 10^5$ photons; for 3 bits of grayscale resolution with these parameters, $C(R) = F \times 1.236 \times 10^4$ photons. Recall that $N = C(R) \times \rho$, so for $\rho_{avg} = 10\%$, the signal level per pixel required for the grayscale task is $1/10^{th}$ of the value of $C(R)$ given by Eq. (6.51). For the InGaAs LMAPD characterized by $k = 0.2$, we have $F = 2.44, 3.52, 4.55$, and 5.56 for $M = 5, 10, 15$, and 20, respectively, as shown in Table 6.3. Although operation of the APD at higher gain is beneficial from the standpoint of the 3D imaging task, it results in worse grayscale imaging performance due to excess multiplication noise. This is a familiar result for LMAPDs, which are typically used in systems where signal shot noise is dominated by amplifier noise, or where accurate measurement of signal amplitude is less important than discriminating signal pulses from noise.

Table 6.3 Required number of photons to achieve a certain grayscale level for an InGaAs LMAPD camera.

Gain M	3 Bits N	6 Bits N
1	1237	79,171
5	3020	193,000
10	4360	279,000
15	5630	360,000
20	6880	440,000

Equation (6.51) shows that the specified 6-bit grayscale resolution for $5\% \le \rho \le 15\%$ is not practically achievable in a single laser shot at low or moderate signal return levels since the required signal level is on the order of 69,000 photons/pixel. The requisite resolution is reached at $N = 193{,}000$ photons for $M = 5$; 269,000 photons for $M = 10$; 360,000 photons for $M = 15$; and 440,000 photons for $M = 20$. For three bits of grayscale resolution, Table 6.3 shows a significant reduction in the required number of photons compared to six bits of grayscale resolution. As we can see for the grayscale case, which is shot noise limited, gain loses its appeal. With no gain ($M = 1$), we use the fewest number of photons. Note that the required number of photons per pixel in this grayscale case in not dependent on which scenario we chose, the large-DAS or small-DAS case. The power required will be different for each scenario due to the other link budget considerations. The number of photons required in a grayscale case of LMAPDs is also independent of whether we are doing foliage poke through or not, except for the factor of 1.6 due to some of the photons being blocked from hitting the final reflector, which will be taken into account when we calculate energy.

If the signal return amplitude measurements from S laser shots are averaged, the standard error of the mean (SEM) amplitude measurement will drop as $S^{-1/2}$ such that, in the shot noise limit, the laser energy per shot can be cut by a factor of S. However, evaluated in terms of the total number of photons required to perform the measurement, no advantage is gained by increasing the number of laser shots. Although this analysis shows that fine resolution of reflectance bins requires strong signals, one advantage of an LMAPD-based sensor is the ability to collect several bits of grayscale data per pixel in a single laser shot. LMAPDs could be useful in applications for which accumulation of a large number of laser shots on the scene is not practical but target discrimination is aided by less-accurate reflectance measurements than analyzed here.

6.3.1.2 HgCdTe LMAPD cameras

Linear-mode HgCdTe electron-injection APDs (e-APDs) have been demonstrated in APD arrays fabricated by Leonardo DRS, Raytheon, CEA/Leti (France), Selex ES (UK), and others.[11] These APDs exhibit a deterministic gain process that results in an excess noise factor near 1 that is independent of gain. Gains up to 1900 with low dark current have been demonstrated in photon-counting FPAs.[12] HgCdTe APD cameras as large as 256×256 pixels for 3D imaging and 1024×668 pixels for gated 2D imaging have been demonstrated. Current HgCdTe LMAPD FPAs have 100-MHz bandwidths that are preamplifier-bandwidth limited or minority-carrier diffusion-time limited. However, the fundamental bandwidth that is set by carrier transit times across the multiplication region is quite high. A bandwidth of 600 MHz that was system resistor–capacitor (RC) time-constant limited has been measured at an APD gain of 3500 (gain bandwidth = 2.1 THz).[10] The modeled, fundamental, carrier-transit-time–limited bandwidth is greater than 10 GHz.[13]

Linear-mode HgCdTe e-APDs as fabricated at Leonardo DRS are frontside-illuminated, cylindrical, p/n–/n+ HgCdTe homojunction photodiodes in the HDVIP® configuration,[14] as shown Fig. 6.17. The architecture for the APD is the same as is used for production, non-APD FPAs. The cylindrical junction is created around a small hole (or via) in a thin passivated p-type HgCdTe membrane that is epoxied to a silicon readout. Metal is deposited in the via to form the contact from the n^+ surface to the input pad on the silicon readout that is under the diode. The array is then antireflection (AR) coated. APD operation is achieved in this structure by increasing the reverse bias to create a high-field multiplication region on the n side of the junction. The p side of the junction becomes the absorption region for the APD.

For typical unit cell geometries, the diffusion lengths of the holes and electrons are greater than the lateral dimensions of the n and p regions. The p (positive) side contains an excess of holes, while the n (negative) side contains an excess of electrons. Thus, at low bias, both the n and p sides of the junction contribute to the photo signal, and the optical fill factor is the area outside the via normalized to the unit cell area, which equals the pitch squared.[15] At high bias, the fill factor is given by the ratio of the area of the p-absorption region to the pitch squared and is typically greater than 60% without a microlens. Microlens arrays have been developed for high–f-number systems that provide 100% fill factor. High–f-number systems have a long focal length compared to the diameter of the optical system.

Many HgCdTe cameras in various formats have been delivered, but we are not aware of a HgCdTe APD camera available as a commercial product.

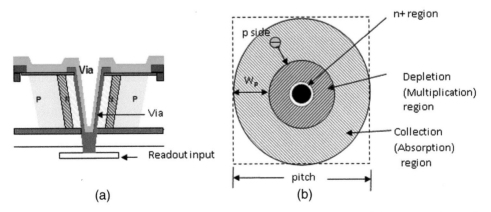

Figure 6.17 Leonardo DRS's frontside-illuminated APD architecture: (a) cross-section and (b) top view (reprinted from Ref. 50).

While you certainly can buy large-format, flash-imaging cameras from Leonardo DRS or Raytheon, these will typically be custom, single-camera purchases, likely to cost on the order of $500K or even more, and possibly associated with some development you specify.

A 5-μm–cutoff HgCdTe APD FPA can be actively operated in any spectral region from about 360 nm in the UV to 5 μm. The MWIR-cutoff HgCdTe APD FPAs need to be cooled to near 80–110 K, which is a disadvantage compared to InGaAs Geiger-mode and linear-mode APDs. The required HgCdTe detector biases, however, are conveniently less than 15 V and are compatible with current 0.18-μm CMOS ROICs. Shorter-wavelength–cutoff HgCdTe APD FPAs have been demonstrated that operate at higher temperatures and even in the thermoelectric-cooler (TEC) temperature range, but these are less mature. Low-noise, LWIR-cutoff HgCdTe APDs have also be demonstrated.

Diffusion dark current is typically negligible compared to tunnel dark current and background photocurrent at 80–110 K in 4.2- to 5-μm–cutoff HgCdTe FPAs. The major source of dark current at the higher gains (higher APD biases) is a bias-dependent dark current that is thought to be due to indirect tunneling processes in the multiplication region. Because this dark current is generated in the multiplication region, it is likely not to be fully gained, and, indeed, noise measurements indicate this.[16] A simplified manner of handling the dark current in the modeling is to use measured dark current at the bias that achieves the required APD gain. This dark current is then divided by the gain to give a gain-normalized dark current. A worst-case, upper-limit dark current can be estimated at any intermediate gain by multiplying this gain-normalized dark current by the gain. Gain-normalized dark current as low as 0.2 fA (2×10^{-16} A) has been measured on 64 μm × 64 μm pixel, photon-counting APDs at a gain of 1100–1200, but typically the gain-normalized dark current for 64 × 64 μm pixel photon-counting APDs is <3.2 fA (< 2×10^4 electrons/s).

The grayscale calculations for HgCdTe are the same as for InGaAs, except that when we are using gain, we would have $F = 1.3$, and we assume a QE of 65%. Since the excess noise does not change with gain, we can simplify the list of required photons (Table 6.3) to those shown in Table 6.4. As it turns out, gain does not help us with grayscale.

Table 6.4 Required number of photons to achieve a grayscale level using HgCdTe.

Number of bits	Photons
3	1608
6	102,921

6.3.1.3 Summary of the advantages and disadvantages of LMAPDs for direct detection

The advantages of InGaAs LMAPD cameras for direct detection are as follows:

- They are thermoelectrically cooled.
- They are commercially available from at least two sources, Voxtel and ASC.
- They are moderately priced.
- 3D images can be formed on a single pulse.
- 3D images can be formed quickly and with simple processing.

The disadvantages of the InGaAs LMAPD cameras are as follows:

- Gain is relatively low (≥ 5) due to excess noise and breakdown issues.
- A complex ROIC is required.
- Because gain is relatively low, it is necessary to keep track of all noise sources.
- They need a relatively high–energy-per-pulse laser.

HgCdTe LMAPD cameras have $k = 0$, meaning that essentially all of the carriers generated during an avalanche are electrons. This allows for very high gains. The advantages of these cameras for direct detection are as follows:

- They are very sensitive, while retaining linear gain.
- They require very low energy for mapping in many cases.
- 3D images can be formed on a single pulse.
- 3D images can be formed quickly and with simple processing.
- They can perform passive and active imaging from the visible to the mid-IR. A day/night passive imager can be inherently co-bore sighted with an active imager.

The disadvantages of HgCdTe LMAPD cameras for direct detection include:

- They are not commercially available, even though you can buy them from Leonardo DRS, Raytheon, or possibly others, so they are more expensive at this time.
- They need to be cooled to near 100 K so require a soda-can–sized cooler.
- They require a complex ROIC.
- They require high–energy-per-pulse lasers.

6.3.2 Direct-detection GMAPD LiDAR camera

LiDAR systems using arrays of GMAPDs were first proposed by Marino[17,18] and demonstrated by MIT Lincoln Laboratory.[19] Development work has continued to advance the technology for Geiger-mode LiDAR components, systems, data processing, and data exploitation in many research groups.[20] This discussion relies on previous work by Fouche,[21] who analyzed signal requirements in the presence of background noise. Recent modeling by Kim et al.[22] provides a detailed description of example system behavior. We restrict our discussion to commercially available Geiger-mode cameras.

Here we consider commercial framing cameras with up to 186-kHz frame rate for the 32×32 pixel format, or up to 110 kHz for the 32×128 pixel format. An asynchronous-readout 32×32 pixel format is also now commercially available and is capable of even higher readout rates limited only by the dead time between detections—the time period after an avalanche during which the detector is not activated to receive a photon. In GMAPDs, the detector is biased above the breakdown voltage, so a photoelectron generated in the absorber region will lead to a large avalanche, often resulting in a voltage fluctuation on the order of 1 V. Whether one photon or many photons hit the detector, the same large avalanche occurs, as is depicted in Fig. 6.18. A dead time of 500 ns to 1 µs occurs after each triggered event. This dead time can block detection of photons arriving later unless the probability of avalanche is kept low. A research thrust is to reduce the amount of dead time after an avalanche. To avoid after-pulsing, GMAPDs are not equipped to detect photons during the dead time.

GMAPDs cannot measure the signal intensity, or grayscale, on each pulse because they always avalanche to the same level. GMAPD flash imagers run at high repetition rates because the probability of detection (avalanche) on a single pulse is often purposely kept low. A target is identified by coincidence processing (looking for multiple returns coincident in the same range bin), which often requires as many as 50 to 100 laser pulses to obtain 2–4 coincident returns in the same range bin.

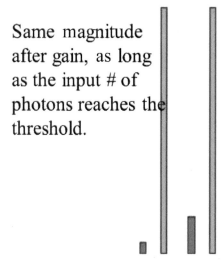

Same magnitude after gain, as long as the input # of photons reaches the threshold.

Figure 6.18 A pictorial description of the GMAPD avalanche.

Four possible GMAPD detection events can trigger the receiver: (1) detection of a desired target photon, (2) detection of an undesired foreground clutter photon (such as from backscatter from fog), (3) detection of undesired background radiation (such as the sun), or (4) undesired detection of a dark electron. Crosstalk can also trigger a GMAPD, where crosstalk is defined as one detector firing based on a photon hitting a nearby detector. If we send out many laser pulses, we will get coincident returns (returns in the same range bin) for reflection from a target or from fixed foreground objects; however, returns from dark current, background, fog, snow, and rain will provide distributed returns with very low probability of range coincidence. A Geiger-mode receiver temporarily over-biases an APD detector for the range gate time during which we expect return signal photons. This is called the gate time, or the time during which a GMAPD detector might avalanche if a photon arrives.

GMAPDs have a very large avalanche gain for any photon hitting a detector, as shown in Fig. 6.19, which is a schematic depiction of a GMAPD device structure.[23]

Photon absorption occurs in a quaternary InGaAsP layer ($E_g \sim 1.03$ eV) that is optimized for detection of single photons at 1.06 µm. The absorption layer is spatially separated from a wider-bandgap InP region ($E_g \sim 1.35$ eV) in which avalanche multiplication occurs. A primary goal in designing the separate absorption and multiplication (SAM) region structure is to maintain a low electric field in the narrower-bandgap absorber (to avoid dark carriers due to tunneling) while maintaining a sufficiently high electric field in the

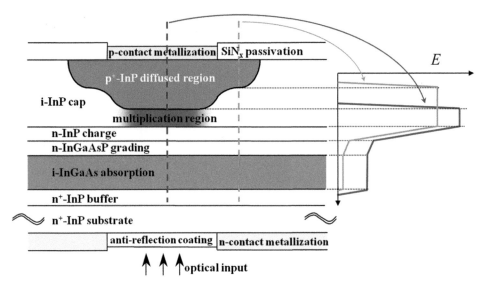

Figure 6.19 Schematic illustration of a diffused-junction planar-geometry APD structure. The electric field profiles at the right show that the peak field intensity is lower in the peripheral region of the diffused p–n junction than in the center of the device (reprinted from Ref. 50).

multiplication region (so that impact ionization leads to significant avalanche multiplication). The creation of a single electron–hole pair by photoexcitation in the absorber layer results in the injection of the hole into the high-field InP multiplication region in which impact ionization results in avalanche gain. With the GMAPD biased above the threshold voltage in its armed state, the resulting avalanche gives rise to a macroscopic current pulse that is sufficiently large to be sensed by a threshold detection circuit contained in the ROIC. Unlike LMAPDs operated below the threshold voltage, the GMAPD detection process is inherently digital, and with appropriately designed detectors and threshold circuits, the detection process is noiseless.

InGaAs light-absorption layers respond in the short-wavelength infrared (SWIR) region and are typically thermoelectrically (TE) cooled. A single-photon detection efficiency (SPDE) of 25%, a dead time of nominally 1 µs following breakdown, and a dark count rate (DCR) of about 6 kHz at 225 K are typical of 25-µm–diameter GMAPD pixels for 1-µm operation. The 128×32 pixel format arrays of 18-µm GMAPD pixels have been reported to operate with 32.5% SPDE and 5-kHz DCR at 253 K thanks to use of a wider-bandgap InGaAsP absorption layer optimized for 1064-nm signal detection. Interframe timing jitter of the 1064-nm–sensitive 128×32 detector format GMAPD array was reported to be about 500 ps and may have been dominated by clock signal distribution issues in its ROIC rather than the fundamental timing performance of the GMAPD pixels themselves. Timing jitter for 32×32 detector format arrays of 1550-nm–sensitive pixels was reported to be in the 150- to 200-ps range.

A higher reflectance area will have a higher probability of return, causing more coincident avalanches to trigger at that range. Some value can be had from multiple returns at the same range in adjacent pixels, depending on how flat the target is. There is a bias toward nearer ranges because of dead time, but this bias is slight if the probability of detection on a single pulse is kept low. This is a major reason that GMAPDs are often run with a probability of detection of less than one on a single pulse. For a given range target, and given transmit and receive aperture diameters, you need to set the emitted laser energy at the right level, or you will not achieve the right probability of an avalanche; therefore, GMAPD-based LiDARs have a narrow dynamic range. GMAPD-based flash imaging is a highly efficient method of 3D mapping an area.

For the case with foliage poke through, we set the average number of photons per pixel to be 0.8 photons returned for the expected range and reflectivity of the target, or a 20% probability of detection per pulse given a photon detection efficiency (PDE) of 25%. With GMAPDs, there is crosstalk between detectors when a photon emitted during breakdown of one pixel triggers breakdown in an another pixel. The noise due to crosstalk tends to be concentrated in the range region where most of the detections occur. Even there, crosstalk noise is much smaller than noise due to background light. For the cases analyzed in this chapter, GMAPD flash-imaging LiDARs tend to be designed to run at high frame rates, and many samples are used to capture the necessary number of photodetection events to achieve the signal level requirements. If laser pulse energy is lowered, the number of photoelectrons generated per pulse is also lowered, along with the probability of a pixel firing. These design parameters provide the technical benefit of keeping peak laser intensity low since each pulse is weak while maintaining high average power. This means that when we calculate the energy required to 3D image a region, the main parameter we will vary is the number of pulses, not the energy transmitted per pulse.

A key advantage of Geiger-mode imaging LiDAR is the relative simplicity of the receiver and data-acquisition electronics compared to wide-bandwidth, linear-mode receivers. A disadvantage is that it takes more processing to make an image because we need to use coincidence processing. In coincidence processing, we only declare a detection if we have multiple returns from the same range in a given detector. This can cause image blurring if there is motion in the scene over the time period during which coincidence processing occurs. Fortunately, the frame rate of a GMAPD camera is high, so not a lot of motion occurs while forming an image, but the camera is not performing true flash imaging. It takes the time needed to obtain many frames of data before coincidence processing can occur.

As previously mentioned, if we send out many laser pulses, we will get coincident returns (returns in the same range bin) for reflection from a real target or from fixed foreground objects. Fog, snow, and rain will provide distributed returns over their extent so will not provide coincident returns except at a low probability. If the fog or clouds are close enough to the target when an avalanche occurs, the dead time will prevent avalanching on return from the target. A problem with GMAPDs is that once the avalanche occurs, there is a dead time of 500 ns to 1 μs. For at least 500 ns, and more likely 1 μs, the GMAPD cannot be armed without a high a probability of after-pulsing (unless the arm times are very short, thus reducing the chance of an after pulse).

With GMAPDs there is no ability to measure the signal intensity per pixel (or grayscale) on a single pulse. Also, there may be some crosstalk between detectors, and dark current can be an issue. GMAPD flash imagers tend to run at high repetition rates because you do not need, or want, a lot of energy per pulse to obtain a response. You keep energy per pulse and probability of detection low on a single pulse, and then use a number of pulses. This low-energy requirement can allow us to use 1.064-μm lasers without eye hazard because of low single-pulse intensities. You can develop an effective grayscale or an effective range profile by using multiple pulses. If you have multiple pulses returned from a given location and keep the probability of detection low, then a higher reflectance area will have a higher probability of return, causing more events to trigger. Also, if you have pixels with mixed range returns, you can map the returns as a function of range. If the probability of triggering is low for any event, you will get events triggering at various ranges. Again, there is a bias toward nearer ranges because of dead time after triggering, but this bias is slight if the probability is low for triggering an individual event.

6.3.2.1 The effect of a bright sun background on GMAPDs

Solar background is detrimental to GMAPD performance in two ways: blocking and noise, blocking being the more important for this analysis. Therefore, one of the first issues to address is whether background from the sun will affect either blocking or noise. If the GMAPD undergoes an avalanche before the signal photons arrive, the detector is "blocked" and is unable to detect the signal until after the dead time. On the other hand, noise can cause the system to erroneously declare a surface to be present. Given a background photon rate per pixel of γ, a PDE of 25%, and a gate width W during which the APD is sensitive, the mean number of background photoelectrons generated in the APD by the sun before the signal occurs is

$$N_b = \gamma(PDE)W\left(\frac{2}{c}\right), \tag{6.52}$$

where c is the speed of light, and the factor of two accounts for the round trip. The Poisson distribution indicates that the probability of the APD not undergoing breakdown before the signal arrives (i.e., the probability of zero photoelectrons) is

$$1 - P_B = \exp(-N_b), \tag{6.53}$$

where P_B is the probability of blocking. To keep the blocking loss below $P_B = 0.2$, the number of background photoelectrons must be kept below

$$N_b = -\mathrm{Ln}(1 - 0.2) = 0.22. \tag{6.54}$$

For a gate width $W = 100$ m, the background must be below $\gamma = 1.33$ MHz. Clearly, dark count rates, which are typically 1–10 kHz, can be neglected.

The background photon rate can be limited by introducing attenuation on the receiver, reducing the aperture, or increasing the focal length and therefore reducing the pixel DAS. The GMAPD community prefers to increase the focal length while maintaining the aperture diameter to reduce blocking loss. The disadvantage of decreasing DAS instead of aperture diameter is that we then must scan more locations to develop the FOV required by the scenario. This will most likely increase collection time.

The sun can block the detector. Table 6.5 shows a high probability of the detector having avalanched from sun photons and therefore being blocked from a signal photon causing an avalanche. For the gate width $W = 100$ m, we

Table 6.5 Probability of avalanche from background sun photons.

Photons per ns	Range bin width (ns)	# of photons per range bin	QE	Probability	# of bins	Range window width (m)	Probability of not having avalanched after # of bins
0.0931	16.67	1.552	25.00%	36.0%	20	100.00	0.00%
0.00238	3.33	0.0079	25.00%	0.2%	200	200.00	63%
0.0149	16.67	0.248	25.00%	6.9%	20	100.00	24%
0.0149	16.67	0.248	25.00%	6.9%	40	200.00	6%
0.00373	16.67	0.062	25.00%	1.8%	20	100.00	70%
0.00373	16.67	0.062	25.00%	1.8%	40	200.00	49%
0.00134	16.67	0.022	25.00%	0.6%	40	200.00	77%

can either reduce the aperture size to 2.5 mm in diameter, or reduce the DAS to 0.25 mrad in order to avoid sun blocking. The smaller-DAS case can use a narrower filter width, which is why decreasing the DAS results in lower energy than decreasing the receive aperture size; and, of course, the reduced DAS provides higher resolution. Innovative processing will also provide a significant advantage for reducing the DAS compared to reducing the receive aperture diameter.

The next section presents coincidence processing, which is used by GMAPDs to achieve the required 90% probability of detection. If we reduce the DAS by a factor of 5 in each dimension, each of our 0.5×0.5 m pixels is made up of 25 0.1×0.1 m pixels. For surfaces that are smoothly varying, we can use these 25 samples to do coincidence processing, resulting in the need for as much as $25\times$ fewer pulses. This will reduce the required energy for mapping the area.

6.3.2.2 Coincidence processing for detection

When using GMAPDs in a foliage poke-through scenario, we keep the probability of detection from a single pulse low because of the dead time after an avalanche (e.g., $P_{\text{det}} = 0.2$). This preserves our ability to see objects farther in range than the initial return. Sometimes people use even lower probability of detection, such as 0.1. As a reminder, a 0.1 probability of detection requires 0.4 photons to be returned per pulse if the QE is 25%. If we do not have mixed pixels with multiple range returns, we can allow the probability of detection to increase. We want to determine the number of pulses N_{p} that must be transmitted to cause a GMAPD pixel to fire on M pulses scattered from the surface of interest (we anticipate that M will be a minimum of two or three detections from the surface of interest). This coincidence detection will determine a real return from a physical object, as compared to a random false return. A judgment call must be made as to how many coincident returns are required to have a certain probability that the returns are from a real object. We rely on the fact that the noise is randomly distributed in time, whereas returns from real objects only occur at the range of an object. We ignore nonuniform detector illumination and sensitivity. Backscatter from fog would be similar to noise, in that it would be randomly distributed.

The probability P_o of detecting a photon backscattered from the object of interest can be expressed as a conditional probability:[24]

$$P_o = P(o|\bar{n})P(\bar{n}), \tag{6.55}$$

where $P(o|\bar{n})$ is the probability of detecting a photon backscattered by the object, given that a photon scattered by some intervening obscurant (or dark count) has *not* been detected; and $P(\bar{n})$ is the probability of *not* detecting a photon from an intervening object. Since detecting a photon from an intervening obscurant is a binary event (it either is detected or it is not detected), the probability of *not* detecting a photon from an intervening obscurant is just one minus the probability of detecting that photon. Hence, Eq. (6.55) can be rewritten as

$$P_o = P(o|\bar{n})(1 - P_n), \tag{6.56}$$

where P_n is the probability of detecting a photon from an intervening obscurant or a dark count. As indicated earlier, the laser radar parameters can be adjusted so that the average values for $P(o|\bar{n})$ and P_n on a single pulse will be much less than one. Multiple pulses will then be required to achieve high probabilities of detection. If the probability of detection is increased for the case with low reflectivity in the foreground, fewer pulses will be required, but more laser energy will be required per pulse.

The value for P_o can be calculated once the parameters of the LiDAR system are specified. However, some insight can be obtained without considering a specific system configuration. The probability of detecting a specified number of photons M backscattered from the object of interest out of N pulses can be described by the binomial distribution as follows:

$$P(M \text{ out of } N \text{ pulses}) = \frac{N!}{M!(N-M)!}[P_o]^M \times (1 - P_o)^{N-M}, \tag{6.57}$$

where P_o is the probability of detection on a single pulse. The results of Eq. (6.57) can be seen in Fig. 6.20 for a probability of detection of 0.1, or 0.4 photons returned per pulse on an average, with a 25% quantum efficiency.

An individual probability of 0.1 would result in 38 and 52 pulses for a level 2 and level 3 of pulse coincidence, respectively. From the plot in Fig. 6.21, with a design probability per pulse of 0.2, 9 pulses will provide a level 2 of pulse coincidence with a 90% probability, and 14 pulses per detector should provide 90% probability of coincidence detection with 3 pulses. If we receive more coincident pulses returning from the same range, then there is a higher probability that a physical hard target at that location caused the

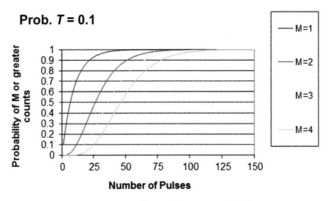

Figure 6.20 Coincidence probability.

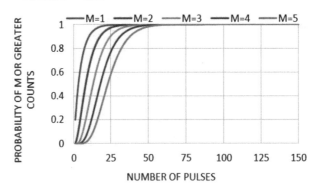

Figure 6.21 Probability of *M* out of *N* detections.

coincident returns. Each avalanche resulting from scatter at a certain range occurs at a random location within the pulse width, so in this case, we are using the entire pulse width, not just its leading edge.

One interesting result involves the large-DAS case, where we will reduce the DAS in both directions by a factor of 5 for bare earth and by a factor of 10 for foliage poke through, resulting in 25 and 100 times as many samples, respectively. If we aggregate 25 small detections into one $0.5\,\text{m} \times 0.5\,\text{m}$ ground sample distance (GSD), we can use a single pulse to obtain 15% probability of level 2 pulse coincidence; in the case of aggregating 100 small-area detections, we can go down to 4% probability of detection as a design criteria and still use only a single pulse.

To do this we recast Eq. (6.56) in the following manner:

$$P_{\text{o}} = P(o|\bar{n})[1 - rP(o|\bar{n})], \qquad (6.58)$$

where *r* is the ratio $P_n/P(o|\bar{n})$, which is the ratio of the strength of scattering from the intervening obscuration to the strength of the return from the target of interest. Equation (6.58) is valuable for the case of specific obscurants, such as intervening foliage, or camouflage. In this form, a value for $P(o|\bar{n})$ can be specified (controllable through the parameters of the laser radar imaging system to provide sufficient power to provide a given probability of detection) and the relative strength of backscatter between the intervening obscurants and the object can be treated as a parameter.

Since we desire to maximize the number of detections from the object of interest rather than the obscurations/false counts, we need to maximize the value of P_o given the constraint that $rP(o|\bar{n}) < 1$, where r is the ratio of near-range–reflected light to target-reflected light. As a reminder, P_o is the probability of detecting a photon from the object of interest, whereas $P(o|\bar{n})$ is the probability of detecting a photon from the object of interest with no obscuration. For our foliage poke-through example, we have twice as much near-range reflection as target reflection, with the last return considered as the target. In this case, $r = 2$. Two-thirds of the return flux comes from the foreground surfaces, and one-third comes from the final surface. We maximize the probability of detecting a photon from the object of interest by differentiating Eq. (6.58) with respect to $P(o|\bar{n})$ and setting the derivative equal to zero. We find that the maximum value occurs for

$$P(o|\bar{n})_{\max} = 1/2r, \qquad (6.59)$$

which can guide where we set our design probability of detection. With our case of $r = 2$ for foliage poke through, we want a design with $P_{\det}$ of 0.25, or 1 photon received from the target with a 25% PDE, not much different from our case without foliage. We note that the expression for $P(o|\bar{n})_{\max}$ is valid for $r \geq 0.5$. Traditionally, GMAPD LiDARs are designed with 0.1–0.2 probability of avalanche from the target, or 0.4–0.8 photons with a PDE of 25%.

To measure grayscale using GMAPD, multiple pulses are transmitted, and the grayscale is built up one photodetection at a time. We compute the number of samples that must be transmitted to achieve the required number of photodetections. We use the term samples because we can multiply the number of pulses by the samples per pulse to obtain the number of samples. The required number of photodetections is determined by the need to have the gray-level separation large enough such that fluctuation in the number of detections is smaller than the gray-level separation.

Since the mean probability of detection P_o on any given pulse is less than one, there will be a fluctuation in the number of detections that will be obtained for a given number of transmitted pulses. As discussed above, the number of detections for a given number of transmitted pulses is a binomial distribution shown in Eq. (6.57). For a binomial distribution, the mean number of detections out of N pulses is NP_o. The variance in the number of detections is $NP_o(1 - P_o)$. To measure N_g number of gray levels ($N_g = 2^{Nbits}$), we need N_g separations, each of which is $3.34\times$ the standard deviation. The factor of 3.34 is used so that approximately 90% of the probability distribution is contained within the gray-level separation. Hence, we need

$$N_p = \frac{11.2 N_g^2 (1 - P_o)}{P_o}. \tag{6.60}$$

As specified earlier, we have assumed a variation in reflectivity from 0.05 to 0.15, or a 10% variation in reflectivity, but the percentage reflectivity change is not a factor in the required number of pulses to obtain a given number of gray levels.

6.3.2.3 Summary of the advantages and disadvantages of GMAPDs for direct detection

GMAPD cameras operate with a low probability of return on a single pulse, but require coincident returns from the same range. They require low–energy-per-pulse lasers with a higher repetition rate. We see that the GMAPD cameras do well in both scenarios when doing bare earth 3D mapping and 3D imaging through trees. The large-DAS scenario does not create a significant energy use issue because of using many coincident samples from a metapixel. In grayscale situations, the GMAPD cameras do use somewhat more energy.

The advantages of GMAPD cameras for direct detection include:

- They are TE cooled.
- They use low–energy-per-pulse, high–repetition-rate lasers, which are easier to obtain because laser diodes are cw and due to the damage thresholds of fiber lasers.
- They can passively image in the near IR.
- They have little noise, so their performance can be easily predicted.
- They are are commercially available from at least two sources, Princeton Lightwave (now ARGO AI) and Spectrolab, a Boeing Co., and are moderately priced.
- Their ROICs are very simple.

The disadvantages of GMAPD cameras for direct detection include:

- There is a dead time of 400 ns to 1 µs after an avalanche. The probability of avalanche must be kept low to avoid seeing only a single range in a pixel with more than one return in range.
- Due to this blocking issue, high background (e.g., a bright sunlight) can be an issue, requiring smaller apertures or increased resolution; however, the innovative processing associated with using multiple samples in a metapixel has mostly mitigated this.
- Forming the image requires significant processing due to both coincident processing and removal of motion.
- The dynamic range is narrow in order to have the right number of return photons.

6.4 Silicon Detectors

Silicon detectors are the least expensive and most mature detectors. Unfortunately, they are in a wavelength region where the eye is sensitive, and lasers in this region are limited to lower powers if they are to remain eye safe. Figure 6.22 shows the response of two types of silicon detectors: normal silicon detectors and hybrid detectors with enhanced response in the near-IR spectral region.[25]

A number of the more-interesting LiDAR applications are in the near-IR spectral region; e.g., many auto LiDARs operate in that region. 905 nm is a very popular auto LiDAR wavelength. T. Baba et al.[25] discuss a Si-based 100-μm–pitch 32 × 32 pixel single-photon avalanche diode (SPAD) sensor. A SPAD array in silicon detectors is essentially the same as what we have been calling a GMAPD array. The major difference is that the dead time after a detection can be an order of magnitude shorter in a SPAD array. Unfortunately, since these arrays are in the visible spectral region, there is even more sunlight as background, so blocking loss can still be an issue.

Macro-pixels, or multipixel–photon-counter (MPPC) arrays, are used with SPADs, just as they are used in GMAPDs.[26] This can provide many detectors in an individual macro-pixel, allowing both a shorter quenching time and an effective method of providing grayscale. Using the macro-pixel approach, S. Adachi et al.[26] measured time over threshold (TOT), which is a method of compensating for range walk. If you measure TOT and know when it occurs, the center of the TOT will be the most likely location of the peak.

A photomultiplier tube (PMT) is another detection device used in the visible region. A PMT is like a SPAD but is a cold vacuum tube. Operation is based on the extrinsic photoelectric effect (from a photocathode) and

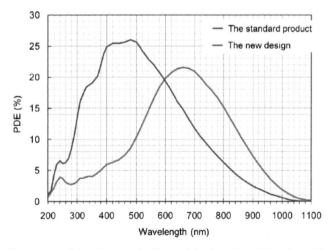

Figure 6.22 Response of two types of silicon detectors (reprinted from Hamamatsu with permission).

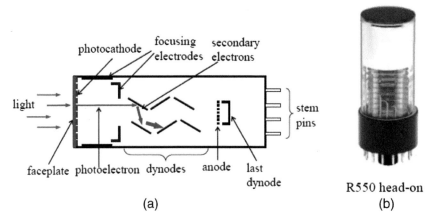

Figure 6.23 Photomultiplier tube: (a) schematic diagram and (b) photograph (reprinted from Hamamatsu with permission).

secondary electron emission from dynodes. It has high intrinsic gain ($\sim 10^6$). As solid state alternatives have grown, the use of PMTs has decreased, but they still may be the best option for many applications. From an analysis point of view, they are also similar to GMAPDs in that they have a large avalanche regardless of how many photons cause the avalanche. Figure 6.23 shows a diagram and a photograph of a PMT.[27]

6.5 Heterodyne Detection

When the return signal interferes with an LO, the resulting optical intensity can fluctuate in time or in space. For spatial heterodyne detection, also called digital holography, the LO is at the same frequency as the signal but is offset in angle to create spatial fringes. For temporal heterodyne detection, we align the LO and the return signal as shown in Fig. 6.24. We have fluctuations near direct current (DC) and at the difference and sum frequencies of the two fields. For optical frequencies, the higher-frequency power oscillations are well beyond the maximum frequency response of detectors. The only detectable power fluctuations are near DC and at the difference frequency.

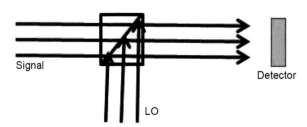

Figure 6.24 LO insertion for temporal heterodyne detection (reprinted from Ref. 49).

The electric field for the signal is given as

$$\mathbf{E}_{\text{sig}} = E_{\text{sig}} \exp(k_{\text{sig}}\mathbf{r}_{\text{sig}} - j\omega_{\text{sig}}t), \qquad (6.61)$$

the electric field for the LO is given as

$$\mathbf{E}_{\text{LO}} = E_{\text{LO}} \exp(k_{\text{LO}}\mathbf{r}_{\text{LO}} - j\omega_{\text{LO}}t), \qquad (6.62)$$

and the intensity or power is given as

$$I = (\mathbf{E}_{\text{sig}} \times \widehat{\mathbf{E}}_{\text{sig}}) + (\mathbf{E}_{\text{LO}} \times \widehat{\mathbf{E}}_{\text{LO}}) + (\mathbf{E}_{\text{sig}} \times \widehat{\mathbf{E}}_{\text{LO}}) + (\mathbf{E}_{\text{LO}} \times \widehat{\mathbf{E}}_{\text{sig}}). \qquad (6.63)$$

The electrical power measured in a heterodyne receiver is linearly proportional to the optical power received because the third and fourth products on the right side of Eq. (6.63) multiply the LO electric field by the return signal electric field. The return signal term therefore is linearly proportional to the intensity when the first two terms in the equation are ignored.

To measure the SNR in this case, the factor of two is eliminated from the denominator because of the factor of two in the signal times LO power. The LO adds shot noise, which is accounted for in the denominator:

$$SNR = \frac{\langle i_s^2 \rangle}{\langle i_n^2 \rangle} = \frac{\eta_{\text{h}} G^2 \mathfrak{R}^2 P_{\text{LO}} P_{\text{s}}}{eBG^2 F\mathfrak{R}[P_{\text{s}} + P_{\text{LO}} + P_{\text{bk}} + P_{\text{dkb}}] + 2eBi_{\text{dks}} + 4ktB/R_{\text{L}}}, \qquad (6.64)$$

where the LO terms have been added to Eq. (6.31), which describes direct-detection SNR. The LO can be increased to dominate other noises. Recently, people have used APDs with heterodyne detection, reducing the SNR-generated need for a high-power LO. The ideal SNR for heterodyne detection is given by

$$SNR = \frac{\eta_{\text{D}}\eta_{\text{h}}P_{\text{s}}}{h\nu B}. \qquad (6.65)$$

Once again, the SNR is directly proportional to the power received and therefore to the number of received photons.

6.5.1 Temporal heterodyne detection

For temporal heterodyne detection, we keep only the time terms of the heterodyne equations. Since the spatial terms are the same in the LO and signal, on interference they sum to zero, so only the temporal frequency differences remain. Equation (6.61), repeated here,

$$\mathbf{E}_{\text{sig}} = E_{\text{sig}} \exp(k_{\text{sig}}\mathbf{r}_{\text{sig}} - j\omega_{\text{sig}}t)$$

becomes

$$\mathbf{E}_{\text{sig}} = E_{\text{sig}} \exp(-j\omega_{\text{sig}}t), \tag{6.66}$$

and Eq. (6.62), repeated here,

$$\mathbf{E}_{\text{LO}} = E_{\text{LO}} \exp(k_{\text{LO}}\mathbf{r}_{\text{LO}} - j\omega_{\text{LO}}t)$$

becomes

$$\mathbf{E}_{\text{LO}} = E_{\text{LO}} \exp(-j\omega_{\text{LO}}t); \tag{6.67}$$

so when they interfere, we have

$$I = 2E_{\text{sig}}E_{\text{LO}} \exp[-j(\omega_{\text{sig}} - \omega_{\text{LO}})]. \tag{6.68}$$

We align the LO with the return signal. We generally offset the LO frequency more than any Doppler shift we expect in the return signal. For the well-designed heterodyne detection receiver (with sufficient LO power), the SNR is proportional to the number of photons received even when the signal is very weak, whereas, for the direct-detection receiver, the signal hitting the detector must be strong enough that its shot (photon) noise dominates all other noise. If a coherent receiver is truly shot noise limited, and if the heterodyne mixing efficiency is unity, then the coherent SNR will always be greater than or equal to the direct-detection SNR (assuming that the detectors have the same quantum efficiency). Heterodyne efficiency is the mixing efficiency between the LO and the return signal.[28] To have a 100% heterodyne mixing efficiency, the LO and the return signal have to be spatially matched. This is not to say that the probability of detection and probability of false alarm are always better for coherent detection, as those depend on the statistical fluctuations of the signal and noise (primarily signal).

6.5.2 Heterodyne mixing efficiency

In order to only keep the time portions of the signal and LO, we need to ensure that the signal and the LO overlap. If perfect overlap does not occur (and it never does), then we have some loss, which we refer to as heterodyne mixing efficiency. We want to match the quadratic spatial phase terms in the LO and in the signal beams; otherwise, the phases will interfere to create rapidly varying spatial fringes that will be difficult to sample.

Figures 6.25 and 6.26 show that, if we have a detector that is much larger than the fringe spacing, many fringes are contained in one detector, and the average value becomes close to zero. If we have finer detector spatial sampling, or fringes with lower spatial frequency content, we will see peaks and valleys across the detector array. With temporal heterodyne detection, we want to focus on the temporal portion of the beat between the LO and signal,

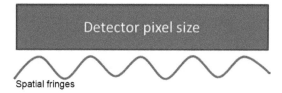

Figure 6.25 Diagram of very low-efficiency heterodyne detection (reprinted from Ref. 49).

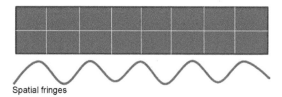

Figure 6.26 Diagram of well-sampled heterodyne detection (reprinted from Ref. 49).

so these spatial fringes are at best a distraction, and at worst they zero out the beat between the LO and signal.

For temporal heterodyne detection, we need a careful design of the LO such that it illuminates the detector array in a manner that matches the return signal, or we will lose heterodyne mixing efficiency. The key is to spatially match the LO and signal. A powerful technique to match the LO with the signal return is to back propagate the LO through the receive optics and the signal transmission path to determine what shape the LO should have in order to provide an exact signal match.

6.5.3 Quadrature detection

In quadrature detection, the IF signal—the beat frequency between the LO and the signal—is divided into two parts. One version of the IF signal is delayed by 90 deg in phase from the other version. This allows us to separate the magnitude and the phase portions of the returned signal. The most direct way to measure the two components of the complex signal would be to direct the signal to two detectors, each with its own reference beam. If the two reference beams are the inphase (I) and the quadrature (Q) components of the transmitted beam, the two detector outputs will be the real and the imaginary components of the signal, respectively. This is the technique commonly used in radar when the wavelength is long. Short optical wavelengths make it impractical to maintain the phase relationship between the references because path lengths must be held constant to a fraction of a wavelength. If, however, we mix the circularly polarized reference with a signal beam that is linearly polarized at 45 deg, we can separate the real part from the imaginary part of the interference term by measuring the intensity in both its horizontally and vertically polarized components.

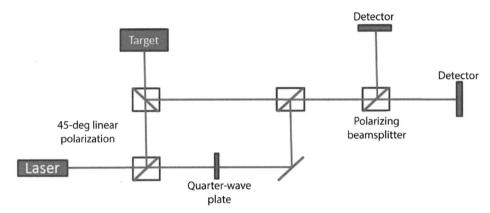

Figure 6.27 Quadrature detection schematic.

Figure 6.27 depicts a laser with 45-deg linear polarization. One leg goes through a quarter-wave plate and becomes circularly polarized. It is then beat against the 45-deg linearly polarized laser beam. This creates two beams with a 90-deg phase that can be separated by a polarizing beamsplitter. Quadrature detection has been applied across a wide range of disciplines including radar, sonar, magnetic resonance imaging, general detection, and LiDAR.[29,30] Inphase and quadrature (IQ) detection in LiDAR is important for many reasons. Two of the most important of these are avoiding the signal fades associated with interferometric detection and discerning the direction of the phase rotation associated with Doppler, SAL, and micro-Doppler signatures. In addition, IQ detection effectively doubles the sampling rate either by providing enhanced sampling or, conversely, by lowering the system's overall sample rate requirement.

Combining the inphase I and quadrature fields, and detecting with a square law detector, the inphase component is[31]

$$I = \left| \frac{1}{2}S + \frac{1}{2}LO \right|^2 = \frac{1}{4}(S^2 + LO^2 + 2S \cdot LO). \qquad (6.69)$$

Similarly, the out-of-phase, or quadrature, component Q is

$$Q = \left| \frac{1}{2}S + \frac{1}{2}e^{i\pi/2}LO \right|^2 = \frac{1}{4}[S^2 + LO^2 + S \cdot LO\, e^{i\pi/2} + S \cdot LO\, e^{-i\pi/2}], \quad (6.70)$$

where the LO leg has been delayed by the exponential phase factor.[31] In practice, the alignment, splitting ratios, and especially the required delay are hard to establish and maintain.

6.5.4 Carrier-to-noise ratio (CNR) for temporal heterodyne detection

For heterodyne detection, the conversion from received optical power to electronic power is linear because the surviving term after interference is a term that multiplies the LO times the return signal, thereby having only one factor of returned signal. Received optical power converts to current, which has to be squared for direct detection to get electronic power, as used in the SNR calculations; however, in this case, instead of squaring the terms, we multiply the received signal by the LO. The carrier-to-noise ratio (CNR) is the SNR on this beat frequency between the signal and the LO. This refers to the mean electrical power at the IF (difference frequency) compared to the noise power at the IF. The following equation shows the intensity, which is the real portion of this equation:

$$I = \text{real}[2E_{\text{sig}}E_{\text{LO}}\exp(k_{\text{sig}}\mathbf{r}_{\text{sig}} - k_{\text{LO}}\mathbf{r}_{\text{LO}})]. \qquad (6.71)$$

CNR comes from the RF community, where it is generally accepted to be a pre-detection measurement, i.e., one made at the RF carrier frequency. CNR refers to the mean electrical power at the IF carrier frequency (difference frequency). In cable industry vernacular, SNR is generally accepted to be a pre-modulation or post-detection measurement, i.e., one made on a baseband signal such as video or audio. For a LiDAR, the term SNR often refers to the signal-to-noise ratio after speckle averaging or other techniques have been performed to make the signal more detectable.

In the microwave or RF community, the term CNR is a pre-detection measurement performed on RF signals. CNR, often given in units of decibels, is the ratio of the level of the carrier to that of the noise in a receiver's IF bandwidth before any nonlinear process such as amplitude limiting and detection takes place.

6.5.5 Spatial heterodyne detection / digital holography

A major advantage of spatial heterodyne detection is the ability to use low-bandwidth framing cameras that are available in large formats. Visible cameras have very large formats. Figure 6.28 shows the tilted LO required for spatial heterodyne detection. At 1.5-μm, camera formats are smaller, but

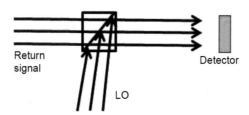

Figure 6.28 LO insertion for spatial heterodyne detection (reprinted from Ref. 49).

512×640 pixels is a very common format, and megapixel cameras are becoming available. We can consider a simple case of a plane wave signal (something viewed far away) and a plane wave LO in the pupil plane. The temporal part of the beat frequency does not matter in this case. We have both the LO and the return signal at the same frequency, except for velocity considerations, which we will initially ignore. We assume that scattering is off of a point reflector, and that the LO is emitted at a point the same distance from the aperture as the point reflector but offset slightly in the cross-travel direction so it hits the detector array at a different angle. We would like to calculate the spatial variation in x of the LO and return signal interfering with each other. To get exactly the right curvature, we put the LO the same distance from the center of the imager as the distance to the point reflector; therefore, if the LO is emitted as a point radiator slightly to the side, it would have to be slightly closer in the z direction, but this is a minor change that we will ignore. Let us also assume that the z displacements are significantly larger than the x displacements.

Only one term oscillates in x; the other term is a constant. We can see that the oscillation is a product of x and x_0, so the farther away from zero we place the LO the more rapid the oscillation in spatial frequency. As we get farther off axis, the fringes oscillate more rapidly. We have to size the detector sampling to see these fringes.

6.5.5.1 SNR for spatial heterodyne detection

For spatial heterodyne detection, we keep only the space terms of the heterodyne equations. Eliminating the time terms, Eq. (6.61), repeated here,

$$\mathbf{E}_{\text{sig}} = E_{\text{sig}} \exp(k_{\text{sig}} \cdot \mathbf{r}_{\text{sig}} - j\omega_{\text{sig}}t)$$

becomes

$$\mathbf{E}_{\text{sig}} = E_{\text{sig}} \exp(k_{\text{sig}}\mathbf{r}_{\text{sig}}), \tag{6.72}$$

and Eq. (6.62), repeated here,

$$\mathbf{E}_{\text{LO}} = E_{\text{LO}} \exp(k_{\text{LO}}\mathbf{r}_{\text{LO}} - j\omega_{\text{LO}}t)$$

becomes

$$\mathbf{E}_{\text{LO}} = E_{\text{LO}} \exp(k_{\text{LO}}\mathbf{r}_{\text{LO}}). \tag{6.73}$$

So, when they interfere, we have Eq. (6.71), repeated here:

$$I = real[2E_{\text{sig}}E_{\text{LO}} \exp(k_{\text{sig}}\mathbf{r}_{\text{sig}} - k_{\text{LO}}\mathbf{r}_{\text{LO}})].$$

The electrical power measured in a spatial heterodyne receiver is also linearly proportional to the optical power received because only the terms with signal times LO are at the IF. This is identical to temporal heterodyne detection from a SNR point of view, so the same SNR equation [Eq. (6.64)] is appropriate and is repeated here:

$$SNR = \frac{\langle i_s^2 \rangle}{\langle i_n^2 \rangle} = \frac{\eta_h G^2 \mathfrak{R}^2 P_{LO} P_s}{eBG^2 F \mathfrak{R} [P_s + P_{LO} + P_{bk} + P_{dkb}] + 2eB i_{dks} + 4ktB/R_L}.$$

If necessary, the LO can be increased to dominate other noises. (Figure 6.29 shows graphically the relative amplitudes when using a high-power LO.) Recently, people have used APDs with heterodyne detection, eliminating the need to use a high-power LO. Once noises have been minimized either by a high-power LO or an APD, the ideal SNR has been reached, so a strong LO is not needed.

Once again, the SNR is directly proportional to the power received; therefore, the number of received photons in this case is given in Eq. (6.65), which is repeated here:

$$SNR = \frac{\eta_D \eta_h P_s}{h\nu B}.$$

6.5.6 Receivers for coherent LiDARs

For a coherent LiDAR, the main method of reducing noise is usually by using a high-power LO to beat with the return signal. If P_{LO} becomes large, the other terms in the denominator of the SNR equation are not relevant. Therefore, in general, gain, such as an APD, has not been used for coherent detection, although it can increase sensitivity. Radar researchers will need to get used to the fact that the signal and the LO beat against each other by each illuminating the detector, rather than after detection, but otherwise coherent LiDAR detection will be similar to radar detection. The beat frequency for temporal heterodyne, where the LO and the signal are spatially aligned but usually offset in frequency, is shown in Fig. 6.29. For a high-power LO to reduce noise, one of the important receiver characteristics is that it is AC coupled to the receiver. If the receiver is DC coupled and the LO is large, the dynamic range of the receiver will be significantly reduced. The figure shows a small AC signal on top of a large DC bias, similar to the high-power–LO situation. If a high gain receiver is used, from a noise mitigation point of view, a high-power LO is not required. In a large array, this can be important when reducing total required LO power.

6.5.6.1 Acousto-optic frequency shifting

Acousto-optical (AO) devices are commonly used to offset the LO from the emitted signal in temporal heterodyne LiDARs. AO devices can beat the EM signal against a diffraction grating set up by the acoustic signal. We can also

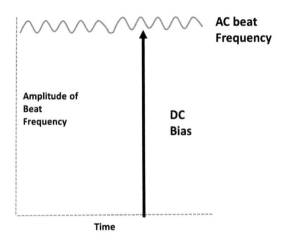

Figure 6.29   The effect of a high-power LO.

think of this as phonons beating against photons, which can shift the frequency and associated wavelength of an optical signal. We stated that for temporal heterodyne LiDAR we would like to have an offset in frequency so that we can determine the direction of the object velocity by seeing whether the Doppler shift is positive or negative.

An acousto-optic modulator (AOM) (see Fig. 6.30) consists of a piezoelectric device that creates sound waves in a material such as glass or quartz. A light beam is diffracted into several orders. By vibrating the material with a pure sinusoid and tilting the AOM such that the light is reflected from the flat sound waves into the first diffraction order, up to 90%

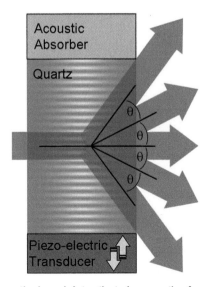

Figure 6.30   An acousto-optical modulator that changes the frequency of the LO (adapted from Ref. 49).

deflection efficiency can be achieved. An AO device can deflect light, but it also can be used to shift the frequency of the light. Light is deflected by

$$\sin \vartheta = \frac{m\lambda}{2\Lambda}, \tag{6.74}$$

where λ is the wavelength of the EM wave, Λ is the wavelength of the acoustic wave, and m is the order of the deflection, which in a thin crystal is ± 1.0. For heterodyne detection, we need to use the AO to frequency shift the beam without deflecting it, or we need to compensate for the beam deflection such that the LO and return signal both hit the target and overlap with high efficiency.

6.5.7 Geiger-mode APDs for coherent imaging

Existing GMAPDs are framing cameras operating at up to 186 kHz frame rate for the 32×32 pixel format, and operating at up to 105 kHz for the 32×128 pixel format; however, a GMAPD camera that became available at the end of 2018 provides asynchronous read out. Each detector fires and is independently reset as soon as possible. The dead time is 400 ns to 1 μs. Ideally, the dead time will be 400 ns, but to be safe, we will count on no more than 1 μs. That means that each detector can continuously respond at a 1- to 2.5-MHz frame rate. This FPA will have 32×32 detectors. As we group detectors together, it will be possible to respond at a higher rates, but with fewer effective detectors. For example, a 4×4 superpixel array of detectors could respond at a 16- to 40-MHz frame rate. The full detector array could respond at 0.9 GHz and is limited by the readout electronics, not the dead time per detector. If it was limited by detector dead time, then its response would be 1–2.5 MHz. Then it would, however, be acting as a single superpixel.

Geiger-mode APDs are ideal both due to their sensitivity, resulting in only needing a low-power LO, and because they inherently capture a full waveform. They have a simple digital readout circuit. The main drawbacks to using GMAPDs for temporal heterodyne coherent imaging are background noise/backscatter and narrow dynamic range. Usually background noise/backscatter is not an issue with coherent imaging, but it will be for GMAPDs. A narrow filter will be required, and applications with high background or backscatter, as well as applications with a highly variable scene, will be problematic. A 1-nm filter, which is very narrow, is 126 GHz at a carrier wavelength of 1550 nm. With a 100-MHz to (maybe) 2- or 3-GHz detector, we normally would not bother with a filter; however, when using GMAPDs, photon blocking is an issue.

With these cameras, we have to use a low-power LO to perform temporal heterodyne sensing,[32] with the LO and the return signal having about the same amplitude. The beat between them modulates the probability of an avalanche. A disadvantage of the weak-LO approach is that the signal amplitude and the LO amplitude need to be approximately matched. GMAPD coherent LiDAR will have a narrow dynamic range. A method will need to be set up to make sure that the LO matches the return signal in amplitude. This could be an issue for scenes with significant differences in reflectivity. Often foliage has a higher reflectivity than manmade objects. A weak LO does not increase the sensitivity of the detector, but in the case of the GMAPD, the cameras are already very sensitive.

We can obtain a commercial 32×32 pixel framing camera. There is also an ITAR-restricted 32×32 pixel asynchronous readout camera that can frame at a higher rate. The next step would be to increase that to a 128×128 pixel asynchronous readout GMAPD camera, or some other larger-format camera; even a 128×32 pixel asynchronous readout camera would be a significant advance. In addition to working to reduce dead time, this increase in camera format would make a nearly ideal GMAPD for flash imaging.

More pulses will be required for coherent detection, for which we need to determine a pattern of pulse positions. Figure 6.31 shows the pulse pattern we would like to detect. The approach to coherent detection with GMAPDs involves setting the LO to about the same intensity as the return signal, and watching the avalanche occurrence rate. When the peaks of the beat frequency occur, we have a higher probability of detection. The probability of detection varies as the intensity of the changes. Figure 6.31 is just a representation showing higher probability where we have the peaks in intensity. There will be a certain constant background probability due to dark current, as well as background radiation, possibly from the sun. The laser returns that beat against the LO will be a varying signal we wish to detect compared to the approximately constant background. We will see an IF return that is the sum of the beat between the LO and the return signal, and the beat between the LO and the constant background signal. We need to distinguish the beat against the signal from the beat against the constant background. Of course, the background will vary statistically.

While direct detection uses coincidence detection, for coherent detection, we need to detect the pattern of more-dense and less-dense returns. Let us

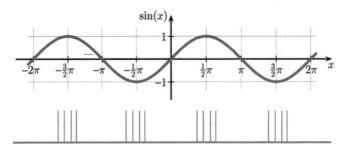

Figure 6.31 Pulse position modulation detection (top graph reprinted from Wikipedia page on sine curve under GNU Free Documentation License).

assume that to detect the pattern we will need at least 10 avalanches. If we design for 0.4 photons returned, we have a 10% probability of detection, and we would like to detect 10 pulses. That means that we will need to transmit 100 pulses. For the case of 0.8 photons, we have a probability of 20%, so we would need 50 pulses. The above assumes a 25% QE, but GMAPDs often have a somewhat higher QE. Of course, we may need 20 avalanches to detect the pattern instead of 10. Figure 6.31 shows a simple sine wave pattern, but the pattern may be more complex. In the case with 0.4 photons returned, for direct detection, we will need 42 pulses. For coherent detection, we will need 2–3 times as many pulses.

6.5.8 A p–i–n diode or LMAPD for coherent imaging

As an alternative to a GMAPD camera, we can use a p–i–n diode with no amplification or a LMAPD. There are two basic types of LMAPDs. There are HgCdTe LMAPDs, which have $k = 0$, where k is the ratio of holes to electrons generated during the avalanche. With $k = 0$, the excess noise generated by an FPA is much less then with a value of k above zero, and avalanche breakdown does occur. HgCdTe cameras have low excess noise, about $F = 1.3$, as a result of $k = 0$. This is actually lower than the McIntyre theory, which would place the excess noise factor around 2. Excess noise is noise added due to the amplification process. When you only amplify the number of electrons, not the holes, you can keep excess noise low. The disadvantages of $k = 0$ APDs are that they need a physically larger cooler than a TE cooler because they operate near 100 K, and they are not commercial available, so cost more. On the positive side, they can have a gain of 1000 or more, so are sensitive and use a lower-power LO.

One of the issues with LMAPDs is you need to develop a fast ROIC to read out the signal and record it behind each detector. While you can do better than that it is likely you might have on the order of 100-MHz bandwidth. In that case, you might want to do linearly frequency–modulated (LFM)-based stretch processing.

When measuring phase and amplitude, Fig. 6.32 can help explain the measurement related to noise. The circles represent a noise distribution in intensity and phase with a unit intensity of 1. Figure 6.32(a) has a signal of 1 unit as well. In reality, the noise intensity varies because it is a probability distribution, but we can ignore that. We have picked a 1-unit intensity signal at a 45-deg angle in phase. What we will actually measure is a value on the circle, so we can see that the intensity could vary from 0 to 2, and the phase could be significantly away from 45 deg. If instead we had an intensity signal of 3 units, still at 45 deg, we would have the situation depicted in Fig. 6.32(b). Now the intensity we will measure could vary from 2 to 4, and we can see that the uncertainty in the phase is significantly reduced. Each time we take a measurement, the noise vector could be any in direction around the circle.

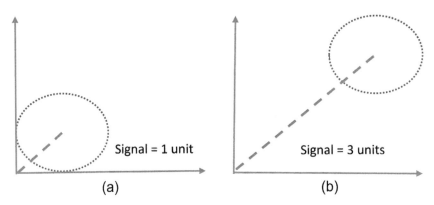

Figure 6.32 Measuring phase and amplitude in the presence of noise. The intensity signal is depicted with a phase of 45 deg by the dashed line: (a) signal is 1 unit and (b) signal is 3 units.

Ignoring noise intensity variation, we can see that multiple samples will increase our ability to measure the intensity and phase of the signal.

6.5.9 Sampling associated with temporal heterodyne sensing

For heterodyne sensing, we need to be able to sample the IF waveform. Therefore, for temporal heterodyne sensing, we need multiple samples across the wavelength of the IF signal. For example, let's assume that the IF is 100 MHz. This means that a sine wave will be 20-ns long on return because of the two-way path of a signal bouncing off of a target. To accurately sample that sine wave, we would like 4 or 5 samples within 20 ns. We therefore need a laser repetition rate and a camera sample rate of no more than 4 or 5 ns. Accurate measurements will require a train of samples at least 10 samples long, but preferably 20 samples long, so would need an 80- to 100-ns–long sampling interval.

6.6 Long–Frame-Time Framing Detectors for LiDAR

Two forms of LiDAR can use long–integration-time framing cameras. One is optical time of flight (OTOF) LiDAR, a polarization-based LiDAR in which a Pockels cell is used to provide timing information to measure range. The second is spatial heterodyne LiDAR, a form of digital holography, in which spatial phase is measured using an angularly offset LO at the same frequency as the return signal, as shown in Fig. 6.24. These applications use visible CCD or CMOS cameras, or they can use near-IR, MWIR, or LWIR cameras, sometimes called FLIRs, or forward-looking infrared cameras. 2D cameras for LiDAR will probably be in the visible spectral region or the near-IR region at about 1.0 or 1.5 μm. These cameras usually frame at 30 or 60 Hz. They do not have any high-speed circuitry like most LiDAR cameras, but are very useful for OTOF and spatial heterodyne LiDAR.

One of the major benefits of these forms of LiDAR is the ability to use these widespread, affordable cameras with large format sizes. In the visible spectral regime, you can buy ten-megapixel cameras for hundreds of dollars. In the near-IR region, cameras are more expensive, but still inexpensive and with larger formats compared to high-bandwidth cameras. Megapixel, near-IR framing cameras are available. Because of their large format size and due to the limited emphasis on single-photon sensitivity, it is likely that these LiDARs will be short range in order to have enough laser power to illuminate the large-format image. The visible cameras are silicon and are very widespread. The near-IR cameras are likely InGaAs, although they could be HgCdTe. The main disadvantage of HgCdTe is that it needs to be cooled.

6.7 Ghost LiDARs

Ghost-imaging LiDAR is a modality requiring comparison of the information from two channels (like imaging with coherent illumination, such as holography or optical coherence tomography).[33] The idea is to reconstruct a fine-resolution image using phase retrieval. This process was described in 1990.[34] An array of point-size detectors samples the intensity of the optical field in the large-aperture plane. A beamsplitter after the small-aperture telescope allows for the detection of intensity simultaneously in two planes [Fig. 6.33(a)]: the usual focal plane, where there exists a diffraction-limited image of the object, and a Fourier transform of its small-aperture plane. Figure 6.33(b) is a block diagram depicting how these three intensity measurements are used to retrieve the phase over the large aperture and reconstruct a fine-resolution image.

The term ghost imaging, which was coined soon after the initial experiments, emphasizes the fact that neither photocurrent alone is sufficient to derive the target image; only by correlating the two photocurrents do you have enough signal to make an image. Due to an intensive search by theoreticians for the best solution, ghost imaging was called "a theme with variations."[35]

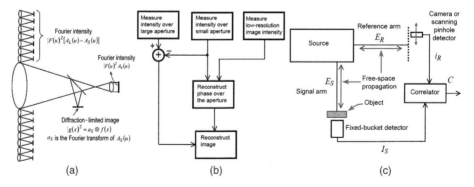

Figure 6.33 (a) Principle of phase retrieval using a low-resolution image; (b) reconstruction of a fine-resolution image from the intensity measurements; (c) ghost-imaging architecture (reprinted from Ref. 33).

Many ghost-imaging experiments have been reported—both for transmission[36–39] and standoff reflection.[40–42] These experiments suggest using two correlated optical beams traversing distinct paths and impinging on two spatially separated photodetectors: one beam interacts with the target and then illuminates a single-pixel detector that provides no spatial resolution (reference channel), whereas the other beam traverses an independent path and impinges on a scanning pinhole detector or a high-resolution camera (signal channel) without any interaction with the target. The image is obtained by correlating the output photocurrents from these photodetectors. Figure 6.33(c) shows an example of the architecture for the case of image transmission.

Subsequent theoretical work addressed the impact of atmospheric turbulence[43] and turbulence combined with laser speckle[44] on ghost-imaging performance. These studies and related experimental work[41] delineate the conditions under which ghost imaging's spatial resolution becomes limited by turbulence, rather than by diffraction.

A conventional LiDAR that flood illuminates a FOV has its spatial resolution's turbulence limit set by the atmospheric coherence length in the receiver's entrance pupil. On the other hand, a ghost imager for the same environment has its spatial resolution's turbulence limit set by the atmospheric coherence length in its structured-illuminator's exit pupil.[42] Therefore, for a bistatic configuration, when the atmospheric coherence length in the transmitter's exit pupil is significantly larger than the one in the receiver's entrance pupil, a ghost imager would yield better turbulence-limited spatial resolution compared to a conventional LiDAR.

6.8 LiDAR Image Stabilization

Did you ever try to take a picture with high magnification and no image stabilization? You need a steady hand to hold the camera, or else or your picture will be blurred. This is similar to the angle/angle portion of a LiDAR image. One difference is that most LiDAR images are captured in nanoseconds, so image blur only occurs when pixels captured at different times are stitched together. You will notice less blur in your camera as the shutter speed is increased. A typical "shutter speed" for a LiDAR will be nanoseconds. The main issue with LiDAR stabilization has to do with stitching together pixels captured at different times. In order to stitch together the returns from images taken at different times, you need to know exactly where each image was taken to a fraction of that image size. If the snapshot at a given time is a single pixel, you must stabilize to a fraction of the angular extent of one pixel. If the snapshot at a given time is a snapshot of many pixels, you only have to stabilize to a fraction of the angular extent of that snapshot.

You can see from Fig. 6.34 that, if you compare imaging 1 pixel at a time to imaging a 6 × 6 array of pixels, the array case need not be stabilized to the

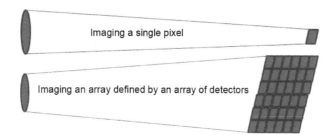

Figure 6.34 Comparison of scanning detector (top) and flash detector (bottom) options (reprinted from Ref. 49).

same level. If the full image is taken on one snapshot, no stabilization is required except for the ability to be pointed at the target when the LiDAR image is taken. One of the major benefits of flash imaging with LiDAR is the release in stabilization requirements. Your camera takes the full image at one instant but has a longer shutter time, so things may move during the detector integration time. LiDAR imaging stops this motion.

6.9 Optical–Time-of-Flight Flash LiDAR

LiDAR traditionally has focused most heavily on using high-bandwidth, time-sensitive detectors. An alternative is to use integrating sensor arrays with no time sensitivity and to modulate the received light in time, thereby mapping time onto intensity. The Air Force looked into this in the early 1990s before any high-bandwidth FPAs were available, with a concept at the time called LIMARS (Laser Imaging and Ranging System).[45–48] A diagram of this polarization-based concept to measure range resolution is provided in Fig. 6.35. Two companies, TetraVue and General Atomics, have pursued this form of 3D imaging. TetraVue now has investment from Continental and is pursuing this

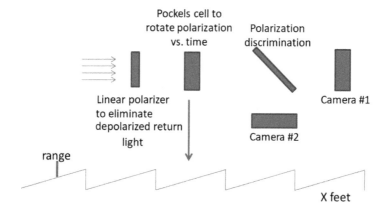

Figure 6.35 OTOF camera layout (reprinted from Ref. 49).

LiDAR approach for auto LiDAR. The biggest benefit is that this form of imaging can leverage commercial, high-resolution CMOS and CCD arrays for 2D digital imaging at high resolution and high precision. These commercial FPAs are very capable in the visible spectral region and are being developed for the near-IR region. These arrays do not have any gain; however, due to their maturity level, they have reduced noise and are relatively sensitive. The major need for gain is to compete against noise, so if we reduce noise, we will not need as much gain.

In this approach, an illumination pulse fills the scene, and the camera lens focuses some part of the reflected signal onto the digital imager through a modulator. The single modulator modulates the entire optical field simultaneously, causing a change in transmitted light for each pixel of the integrating CMOS or CCD array as a function of when the light incident on that pixel passed through the modulator. A second array can be used to obtain the relative reflectivity of the surface so that the derived range is independent of surface reflectivity. In this way, the traditional problem of parallel precision time measurement (with 1-GHz to 100-GHz effective bandwidths) becomes a parallel precision intensity measurement problem, which is where modern CCD and CMOS arrays do well. If two arrays are used, then it is critical to accurately align the two arrays because the ratio of intensity or corresponding pixels in the two arrays is used to measure range.

Assuming accurate alignment between the two arrays, the range precision is determined by the range noise, which is related to the RMS uncertainty in the intensity measurement of each simple pixel. For this reason, grayscale measurements are inherent in this LiDAR modality. For high-precision modalities, the fundamental noise limit is determined by the shot noise of the photon field. For longer-range, lower-precision modalities, the noise limit is the read noise of the sensor. Today, even consumer-grade visible CMOS imagers are achieving read noise levels $<< 10$ e$^-$, and some high-end CMOS sensors approach 1 e$^-$ without external cooling. Since integration times can be short, dark noise is not relevant. The RMS range precision scales as

$$\sigma_R = C \left[\frac{R^2 N_{px}}{D_{tele}^2 \rho E_{illum}} \right]^k \Delta R, \qquad (6.75)$$

where R is the range, N_{px} is the number of pixels in the array, D is the collecting aperture diameter, ρ is the surface reflectivity, E_{illum} is the illuminator pulse energy, and ΔR is the range ambiguity interval. C is a proportionality constant, and k is a constant equaling 1 for read-noise–dominated modalities and equaling 0.5 for shot-noise–dominated modalities.

OTOF LiDAR has advantages such as access to megapixel and larger FPAs for high lateral resolution or high pixel count, simultaneous acquisition of high-bit-depth grayscale imagery, good performance with high-contrast scenes (>100:1) and objects, and millimeter-scale range precision achieved for certain high-signal, small–range-window scenarios. Low-cost imager sensor arrays and electronics are available for $100s or $1000s, especially in the visible and near-IR, and no cooling is required for a 3D camera system. However, OTOF 3D cameras do not detect multiple returns in a single pixel, so mixed pixels give only the average range. Also, the need to use an external modulator such as a Pockels cell can result in more-complex optical designs.

Grayscale imagery is intrinsic to OTOF because of its single-pulse acquisition. The actual available bit depth of the grayscale images for each scenario is determined by ratio of a 100% reflective surface to the dominant noise term (read or shot noise) after binning and some image processing.

6.9.1 Summary of the advantages and disadvantages of OTOF cameras

An OTOF system using low-bandwidth cameras with a medium-bandwidth Pockels cell has the advantage of using commercially available 2D cameras for flash 3D imagery. In the visible or near-IR spectral regions, you can obtain huge-format cameras with tens of megapixels for hundreds of dollars, promising high performance at low cost. Even in the SWIR, you can obtain up to a 1920 × 1080 pixel custom camera, and smaller cameras for as low as $25K, from multiple vendors. These cameras are mature and can have low noise even when operating at room temperature. They are sensitive even without gain. The main disadvantage is that you need a Pockels cell, which is costly. A secondary disadvantage is that if you use two cameras, you must align them carefully. The OTOF cameras show low energy use for 3D mapping with grayscale. That said, it is likely that some of the other sensing modalities will be able to adopt some of the noise-reduction techniques being employed. The OTOF LiDAR has the advantage of using commercially available cameras built for a large commercial market.

If you want high range resolution, you are currently best off using either a GMAPD array or an OTOF imager. HgCdTe arrays are currently probably second in bandwidth/range resolution. All LMAPD cameras have similar challenges in high range resolution.

The range precision for these scenarios is not a challenge for GMAPDs and could be significantly better. GMAPDs typically have an advantage over LMAPDs in terms of inherent timing precision when detecting isolated optical pulses because the current pulses generated by breakdown of a GMAPD pixel are stronger and are of much more uniform amplitude than the current pulses emitted by an LMAPD pixel in response to weak signals. The mean response of an LMAPD pixel to an ensemble of identically prepared input optical pulses is proportional to the optical signal strength, which enables direct measurement of signal amplitude. However, an LMAPD's response to such an ensemble of identical input signals varies stochastically around the mean, limiting the accuracy of a single amplitude measurement and affecting the timing of when the signal crosses the detection threshold of a decision circuit. When using LMAPD pixels, timing jitter is large if the APD's response barely exceeds the detection threshold, and range precision improves for stronger signal returns. Consequently, scenarios that prioritize the best range precision with the least transmitted energy tend to favor GMAPD detectors, whereas scenarios that require penetrating obscurants or collecting reflectance information in a single observation (for instance to 'freeze' a dynamic scene) tend to favor LMAPDs. In this chapter, we have attempted to select scenarios that straddle these respective areas of strength and weakness, but these general characteristics should be kept in mind when considering specific applications.

Problems and Solutions

6-1. A detector has a mean dark count rate of 2×10^4 counts/s. A constant signal optical power of 5×10^{-16} W is incident on this detector. Assume that the QE is 0.6 and that the mean wavelength of the incident light is 1550 nm.

(a) What is the value of the dark current associated with the dark counts?
(b) Calculated the responsivity.
(c) What is the value of the shot-noise current associated with the dark counts and the signal?

6.1 Solution:

(a) We have 2×10^4 counts/s of dark count; therefore, the dark current will be 2×10^4 times the charge of the electron; dark current = dark count $\times$ e = $<I_{dark}>$.

$$B = 2 \times 10^4 \text{ counts/s},$$

$$<I_{dark}> = 2 \times 10^4 \frac{\text{counts}}{\text{s}} \times 1.6\text{E}{-}19 = 3.2\text{E}{-}15 \text{ amps.}$$

(b) The responsivity is given by

$$QE = \frac{R}{\lambda} \times \frac{hc}{e} = 0.6,$$

$$R = \frac{0.6\lambda e}{hc} = \frac{(0.6)(1.550 \times 10^{-6}\text{m})(1.6 \times 10^{-19}\text{A s})}{(6.626 \times 10^{-34}\text{J s})(3 \times 10^8 \text{m/s})} = 0.7481 \text{ A/W.}$$

(c) Total noise power is an integral of the power spectrum over the electrical bandwidth; in Schottky's formula, we have for shot noise (SN),

$$\overline{i_{SN}^2} = 2e\langle I \rangle B.$$

We can suppose that $2B = 1$. Therefore, $e{<}I{>} = \sigma^2_{\text{Shot Noise}}$, where e is the charge of an electron, B is the bandwidth, and $\langle I \rangle = e \times \text{rate} = e\frac{\eta P}{h\nu}$. Therefore, for the dark current and signal, we have

$$(2B)e{<}I_{\text{signal}}{>} = e{<}I_{\text{signal}}{>} = e \cdot e\frac{\eta P}{h\nu} = \sigma^2_{\text{Shot Noise, Signal}},$$

$$(2B)e{<}I_{\text{dark}}{>} = e{<}I_{\text{dark}}{>} = \sigma^2_{\text{Shot Noise, Dark}},$$

where $P = 5 \times 10^{-16}$W.

When $G = 1$ and $F = 1$,

$$\langle i^2_{\text{shot sig}} \rangle = 2eG^2FRP_sB$$
$$= 2(1.6 \times 10^{-19}\,\text{A s})(0.7481\,\text{A/W})(2.57 \times 10^{-15}\,\text{W})(2 \times 10^4\,\text{counts/s})$$
$$= 2.3939 \times 10^{-30}\,\text{A}^2.$$

6-2. For the detector in Problem 6-1,

(a) What is the value of the total shot noise of the current out of this detector?
(b) What is the SNR?

6-2 Solution:

As seen in Problem 6-1, for the dark current and signal, we have

$$(2B)e{<}I_{\text{signal}}{>} = e{<}I_{\text{signal}}{>} = e \cdot e\frac{\eta P}{h\nu} = \sigma^2_{\text{Shot Noise, Signal}},$$

$$(2B)e{<}I_{\text{dark}}{>} = e{<}I_{\text{dark}}{>} = \sigma^2_{\text{Shot Noise, Dark}},$$

where η is quantum efficiency, P is power of the signal, h is Plank's constant, and ν is frequency. Therefore,

(a) $\sigma^2_{\text{Shot Noise}} = \sigma^2_{\text{Shot Noise, Signal}} + \sigma^2_{\text{Shot Noise Dark}} = e({<}I_{\text{Signal}}{>} + {<}I_{\text{Dark}}{>}),$

$$\langle i^2_{\text{dk}} \rangle + \langle i^2_{\text{shot sig}} \rangle = 1.47 \times 10^{-29}\text{A}^2,$$

(b) $SNR = (\sigma^2_{\text{Shot Noise}})^{0.5} = (\sigma^2_{\text{Shot Noise, Signal}} + \sigma^2_{\text{Shot Noise Dark}})^{0.5},$

$$\langle i^2_{\text{sig}} \rangle = (0.7481\,\text{A/W})^2(5 \times 10^{-16}\,\text{W})^2 = 1.3991 \times 10^{-31}\,\text{A}^2,$$

$$SNR = \frac{\langle i^2_{\text{sig}} \rangle}{\langle i^2_{\text{dk}} \rangle + \langle i^2_{\text{shot sig}} \rangle} = \frac{1.3991 \times 10^{-31}\text{A}^2}{1.47 \times 10^{-29}\text{A}^2} = 0.0095.$$

6-3. A noisy detector has a mean dark count rate of 1×10^3 counts/s. A constant signal optical power of 8×10^{-16} W is incident on this detector, and the signal is gathered over 30 ms. Assume that the QE is 0.5 and that the mean wavelength of the incident light is 1 μm.

(a) Plot the PDF of the dark counts and find a threshold such that the probability of false alarm is 5%.

(b) Plot the PDF of the signal + dark counts. Find the probability of detection using the threshold defined in part (a).

6-3 Solution:

(a) We assume that we have only the shot noise and dark noise. As mentioned, the mean dark count rate of the noisy detector is 1000 counts/s, and the constant signal power is $P = 8 \times 10^{-16}$ W. Additionally, $T = 30 \times 10^{-3}$ s, and $\eta = 0.5$, $\lambda = 1 \times 10^{-6}$ m $\rightarrow \nu = c/\lambda = 3 \times 10^{14}$ Hz.

For dark counts, $\lambda T = 30 = \sigma_n^2 = <k_n>$ because noise is assumed to be Poisson distributed. However, $<k_n>$ is high enough to be very close to a normal distribution.

Because $\text{PDF}_{\text{dark noise}}$ is Poisson distributed, it equals

$$\frac{(\lambda T)^k}{k!} e^{-\lambda T} = \frac{\langle k_n \rangle^k}{k!} e^{-\langle k_n \rangle} = \frac{30^k}{k!} e^{-30} \approx \frac{1}{\sqrt{2\pi\sigma_n^2}} e^{-\frac{(k-\langle k_n \rangle)^2}{2\sigma_n^2}} = \frac{1}{\sqrt{60\pi}} e^{-\frac{(k-30)^2}{60}},$$

which is very close to a normal distribution.

$$P_{\text{false alarm}} = \int_{\text{threshold}}^{\infty} \text{PDF}_{\text{dark noise}} dk = \int_{\text{threshold}}^{\infty} \frac{1}{\sqrt{60\pi}} e^{-\frac{(k-30)^2}{60}} dk = 5\% \rightarrow$$

threshold $= 39$.

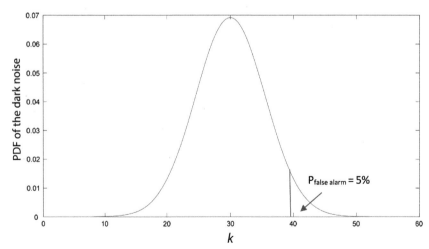

(b) We have a constant signal with $P = 8 \times 10^{-16}$ W on the detector. This means that only the signal-induced noise is shot noise. We have

$$\lambda_s = \frac{\eta P}{h\nu} = \frac{0.5 \times 8 \times 10^{-16}\,\mathrm{J/s}}{6.626 \times 10^{-34} \times 3 \times 10^{14}\,\mathrm{J}} = 201.23\,\mathrm{counts/s}$$

$$\rightarrow \langle k_s \rangle = \lambda_s T = 201.23 \times 30 \times 10^{-3} = 6.04 \approx 6 \rightarrow \langle k_s \rangle = \sigma_s^2 = 6.$$

Because the shot noise is Poisson distributed too, the PDF of the signal is

$$PDF_{\mathrm{signal}} = \frac{(\lambda_s T)^k}{k!} e^{-\lambda_s T} = \frac{\langle k_s \rangle^k}{k!} e^{-\langle k_s \rangle} = \frac{6^k}{k!} e^{-6}.$$

As seen, both the noise and the signal are Poisson distributed. Therefore, their convolution is Poisson distributed, too. We have

$$PDF_{\mathrm{s+n}} = PDF_s \times PDF_n = \frac{36^k}{k!} e^{-36}.$$

Therefore, $\langle k_{\mathrm{s+n}} \rangle$ is 36 and is high enough to be considered as normally distributed:

$$PDF_{\mathrm{s+n}} = \frac{36^k}{k!} e^{-36} \simeq \frac{1}{\sqrt{62\pi}} e^{-\frac{(k-36)^2}{62}}.$$

Therefore, the probability of detection will be

$$P_{\mathrm{detection}} = \int_{\mathrm{threshold}}^{\infty} PDF_{\mathrm{s+n}} dk = \int_{\mathrm{threshold}}^{\infty} \frac{1}{\sqrt{62\pi}} e^{-\frac{(k-36)^2}{62}} dk = 31\%.$$

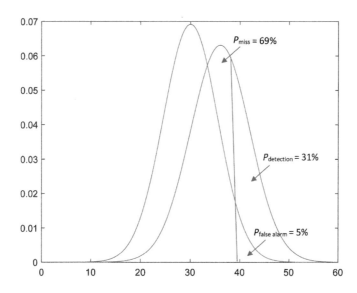

6-4. In Problem 6-3, what can we do to increase the probability of detection? List at least four methods we could use.

6-4 Solution:

As seen in Problem 6-3, although the signal is constant, the probability of detection is still low. This is because the detector has high dark noise, the power of the incident signal is very small, and the exposure time is short, too. To improve the detection probability, we can do the following:

1. Change the detector to one having a higher QE and a lower dark noise.
2. Increase the exposure time of the signal.
3. Increase the incident signal power.
4. Use a lower frequency or a longer wavelength so that a fixed power has more photons.
5. Average several pulses on the detector instead of only one pulse.

6-5. For shot-noise–limited detection, derive an expression for the SNR associated with shot noise in terms of bandwidth, power, quantum efficiency, and frequency. Then show that SNR is related to the number of detected photons. If 100 photons are detected per integration period, what would the SNR be?

6-5 Solution:

Mean shot noise current is calculated as

$$\langle I \rangle = e \times \text{rate} = e \frac{\eta P}{h\nu}.$$

Schottky's formula says that $\sigma_I^2 = 2e\langle I \rangle B$.

Therefore, the SNR would be

$$SNR = \sqrt{\frac{\eta P}{h\nu 2B}} = \sqrt{\frac{\langle k \rangle}{T} \frac{1}{\frac{1}{T}}} = \sqrt{\langle k \rangle}.$$

For $\langle k \rangle = 100$, $SNR = \sqrt{100} = 10$.

6-6. We are interested in developing a laser rangefinder. The goal is to detect the presence of the scattered return pulse energy. Since we want the rangefinder to be inexpensive, we will be using a noisy detector. Assume that the detector is limited by thermal (Johnson) noise, which is zero mean, Gaussian distributed with a variance of 10 photoelectrons.

(a) Plot the PDF of the noise.
(b) Find a threshold such that that the probability of false alarm is less than 0.05.

6-6 Solution:

(a) We have $\sigma_n^2 = 10$, $\langle x_n \rangle = 0$, and the noise is Gaussian distributed. Therefore, the PDF of the noise will be

$$PDF_{noise} = \frac{1}{\sqrt{20\pi}} e^{-\frac{x_n^2}{20}}.$$

The plot will be as follows:

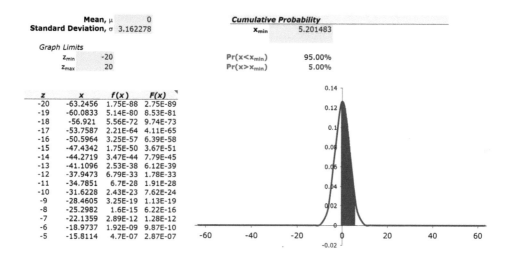

	Mean, μ		0
Standard Deviation, σ			3.162278

Graph Limits		
z_{min}		-20
z_{max}		20

	Cumulative Probability	
	x_{min}	5.201483
$Pr(x<x_{min})$		95.00%
$Pr(x>x_{min})$		5.00%

z	x	f(x)	F(x)
-20	-63.2456	1.75E-88	2.75E-89
-19	-60.0833	5.14E-80	8.53E-81
-18	-56.921	5.56E-72	9.74E-73
-17	-53.7587	2.21E-64	4.11E-65
-16	-50.5964	3.25E-57	6.39E-58
-15	-47.4342	1.75E-50	3.67E-51
-14	-44.2719	3.47E-44	7.79E-45
-13	-41.1096	2.53E-38	6.12E-39
-12	-37.9473	6.79E-33	1.78E-33
-11	-34.7851	6.7E-28	1.91E-28
-10	-31.6228	2.43E-23	7.62E-24
-9	-28.4605	3.25E-19	1.13E-19
-8	-25.2982	1.6E-15	6.22E-16
-7	-22.1359	2.89E-12	1.28E-12
-6	-18.9737	1.92E-09	9.87E-10
-5	-15.8114	4.7E-07	2.87E-07

(b) As shown in the plot,

$$P_{false\ alarm} = \int_{threshold}^{\infty} PDF_{noise} dx = \int_{threshold}^{\infty} \frac{1}{\sqrt{20\pi}} e^{-\frac{x_n^2}{20}} dx = 5\%.$$

Threshold $= 5.201683 \approx 5$.

6-7. In Problem 6-6, assume that the laser pulse comes from a stable laser and that the reflecting surface is specular such that only signal-induced noise is shot noise. The mean value of the number of photoelectrons generated by the return pulse is 10.

(a) Plot the PDF of the number of photoelectrons due to the signal.
(b) Plot the joint PDF of the number of photoelectrons of the signal + detector noise (assume that the signal and noise are independent).
(c) Determine the probability of detection and the probability of missed detections of the returned pulse using the threshold defined in part (b) of Problem 6-6.
(d) Suppose that the requirement is to achieve a probability of detection of 99% while only doubling the amount of power being used. This can be done in two ways. First, the pulse energy could be increased by a factor

of 2. Second, two pulses could be used. Which would you choose? Why? (Be quantitative.)

6-7 Solution:

(a) The PDF of the photoelectrons due to the signal is PDF_s:

$$PDF_s = \frac{\langle k \rangle^k}{k!} e^{-\langle k \rangle} \text{ and } \langle k \rangle = \sigma^2 = 10 \rightarrow PDF_s = \frac{10^k}{k!} e^{-10}.$$

We may suppose that the PDF of the signal is almost close to a Gaussian distribution because $\langle k \rangle$ or σ^2 is sufficiently high. Therefore,

$$PDF_s = \frac{10^k}{k!} e^{-10} \approx \frac{1}{\sqrt{20\pi}} e^{-\frac{(x-10)^2}{20}}.$$

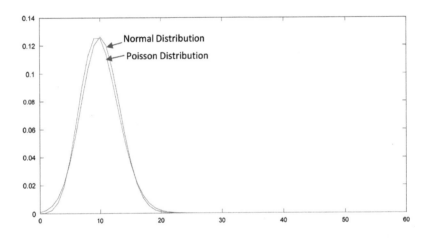

(b) For the signal, we had $PDF_s \approx \frac{1}{\sqrt{20\pi}} e^{-\frac{(x-10)^2}{20}}$.

We have two noise sources: the shot noise associated with the signal and the noise of the detector. We have

$$PDF_n \approx \frac{1}{\sqrt{20\pi}} e^{-\frac{x^2}{20}}.$$

Since the signal and noise are independent, the PDF of the number of photoelectrons of the signal + detector noise PDF_{n+s} is the convolution of PDF_s and PDF_n:

$$PDF_{s+n} = PDF_s \times PDF_n = \frac{1}{\sqrt{40\pi}} e^{-\frac{(x-10)^2}{40}}.$$

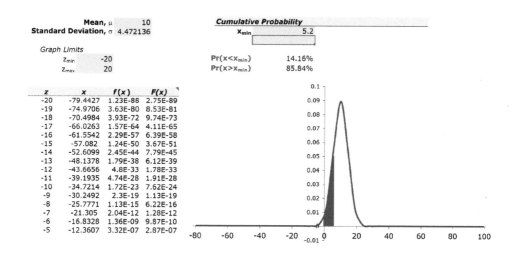

(c) Probability of false alarm:

$$P_{\text{false alarm}} = \int_{\text{threshold}}^{\infty} PDF_{\text{noise}}dx = \int_{\text{threshold}}^{\infty} \frac{1}{\sqrt{20\pi}} e^{-\frac{x^2}{20}}dx = 5\% \rightarrow$$

Threshold $= 5.2$.

$$P_{\text{detection}} = \int_{\text{threshold}}^{\infty} PDF_{\text{s+n}}dx = \int_{5.2}^{\infty} \frac{1}{\sqrt{40\pi}} e^{-\frac{(x-10)^2}{40}}dx = 85.8\%.$$

Probability of missed detection $= P_{\text{miss}} = 1 - P_{\text{detection}} = 14.2\%$.

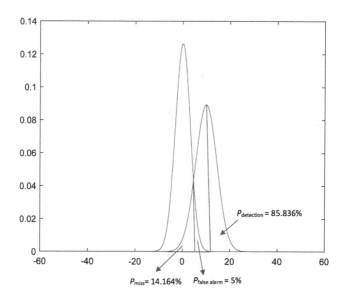

(d) To reach the detection probability of 99%, we can use the following two approaches, both of which are examined:

Approach 1: Using two pulses

Instead of having one pulse, we can have two pulses (1→2). Since we have two independent pulses, we will have the following probabilities:

Pulse 1	Pulse 2
Detected	Detected
Detected	Missed
Missed	Detected
Missed	Missed

Therefore, the probability of detection will be $P_{detection} = 1 - P_{missed} = 98\%$, which is smaller than 99%. This means that even with two pulses, we cannot reach a detection probability of 99%.

Approach 2: Increasing the pulse energy two times

We can have a pulse energy that is two times larger than another option. Therefore, we will have $\langle k \rangle = 20$ instead of 10. By repeating the calculation above, the probability of detection will be $P_{detection} > 99\%$. Therefore, to reach a probability of detection of 99%, it would be preferable to have a twice-larger pulse than to have two pulses.

6-8. Plot the Poisson distribution for 65 photons and 4 detections.

6-8 Solution: $q(k,M) = \frac{M^k e^{-M}}{k!}$. Plotting, we have

```
k=1:65
M=4
plot(M.^k.*exp(-M)./factorial(k))
grid on
```

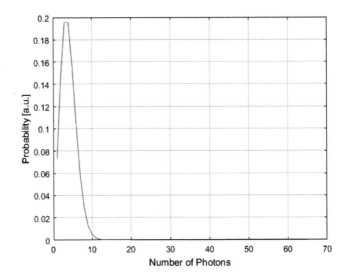

6-9. How many photons will be required per detector for

(a) an InGaAs LMAPD with a gain of 15?
(b) a HgCdTe LMAPD with a gain of 400?

6-9 Solution:

(a)

$$C(R) = 6.916 \times 10^5 \text{ photons} \times F,$$

$$F = M\left[1 - (1 - k)\left(\frac{M - 1}{M}\right)^2\right] = 15\left[1 - (1 - 0.2)\left(\frac{15 - 1}{15}\right)^2\right] = 4.5467,$$

$$C(R) = 3.1445 \times 10^6 \text{ photons}.$$

(b) Since a HgCdTe APD does not obey McIntyre's law and has an approximate value of $F = 1.3$ regardless,

$$C(R) = 8.9908 \times 10^5 \text{ photons}.$$

6-10. How many frames will be required for coincidence detection with a GMAPD having a bit width of 8 and a mean probability of detection of 50%?

6-10 Solution:

$$N_p = \frac{11.2 N_g^2 (1 - P_o)}{P_o}.$$

$N_g = 2^8$
$N_p = \text{ceil}[11.2 * N_g^2 * (1 - 0.5) / 0.5]$
734,004 frames

Notes and References

1. P. Gatt and S. W. Henderson, "Laser radar detection statistics: a comparison of coherent and direct-detection receivers," *Proc. SPIE* **4377**, 251–262 (2001) [doi: 10.1117/12.440113].
2. M. S. Salisbury, P. F. McManamon, and B. D. Duncan, "Optical-fiber preamplifiers for ladar detection and associated measurements for improving the signal-to-noise ratio," *Optical Engineering* **33**(12), 4023–4032 (1994) [doi: 10.1117/12.183406].
3. https://en.wikipedia.org/wiki/Image_intensifier.
4. G. M. Williams, M. Compton, D. A. Ramirez, M. M. Hayat, and A. S. Huntington, "Multi-gain-stage InGaAs avalanche photodiode with enhanced gain and reduced excess noise," *IEEE J. Electronic Devices Society* **1**(2), 54–65 (2013).

5. M. Ren, S. J. Maddox, M. E. Woodson, Y. Chen, S. R. Bank, and J. C. Campbell, "AlInAsSb separate absorption, charge, and multiplication avalanche photodiodes," *Applied Physics Letters* **108**, 191108 (2016).

6. S. J. Maddox, W. Sun, Z. Lu, H. P. Nair, and J. C. Campbell, "Enhanced low-noise gain from InAs avalanche photodiodes with reduced dark current and background doping," *Applied Physics Letters* **101**, 151124 (2012).

7. Voxtel presently offers a prototype 128 × 128 flash LiDAR LMAPD camera with an InGaAs photodiode detector array that is TE cooled. Among others, ASC (recently acquired by Continental) sells 128 × 128 InGaAs LMAPD-based LiDAR cameras.

8. R. J. McIntyre, "Multiplication noise in uniform avalanche photodiodes," *IEEE Trans. Electron. Devices* **ED-13**, 164–168 (1966).

9. R. J. McIntyre, "The distribution of gains in uniformly multiplying avalanche photodiodes: theory," *IEEE Trans. Electron. Devices* **ED-19**(6), 603–613 (1962).

10. S. O. Rice, "Mathematical analysis of random noise," *Bell System Technical Journal* **23**(3), 282–332 (1944) & **24**(1), 46–156 (1945).

11. X. Sun, J. B. Abshire, and J. D. Beck, "HgCdTe eAPD detector arrays with single photon sensitivity for space LiDAR applications," *Proc. SPIE* **9114**, 91140K (2014) [doi: 10.1116/12.2053656].

12. W. Sullivan, J. Beck, R. Scritchfield, M. Skokan, P. Mitra, X. Sun, J. Abshire, D. Carpenter, and B. Lane, "Linear mode photon counting from visible to MWIR with HgCdTe avalanche photodiode focal plane arrays," *Proc. SPIE* **9492**, 94920T (2015) [doi: 10.1117/12.2180394].

13. G. Perrais, S. Derelle, L. Mollard, J.-P. Chamonal, G. Destefanis, G. Vincent, S. Bernhardt, and J. Rothman, "Study of the transit-time limitations of the impulse response in mid-wave infrared HgCdTe avalanche photodiodes," *J. Electronic Materials* **38**(8), 1790–1799 (2009).

14. HDVIP® stands for high-density vertically integrated photodiode.

15. It is assumed that the unit cell is surrounded by other unit cells in a 2D array configuration with a center-to-center spacing defined as the pitch.

16. J. Beck, R. Scritchfield, M. Goodwin, L. Wood, M. Ohlson, M. Skokan, M. Kinch, P. Mitra, C.-F. Wan, J. Robinson, W. Sullivan III, and J. Teherani, "Performance and modeling of the MWIR HgCdTe e-APD," *J. Electonic Materials* **38**(8), 1579–1592 (2009).

17. M. B. R. Marino and R. Spitzberg, "A photon counting 3-D imaging laser radar for advanced discriminating interceptor seekers," in 2nd Annual Interceptor Technology Conference, American Institute of Aeronautics and Astronautics, Washington, D.C. (1993).

18. R. Marino, "Method and apparatus for imaging a scene using a light detector operating in non-linear Geiger-mode," U.S. Patent 5,892,565 (1999).

19. R. Heinrichs, B. F. Aull, R. M. Marino, et al., "Three-dimensional laser radar with APD arrays," *Proc. SPIE* **4377**, 106–117 (2001) [doi: 10.1117/12.440098].

20. M. A. Itzler, X. Jiang, M. Entwistle, K. Slomkowski, A. Tosi, F. Acerbi, F. Zappa, and S. Cova, "Advances in InGaAsP-based avalanche diode single photon detectors," *J. Modern Optics* **58**(3–4), 1–26 (2011).

21. D. G. Fouche, "Detection and false-alarm probabilities for laser radars that use Geiger-mode detectors," *Applied Optics* **42**(27), 5388–5398 (2003).

22. S. Kim, I. Lee, and Y. J. Kwon, "Simulation of a Geiger-mode imaging ladar system for performance assessment," *Sensors* **13**(6), 8461–8489 (2003).

23. M. A. Itzler, M. Entwistle, M. Owens, K. Patel, X. Jiang, K. Slomkowski, and S. Rangwala, "Comparison of 32 × 128 and 32 × 32 Geiger-mode APD FPAs for single photon 3D LADAR imaging," *Proc. SPIE* **8033**, 80330G (2011) [doi: 10.1116/12.884693].

24. E. A. Watson, "New imaging modalities for laser-based systems," *2001 IEEE Aerospace Conference Proceedings* (2001).

25. T. Baba, T. Nagano, A. Ishida, S. Adachi, S. Nakamura, and K. Yamamoto, "Silicon hybrid SPAD with high-NIR-sensitivity for TOF applications," *Proc. SPIE* **10108**, 101080Y (2017) [doi: 10.1117/12.2250165].

26. S. Adachi, T. Baba, T. Nagano, S. Nakamura, and K. Yamamoto, "Development of silicon hybrid SPAD 1D arrays for LIDAR and spectrometer applications," *Proc. SPIE* **10537**, 105371D (2018) [doi: 10.1117/12.2288159].

27. Hamamatsu Photodetectors: Selection Guide, Slawomir Piatek NJIT/Rutgers/Hamamatsu, Slide 34, © 2017 Hamamatsu.

28. J. H. Shapiro, "Heterodyne mixing efficiency for detector arrays," *Applied Optics* **26**(17), 3600–3606 (1987).

29. D. Hogenboom and C. DiMarzio, "Quadrature detection of a Doppler signal," *Applied Optics* **37**(13), 2569 (1998).

30. J.-Y. Lee and D.-C. Su, "Improved common-path optical heterodyne interferometer for measuring small optical rotation angle of chiral medium," *Optics Communications* **256**(4–6), 337–341 (2005).

31. F. Tosco, Ed., *Fiber Optic Communications Handbook*, TAB Professional and Reference Books, Blue Ridge Summit, Pennsylvania, p. 970 (1990).

32. L. A. Jiang and J. X. Luu, "Heterodyne detection with a weak local oscillator," *Applied Optics* **47**(10), 1486–1503 (2008).

33. V. Molebny, P. F. McManamon, O. Steinvall, T. Kobayashi, and W. Chen, "Laser radar: historical prospective—from the East to the West," *Optical Engineering* **56**(3), 031220 (2016) [doi: 10.1117/1.OE.56.3.031220].

34. J. R. Fienup and A. M. Kowalczyk, "Phase retrieval for a complex-valued object by using a low-resolution image," *J. Optical Society of America A* **7**(3), 450–458 (1990).

35. J. H. Shapiro, "Ghost imaging: a theme with variations," *Imaging and Applied Optics 2015, OSA Technical Digest*, Optical Society of America, paper Ct4F.4 (2015).

36. F. Ferri, D. Magatti, A. Gatti, M. Bache, E. Brambilla, and L. A. Lugiato, "High-resolution ghost image and ghost diffraction experiments with thermal light," *Physical Review Letters* **94**, 183602 (2005).

37. G. Scarcelli, V. Berardi, and Y. Shih, "Can two-photon correlation of chaotic light be considered as correlation of intensity fluctuations?" *Physical Review Letters* **96**, 063602 (2006).

38. F. Ferri, D. Magatti, L. A. Lugiato, and A. Gatti, "Differential ghost imaging," *Physical Review Letters* **104**, 253603 (2010).

39. J. H. Shapiro and R. W. Boyd, "The physics of ghost imaging," *Quantum Information Processing* **11**(4), 949–993 (2012).

40. R. Meyers, K. S. Deacon, and Y. Shih, "Ghost-imaging experiment by measuring reflected photons," *Physical Reviews A* **77**, 041801(R) (2008).

41. P. B. Dixon, G. A. Howland, K. W. C. Chan, et al., "Quantum ghost imaging through turbulence," *Physical Reviews A* **83**, 051803(R) (2011).

42. B. I. Erkmen, "Computational ghost imaging for remote sensing," *J. Optical Society of America A* **29**(5), 782–789 (2012).

43. J. Cheng, "Ghost imaging through turbulent atmosphere," *Optics Express* **17**, 7916–7921 (2009).

44. N. D. Hardy and J. H. Shapiro, "Reflective ghost imaging through turbulence," *Physical Reviews A* **84**, 063824 (2011).

45. L. Tamborino and J. Taboada, "Laser imaging and ranging system using one camera," U.S. Patent 5,162,861, 1992.

46. L. Tamborino and J. Taboada, "Laser imaging and ranging system using two cameras," U.S. Patent 5,156,451, 1992.

47. K. W. Ayer, W. C. Martin, J. M. Jacobs, and R. H. Fetner, "Laser IMaging And Ranging System (LIMARS): a proof of concept experiment," *Proc. SPIE* **1633**, 54–62 (1992) [doi: 10.1117/12.59206].

48. M. B. Mark, "Laser imaging and ranging system (LIMARS) range accuracy analyses," WL-TR-92-1053, March (1992).

49. P. McManamon, *Field Guide to Lidar*, SPIE Press, Bellingham, Washington (2015) [doi: 10.1117/3.2186106].

50. P. F. McManamon, P. S. Banks, J. D. Beck, D. G. Fried, A. S. Huntington, and E. A. Watson, "Comparison of flash lidar detector options," *Optical Engineering* **56**(3), 031223 (2017) [doi: 10.1117/1.OE.56.3.031223].

Chapter 7
LiDAR Beam Steering and Optics

There are many ways to steer optical beams. Historically, steering an optical beam meant moving a mirror or a transmissive optical element. Over the last few decades, other approaches have been developed that involve creating an optical path difference (OPD) or a phase difference. We can group beam steering methods into mechanical and nonmechanical approaches. Non-mechanical approaches are sometimes called optical phased arrays (OPAs) because they change the angle of an optical beam based on changing the phase profile of that optical beam. Microwave phased-array radars, which have been around for decades, use this approach of changing the phase of an outgoing or incoming beam. The much-shorter wavelength associated with optical systems makes this approach more difficult. Following the discussion of mechanical and nonmechanical approaches to beam steering, a few relevant optical design parameters for a LiDAR are presented. This topic includes a brief discussion of adaptive optics, as well as how to use some of the presented nonmechanical techniques to make dynamically adaptive optical elements. If it is possible to change a phase profile across an optical element to steer a beam, then it is also possible to change a phase profile to make a lens or another optical element. The future era of flexible optical elements will be an enabler.

7.1 Mechanical Beam-Steering Approaches for LiDAR

7.1.1 Gimbals

Gimbals are mechanically complex and expensive but can steer very accurately over wide angles. Except for obscurations, a three-axis gimbal can point in any direction, and each axis can rotate. Using all three axes, it is possible to hold a LiDAR aperture in the inner ring and point in any direction.

Figure 7.1 shows a schematic diagram of a three-axis gimbal system. It is obvious that the aperture inside the inner ring must be significantly smaller than the size of the outer ring because there has to be room for the various rings to fit inside each other. Typically, a hole that is 2.5–3 times the size of the aperture must be cut in the vehicle on which a gimbal is mounted. Figure 7.2 shows photographs of a couple of operational gimbal systems.[1] There are two interesting ratios to notice in this figure. One is the size of the gimbal compared to the optical aperture. The other is the size of the hole that would be needed to attach the gimbal to a vehicle compared to the size of the optical

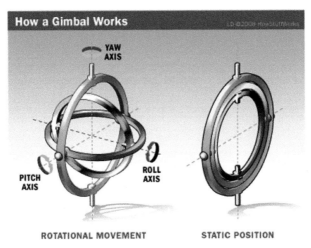

Figure 7.1 A diagram showing how a gimbal works (reprinted from How Stuff Works website: http://science.howstuffworks.com/gimbal.htm).

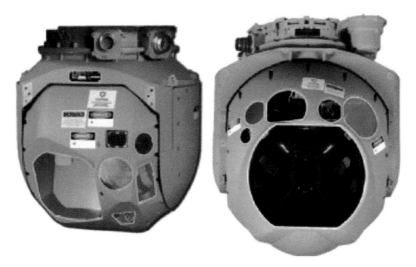

Figure 7.2 Two current optical gimbals: the Raytheon MTS A (left) and the Raytheon MTS B (right) (reprinted from Ref. 1).

aperture. One of the gimbals has a better ratio of clear aperture size to the size of the gimbal size compared to the other gimbal.

An issue with gimbals is how to get electrical or other connections into the inner gimbal, and how to get data out. From Fig. 7.1, it would be hard to run a wire to the inner gimbals without twisting the wire. Slip rings are sometimes used for electrical connections, where two conducting surfaces slide against each other to keep electrical contact as a ring rotates. Wireless systems can be used to get data to and from the gimbal. Rotating fiber optic joints along the axis of a gimbal stage can get data on and off of a gimbal stage.[2] In that case, two fibers are aligned on the axis, and one of them spins, but data can be transmitted. These types of gimbals can be purchased commercially. Many military optical systems currently use gimbal pointing.

Gimbals can usually be pointed and stabilized very accurately, but they tend to cost hundreds of thousands of dollars.[3] Some estimates of the cost of various gimbal are:

- pilotage gimbals: ~14-in. diameter, on the order $75–100K,
- medium-altitude–mission gimbals: ~17-in. diameter, on the order of $100–125K, and
- high-altitude gimbals: ~22-in. diameter, on the order of $150–175K.[4] (Gimbals for small UAVs can be smaller and cheaper.)

Gimbals are a great method of pointing if (1) the response time does not have to be very rapid, (2) precise pointing and stabilization is required, and (3) the cost for the optical system is not a significant constraint. Gimbals will not be used in large-volume production LiDARs but are likely to be used in high-end, limited production. For high-accuracy pointing, the required time to stabilize a gimbal can be a constraint.

7.1.2 Fast-steering mirrors

A fast-steering mirror (FSM) is a flat mirror that can be moved over small angles very quickly. The beam is steered as the mirror is rotated, using angle of incidence equals angle of reflection to steer the beam. The larger the mirror the more time and power it takes to move it quickly, so there is a tradeoff between mirror size and speed.

Figure 7.3 shows a schematic of a FSM and a telescope. FSMs are rotating flat mirrors, but of different sizes. FSMs should be small to be fast, so they should be placed where the optical beam is small. A beam-expanding telescope will normally be required after the FSM. The final steering angle is reduced by the magnification of the beam to the final aperture size:

$$\theta_f = \frac{\theta}{M},$$ (7.1)

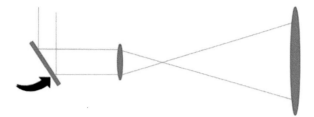

Figure 7.3 A diagram of a fast-steering mirror.

where θ_f is the final steering angle, θ is the angle steered by the steering mirror, and M is magnification. This angle reduction can create a need for a wide-FOV telescope to magnify the beam to the final aperture size. The design of an aberration-free telescope that can accept a wide steering angle can be a limitation.

Fast-steering mirrors can move at ~1 KHz bandwidth so can allow for fast small-angle pointing. There will be a settle time after moving quickly from one angular location to another. The settle time will likely be significantly longer than one over the FSM bandwidth. Even an expensive FSM will have a 6-ms settle time, or more.

Steering flats look just like FSMs; the only difference is that they are bigger and slower. Figure 7.4 is a photograph of a steering flat.[5] One way to drive steering flats is by using galvanometers. A galvanometer is a type of sensitive ammeter, an instrument for detecting electric current. It is an analog electromechanical transducer that produces a rotary deflection of some type of pointer in response to electric current flowing through its coil in a magnetic field.[6]

A couple of galvo scanners are shown in Fig. 7.5. In 2007, the larger scanner in the figure cost $3435 for a single-axis scanner with a 5-cm clear aperture. It can do ±20 deg optical motion, or ±10 deg mechanical motion. It

Figure 7.4 Photograph of a fast-steering mirror (reprinted from Ref. 5).

Figure 7.5 Two galvo scanners (reprinted from Cambridge Technology website: https://www.cambridgetechnology.com/products/galvanometer-xy-sets).

can do a 0.1-deg mechanical step in ∼2.3 ms and a 20-deg mechanical step in ∼14 ms.[7]

7.1.3 Risley prisms and Risley gratings

Risley prisms are two (or sometimes three) rotating wedges or prisms. They steer light by rotating two or three prisms. A typical steer and settle time for a Risley prism is 15 ms.[8] Risley prisms use rotation, a very simple motion, to steer optical beams. The prisms rotate with respect to one another to steer a beam. Control of the pointing can be a little complicated because Lissajous figures[9] are created by the motion. While the relative size of the aperture versus the size of the hole holding the Risley prism beam scanner is an issue, it is possible to increase the fill factor by placing the motor and much of the rotation mechanism in an area where the beam is smaller. Some researchers have also held each Risley prism (or grating) on a shaft in the center. This creates a small obscuration in the center. Another potential issue with Risley-prism–based steering is the zero-crossing issue. A track through zero requires the wedges to rotate by 180 deg. Risley prisms can be a flush, or conformal, steering approach, which is very beneficial for some applications. Any time a steering mechanism is flush with a flat surface, there is a cosine loss in power sent to a given angle, for which the cosine is taken from a perpendicular direction away from the surface. This cosine loss is due to the reduction of the projected area of the aperture in the direction the beam is propagated.

Recently, Risley prisms,[10] and even Risley gratings,[11] have become more popular. For narrowband systems, Risley gratings have an advantage over Risley prisms. Risley gratings are lighter and therefore can move more quickly using the same amount of power to drive them, but they have dispersion. Since Risley prisms and Risley gratings are both moved by motors, being lighter in weight will allow a smaller motor (using less electricity), or will allow faster motion. Dispersion with Risley gratings will not be an issue for LiDAR unless the LiDAR is very high bandwidth, meaning that it has high range resolution. Unless the LiDAR is attempting extreme range resolution, the LiDAR bandwidth should not be broad enough to be an issue. Certain Risley gratings can solve the zero-crossing issue by switching the handedness of the incoming circular polarized light, if a polarization birefringent grating is used, as explained in Ref. 11. Figure 7.6(a) shows an early Lockheed Martin Risley steering stage,[12] and Fig. 7.6(b) shows an OPTRA commercial Risley prism steering device.[13] (OPTRA has since been gone out of business.) It is possible to steer to large angles using Risley devices that use only rotation, a very simple motion. The OPTRA devices, however, had only a little more than 50% linear fill factor. Lockheed Martin has a Risley device with about 70% linear fill factor.[14] Fill factor is defined as the size of the clear aperture compared to the size of the hole required in the vehicle carrying the beam-steering device.

For back-and-forth scanning motion from 0 to 24 deg, it is possible to use two prisms rotating at the same speed in opposite directions. At zero-degree scanning, the wedges are positioned as seen in Fig. 7.7. At 24-deg pointing, the

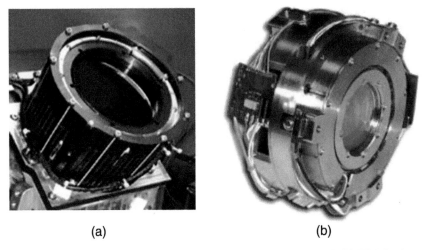

        (a)   (b)

Figure 7.6 (a) Early Lockheed Martin Risley prism scanner and (b) OPTRA Risley prism scanner [part (a) courtesy of D. J. Adams, Lockheed Martin MST and part (b) reprinted from Ref. 5].

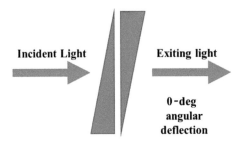

Figure 7.7 Zero-deg pointing with Risleys.

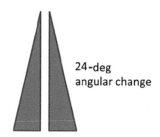

Figure 7.8 24-deg pointing.

Figure 7.9 Risley scan patterns.

wedges are positioned as shown in Fig. 7.8. Two 12-deg Risley prisms will scan from 0 deg to 24 deg and back again, as shown in Fig. 7.9. The scanning will be off to one side rather than curtained around the middle.

A paper[15] by Gerald Marshall, "Risley prism scan patterns," shows the equations for Risley steering as

$$D_x = D_1 \cos(\omega_1 t) + D_2 \cos(\omega_2 t - \phi), \tag{7.2}$$

$$D_y = D_1 \cos(\omega_1 t) + D_2 \sin(\omega_2 t - \phi), \tag{7.3}$$

where D_1 and D_2 are the wedge angles for the two wedges, and ω_1 and ω_2 are the rotation frequencies.[15] To have a horizontal scan of 24 deg, we would use two identical 12-deg wedges rotating in opposite directions. In that case, $D_1 = D_2 = D$, and $\omega_1 = -\omega_2$. For this case,

$$\begin{aligned} D_x &= D \cos(\omega t) + D \cos(-\omega t - \phi) \\ &= D\{\cos(\omega t) + [\cos(-\omega t) \cos(\phi) - \sin(-\omega t) \sin(\phi)]\}. \end{aligned} \tag{7.4}$$

If ϕ is zero, then

$$D_x = 2D \cos(\omega t), \tag{7.5}$$

and, in a similar fashion, D_y is zero for $\phi = 0$.

If ϕ is zero, this device becomes a simple back-and-forth scanner. This is unfortunately a sinusoidal scan with the higher dwell time at the edges of the scan rather than in the middle, where it would be preferable. This is one of the disadvantaged of Risley-prism–based scanning. An engineer would usually spend more of the scan period in the center, and less time in the edges.

7.1.4 Rotating polygonal mirrors

Rotating polygons are shown in Figs. 7.10 and 7.11. Figure 7.10 shows the beam hitting a corner, which is a loss because some of the beam is reflected in one direction, and some in a different direction. The motion of a polygon is simple rotation. Because a polygon has many sides, even slow rotation can scan the angle rapidly. Polygons, however, only scan in one dimension, although it is possible to have some sides tilted at a different angle, providing an elevation step scan. A hexagon, which has 6 sides, will scan over 120 deg because angle of incidence equals angle of reflection. The mechanical scan angle is 60 deg, but the optical scan angle is 120 deg. The beam-steering angle is always twice the mechanical rotation angle. The more sides on the polygon, the smaller the scan angle before the scan jumps to the next scan, but the physically larger the polygon. A polygon has no fly-back period but does have scan loss where the beam hits the corners. As soon as the beam crosses from one flat face to the next, the next scan begins. As you can see, it will take time for the whole beam to traverse each corner. The size of the beam compared to the length of one side on the polygon will provide a scan efficiency measure. An optical designer can design to only use a portion of one side. For example, a 60-deg mechanical scan with a 120-deg optical scan could be designed to only provide 40, 60, or 80 deg of scanning, completely avoiding the losses at the corners. This is a loss of scan efficiency, which is discussed shortly.

A large drawback to polygon scanners is the polygon size. The polygon has to be thicker than the clear beam and can become large for small-angle

Figure 7.10   A rotating polygon mirror with the beam hitting a corner (reprinted from Ref. 50).

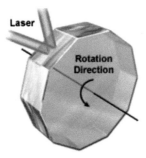

Figure 7.11   A rotating polygon scanning mirror (reprinted from Nikon Instruments Inc: https://www.microscopyu.com/techniques/confocal/resonant-scanning-in-laser-confocal-microscopy).

scanning because the polygon will require many sides. If you limit the number of sides of the polygon, scan efficiency will decrease for scanning over small angles. Another drawback to polygon scanning is the fact that the center of rotation is not on the mirror surface, such as happens with a fast-steering mirror.[16] Because the center of rotation is in a different location than the surface of the polygon, the steered beam will translate back and forth slightly. This translation can cause issues in some applications.

Angle of incidence equals angle of reflection, so the scan rate is twice the mechanical angular rate of movement of the polygon. There is a total of 720 deg of optical scanning for each 360 deg of physical rotation of the polygon. A polygon has a smooth rotating motion, which is an advantage. Figure 7.12 shows a six-sided polygon made up of isosceles triangles. The base of each triangle, measured in degrees, is 360/(number of sides). The height of the isosceles triangles, which is the radius of the polygon, is given by

$$y = \frac{x}{\tan \vartheta}. \tag{7.6}$$

Equation (7.6) can be used to calculate the diameter of a polygon for a given length of a side, which we will call d_{side}. Let us define N_{sides} as the number of sides and d_{beam} as the beam diameter. Then we calculate the scanning efficiency according to

$$S_{\text{eff}} = \frac{N_{\text{side}}}{30} \cdot \frac{d_{\text{side}} - d_{\text{beam}}}{d_{\text{beam}}}. \tag{7.7}$$

This equation was used to calculate the values in Table 7.1.

The entire beam is completely on a given side only for a brief period. For it to be fully on the scan side, we need to subtract twice the diameter of the beam. Some engineers would count when half of the beam is on the side, which would improve efficiency, but at this time, only half of the beam is being deflected in the correct direction. Table 7.1 shows both approaches, but unless the beam is very small, or the polygon is very big, this is not an efficient scanner.

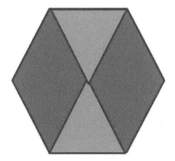

Figure 7.12 Six-sided polygon.

Table 7.1 Various polygon parameters and scan efficiencies.

Side length (cm)	# of sides	Diameter (cm)	Scan efficiency for a 2-cm–diameter beam (full beam)	Scan efficiency for a 2-cm–diameter beam (half beam)	Scan efficiency for a 5-cm–diameter beam	Scan efficiency for a 5-cm–diameter beam (half beam)
5	12	37.32	8.00%	24.00%	0.00%	0.00%
5	18	56.72	12.00%	36.00%	0.00%	0.00%
5	24	75.96	16.00%	48.00%	0.00%	0.00%
5	30	95.14	20.00%	60.00%	0.00%	0.00%
10	12	74.64	24.00%	32.00%	0.00%	20.00%
10	18	113.42	36.00%	48.00%	0.00%	30.00%
10	24	151.92	48.00%	64.00%	0.00%	40.00%
10	30	190.28	60.00%	80.00%	0.00%	50.00%
15	12	111.96	29.33%	34.67%	13.33%	26.67%
15	18	170.14	44.00%	52.00%	20.00%	40.00%
15	24	227.88	58.67%	69.33%	26.67%	53.33%
15	30	285.44	73.33%	86.67%	33.33%	66.67%

A polygon scanner can also be used to create motion in the nonscanned direction. Imagine if each side is tilted at a slightly different angle. It is possible to change the tilt angle enough to change the FOV in the cross-scan direction, or to tilt the angle enough to only interlace the elevation samples. For example, with 100 elevation samples and scanning in azimuth with the polygon, the tilt could be large enough to change elevation by the full 100 elevation samples, or could be only 1/6 of a single elevation sample spacing in angle for a 6-sided polygon. In either case, a 6-sided polygon could increase the number of elevation samples from 100 to 600; however, in one case, the elevation scanning would be a microscan, and in the other case, there would be significant elevation change in angle between each side of the polygon scanner.

7.1.5 MEMS beam steering for LiDAR

Mirrors are dispersion-free and polarization-invariant devices. Micro-electromechanical system (MEMS) devices are normally fabricated in cleanroom facilities. Large single-aperture mirrors are bulky, slow, and consume high power. MEMS mirrors usually are formed in an array, with each mirror being very small, but the array can be large. Sometime MEMS mirrors are individual, in which case they are a type of FSM. MEMS mirrors can steer quickly because each individual mirror has low mass and inertia. One decision when using MEMs mirrors is whether or not to phase up the array of individual small mirrors. If the array is not phased up, the aperture size, from a diffraction point of view, is the size of an individual MEMS mirror. This can be acceptable if high angular resolution is not required. However, to form an aperture that is effectively large, a $2\pi n$ phase shift will be required between the MEMS mirrors, so the wavefront from each aperture is aligned in phase with the other apertures. If this modulo 2π beam steering is performed, MEMS arrays will be dispersive. If the MEMS array is not phased up, it will not be dispersive but will have low angular resolution. A diagram showing a typical structure of a MEMS micro-mirror array is shown in Fig. 7.13, and a photograph of a MEMS array is shown in Fig. 7.14.

A 2π phase shift can be provided either by positioning each micro-mirror in depth, or by adding a phase shift layer before each micro-mirror. A phase shift layer could be, for example, a liquid crystal layer that can adjust the phase hitting each micro-mirror. Since we are in a reflective mode, the phase layer has to be capable of causing a phase shift of up to π. In reflective mode, this means that the phase layer can provide 0 to 2π phase shift since the light passes through the phase layer twice. MEMS-mirror–based steering has to operate in a reflective mode. The need to make the steering elements reflective can increase the size of a steering system because it restricts the optical design choices. There is no fly-back region—a section of the array where the beam is defected at the wrong angle—with MEMS mirrors, such as we have with liquid crystal phased arrays. Fly back will be further

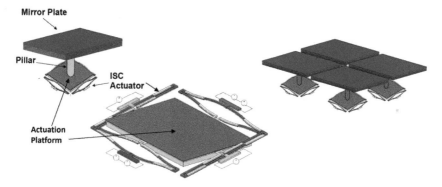

Figure 7.13   Structure of MEMS micro-mirror.

Figure 7.14   A MEMS array (reprinted from Ref. 17).

discussed in Section 7.2.1.4 on liquid crystal OPAs. The method of steering is by tilting the micro-mirrors. There is no digitization loss since the ramp is a single MEMS mirror. Fill factor will need to be addressed to keep high throughput efficiency with MEMS steering. In some structures, the hardware to hold and move a MEMS mirror takes a considerable portion of a steering pixel. A combination of a reflective array of MEMS mirrors with a step steering approach (see Section 7.2.6.1) could be very efficient over wide angles. In a step steering approach, an optical system does not steer continuously. Rather, it steps from steering at one angle to another discrete angle. MEMS arrays can have some shadowing effects. Figure 7.15 shows the tilt of a MEMS array of mirrors with a phase delay layer shown in front of the MEMS array. The phase layer is there to adjust the OPD so that there is a 2π phase shift between the top of one slanted MEMS mirror and the bottom of the next MEMS mirror.

For MEMS mirror arrays, it is critical to have high area fill factors. One way to boost fill factors is to hide actuators behind reflective surface, i.e., to place micro-actuators under mirror plates. Hidden actuators can be easily implemented in thin-film surface-micromachining processes.

MEMS mirror arrays need to have very high reflectivity in order to handle high powers since the thermal path for cooling is very slow. MEMS arrays may have a fill factor issue, resulting in the need for lenslets in front of the MEMS array to increase the fill factor. Another possible issue with MEMS mirrors is the ability to point accurately in a dynamic environment. MEMS mirror pointing structures are usually resonant structures. If a MEMS array is put into a dynamic environment, say on a small UAV bouncing around the sky, it will be very difficult to accurately control the pointing angles. There is ongoing work on closed-loop control to maintain MEMS pointing accuracy in a dynamic environment, but this technology is not yet mature.[18] The author would caution readers to take into account the impact of a dynamic environment when considering using a MEMS mirror array for optical pointing.

Some MEMS devices steer in about 10 ms; others can steer in about 1 ms. A critical consideration concerning MEMS arrays is that each MEMS

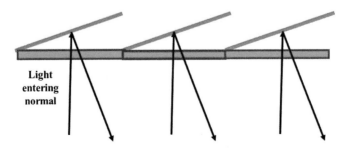

Figure 7.15 Tilted MEMS array with a phase shift layer.

element must both tilt and move in piston if the array is to be phased up so that it acts like a single optical element from a diffraction point of view. The tilt is obviously required to steer the array. The piston is required to phase up one element with the next. Often people promoting the use of MEMS arrays do not understand the need for precise piston control, but this control is critical unless you are willing to use an effective aperture that is the size of a single MEMS element. Characteristics of a typical MEMS device are shown in Table 7.2.

As already mentioned, one of the major potential losses from using a MEMS array is fill factor. The MEMS array shown in Fig. 7.16 has only a 45% fill factor. By using hidden actuators, that fill factor can be increased to about 90%. In addition, one or more lenslet arrays can be used to increase the effective fill factor. An ideal effective fill factor would be >99%. This array does not have high reflectivity. Ideally, reflectivity would be >99%, which should be achievable[20] because getting highly reflective arrays is generally reasonably simple. Gold or silver coatings may be required to obtain the high reflectivity. Steering angles can be about ±10 deg mechanical, or about ±20 deg optical.

Table 7.2 Typical MEMS array characteristics (mechanical angle is one-half of the optical angle).

Parameter	Values
Array size	4 × 4 pixels
Pixel size	2.12 mm × 1.5 mm
Individual-mirror aperture size	1.41 mm × 1.0 mm
Die size	9.5 mm × 7.5 mm
Fill factor	45%
Reflectivity @ 633 nm	72%
Response time (10%–90%)	11 ms
Mechanical angle in the x dimension [deg]	±12 deg
Mechanical angle in the y direction [deg]	±12 deg

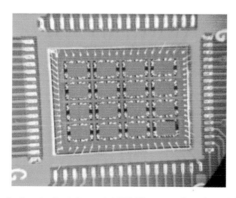

Figure 7.16 A 4 × 4 pixel format MEMS array (reprinted from Ref. 19).

MEMS devices can steer in both azimuth and elevation with a single array. To increase power handling, maximum reflectivity will be required. Devices can currently be made 7.5 mm across or up to 10 mm across.

7.1.6 Lenslet-based beam steering

A microlens array can steer a beam to discrete locations. The array makes use of two or three cascaded arrays. One disadvantage of lenslet-based beam steering is that it has multiple layers. Each layer adds complexity and can reduce steering efficiency due to mundane engineering factors such as absorption, reflection, and scattering. A lenslet array is similar to a phased array, but with a fixed period before a reset.[21] Because most implementations of lenslet arrays use mechanical motion, we include lenslet arrays in this section, but they do have many similarities to OPAs. Also, it is possible to write lenslet arrays by removing mechanical motion from their implementation. By writing a lenslet array, we mean using a birefringent material, such as a liquid crystal, along with an appropriate E-field distribution, to make a writable device that acts as a lens. An advantage of lenslet-array–based steering is that a small amount of motion can cause steering to a significant angle. This is also a disadvantage because small, unintended motion can cause steering as well.

7.1.6.1 All-convex-lens–based steering

A single triplet of the cascaded arrays is shown in Fig. 7.17. Early work on lenslet-based steering used a doublet for steering, but, later, a third lens, a field lens, between the two lenslets redirected light such that in theory all of the light passing through the first lens reaches the last lens. Using a field lens between the other two lenses increases steering efficiency.

The steering angle is given by

$$\tan \vartheta = \frac{\Delta x}{f_1}, \tag{7.8}$$

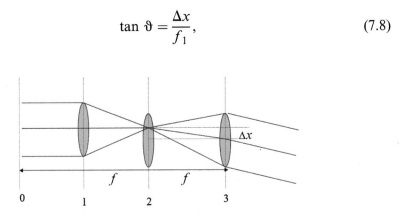

Figure 7.17 A three-lens steering implementation (reprinted from Ref. 34).

where Δx is the amount of decenter of the first microlens with respect to the optical axis of the other two microlenses, and f_1 is the common focal length of all three microlenses in the triplet. Steering is based on changing the diffraction orders, so a phase corrector plate needs to be used to dither the angular location of orders; alternatively, a fine-angle beam steerer must be used to steer between orders. A phase corrector plate imposes a fixed phase delay over some spatial section of the device. If a phase plate is inserted, each phase shift element must be the size of one lenslet. An example of the lenslet array parameters corresponding to various discrete steering angles is shown in Table 7.3.

An f-number of $f/1$ can achieve steering up to 25 deg, so this can be a moderately wide-angle steering approach, as well as moving to high efficiency. It is important to have low f-numbers to steer to large angles.

7.1.6.2 Mixed-lenslet arrays

An alternative lenslet-based steering approach uses a mixed set of lenslet arrays with both positive and negative lenslets, as shown in Fig. 7.18. The mixed-lenslet approach for beam steering has several advantages: (1) it eliminates the foci, thus increasing power-handling capability, (2) it makes the

Table 7.3 Steering angles for various lenslet arrays.

Δx, μm	Steering angle, deg	Focal length, μm	Lenslet diameter, μm	$f/\#$	OPD, μm
67.5	25	145	150	1	19.4
67.5	20	185	150	1.2	15.2
67.5	10	383	150	2.6	7.3
67.5	3	1288	150	8.6	2.2
67.5	1.5	2578	150	17.2	1.1

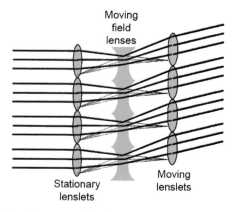

Figure 7.18 A mixed-lenslet approach to beam steering has both positive and negative lenslets (adapted from Ref. 34).

set of lenslets more compact, cutting the depth of the lenslet array by a factor of two, and (3) the lenslets are not required to have a focus.

If lenslets become too small and have a high *f/#*, then the "focus" is in the far field of the lenslet. This means that the lens does not have a quality focus. For mixed lenslets, the focused beam reaches only half the distance to a focus before it is diverged again by the negative lenslet. In that case, a focus is not required, so we eliminate the issue of an imperfect lens. We also increase power-handling capability. The focal length of the positive lenslets is twice the focal length of the all-positive lenslet approach if the steering angle is kept constant. The negative lenslet requires a focal length that is half as large, but it could be implemented using a high–refractive-index fixed material. This approach requires the two positive, outside lenslet arrays to have identical focal lengths and to be placed one focal length apart. This distance is one-half of the separation in the all-positive-lenslet array approach. For mixed-lenslet steering, the center lenslet is moved down one unit, while the positive lenslet on the other side is moved up two units. For a mixed-lens approach, the steering angle is

$$\theta = \arctan\left(\frac{2\Delta x}{f_1}\right), \tag{7.9}$$

where f_1 is the focal length of the positive lenslets. The focal length of the negative lenslet is one-fourth of that of the positive lenslet. Digital simulations show the mixed-lenslet approach having >97% steering efficiency for small angles such as 1.5 deg. The middle lenslet could still clip energy entering the first lenslet; however, the main efficiency issue will be the engineering challenge of loss through multiple layers.

7.2 Nonmechanical Beam-Steering Approaches for Steering LiDAR Optical Beams

To date, most OPA steering uses space-fed, phased-array steering as compared to individual transmit/receive (T/R)-module–based steering, which is often used with phased-array microwave radars. Most of the steering in current OPAs is in one dimension at a time. Traditional radar phased arrays have individual phase elements that are at half-wavelength spacing or smaller. Consider an X-band radar at 10 GHz with a wavelength is 3 cm. Wavelength spacing for individual phase shifters with a 3-cm wavelength would mean that the pitch between individual phased-array elements is 1.5 cm. It is possible to build and address individual T/R modules that are 1.5 cm × 1.5 cm on the radiating surface. Consider a radar aperture that is 75 cm × 75 cm. Making this a square 50 × 50 pixel array of T/R modules that are 1.5 cm on a side would result in 2500 individual T/R module elements. To make this array round instead of square, the number of elements decreases to about 2000 elements.

Now let's consider a similar OPA based on T/R modules. Assume an optical wavelength of 1.5 μm. A half-wavelength pitch means 0.75 μm between elements. Optical apertures usually are not as large as microwave apertures, so let's assume that our optical aperture is 30 cm × 30 cm. To build this aperture will require 400,000 elements on a side, or a total of 160,000,000,000 T/R modules in a square aperture; or about 125,000,000,000 elements in a round aperture that is 30 cm in diameter. This same array would have 400,000 elements in one direction. Two crossed 1D arrays would need 700,000 elements. This is still a large number. To put these numbers in context, however, the iPhone 6 in 2015 had 2,000,000,000 transistors on a chip, whereas a Pentium 4 in 2001 had 2,000,000 transistors.[22] This is a factor of 1000 increase in 14 years. There is rapid progress in addressing larger and larger arrays of transistors on a chip. Of course, those are for large-volume commercial products with millions of units. Also, for a 2D array, each T/R module has to be 0.75 μm × 0.75 μm on the surface of the apertures, which is very small. In today's technology, this is impractical. As a result, the optical community has started working OPAs that steer in one dimension at a time and are space fed. A space-fed array changes the phase of an optical beam passing through it, resulting in steering the optical beam. Figure 7.19 shows two crossed 1D space-fed phased arrays.

7.2.1 OPD-based nonmechanical approaches

7.2.1.1 Modulo 2π beam steering

The index of refraction in a prism is larger than that of air, so light travels more slowly within the prism. The angle of light passing through a prism will be changed because the light moving through the thick portion of the prism will be delayed compared to light traveling through the thin portion. This tilts

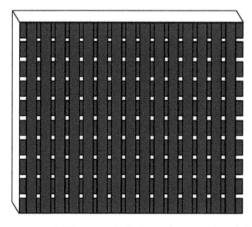

Figure 7.19 Two crossed 1D space-fed phased arrays (reprinted from Ref. 51).

the phase front, so the outgoing phase front is at a different angle than the incoming phase front, thus steering the optical beam. Steering can be accomplished by changing the thickness of the prism. Light could be steered electronically by writing a prism. The problem with doing this is that it is difficult to create an OPD as large as would be required to write a full prism of appreciable width. For example, a 10-cm–wide aperture steering to 30 deg would require 5 cm of OPD on the thick side of the prism.

For a narrow-wavelength band, we can take advantage of the fact that light is a sine wave. With sine waves, it does not matter if there is a 0, 2π, 4π or a $2\pi n$ phase shift from a phase perspective; they are all the same. Therefore, as one moves across the width of the prism, one can subtract $2\pi n$ of phase every time the phase reaches $2\pi n$, resulting in a sawtooth phase profile. The unfolded phase, which is called a modulo 2π phase profile, looks like the phase profile that would result from propagation through a full prism and steers light in the same manner. This is shown in Fig. 7.20. The benefit of using a modulo 2π phase profile is that the required OPD can be small, on the order of one wavelength. The modulo 2π steering approach, however, makes the beam steerer very wavelength dependent (dispersive). Figure 7.20 shows modulo 2π phase shifting to create beam steering.

Modulo 2π beam steering is similar to using a Fresnel lens, except that a Fresnel lens[23] has resets much larger than 2π so is not as dispersive. Many people are familiar with Fresnel lenses, as they were commonly used in overhead projectors. The phase shift profile for agile modulo 2π beam steering is shown in Fig. 7.21, and a grating model for modulo 2π beam steering is shown in Fig. 7.22.

Modulo 2π beam steering provides

- a dynamic writeable blazed-phase grating and
- a modulo 2π phase ramp structure, which gives maximum switching speed by making the OPD change layer a thin as possible.

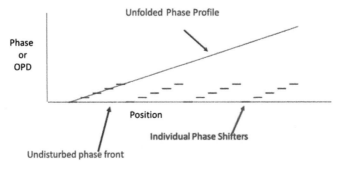

Figure 7.20 An example of the phase profile for a modulo-2π–based beam steerer using digital steps (adapted from Ref. 50).

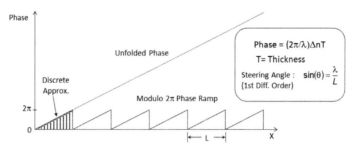

Figure 7.21 Phase shift profile for modulo 2π beam steering.

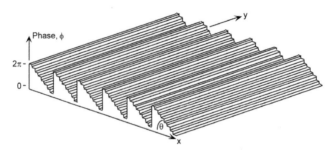

Figure 7.22 The grating model for modulo 2π beam steering.

In Fig. 7.22, each reset period is shown to be made up of q phase shifters that form a quantized, blazed grating with steering efficiency η, which is calculated as

$$\eta = \left(\frac{\sin(m\pi/q)}{m\pi/q} \right)^2 , \qquad (7.10)$$

where m is any integer value but usually equals 1 because in that case the active layer is thinner. According to Figs. 7.21 and 7.22, the deflection angle depends on the reset period and the blazed grating. Therefore, there are two ways to change the angle and steer the optical beam: vary the reset period, or vary the blaze. These two methods are explained in greater detail in the next sections.

Variable-period, fine-beam steering. It is possible to vary the period of a reset to steer an optical beam. As seen in Fig. 7.23, the length of the distance before a reset changes: In case 1, the reset length is $\Lambda 1$ and three electrodes are

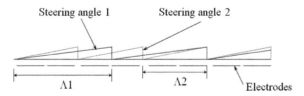

Figure 7.23 Variable-period fine beam steering (reprinted from Ref. 34).

included in one reset. In case 2, the reset length of $\Lambda2$ is the width of two electrodes. Changing the reset period (length) from $\Lambda1$ to $\Lambda2$ leads to a different deflection angle. The steering angle depends on the number of electrodes q in one reset period. Therefore, the change in optical beam angle is

$$\Delta\theta \approx \frac{\lambda N}{Dq^2}, \qquad (7.11)$$

where N is the number of electrodes across an aperture. The phase shifters are individually addressed, and the steering is essentially continuous for q greater than $\sqrt{N}$.

Variable-blaze, large-angle beam steering. Another way to steer an optical beam is to vary the blaze, as shown in Fig. 7.24. This is discrete steering because continuous steering requires a reset length greater than the aperture diameter (Λ > aperture diameter). Therefore, a second device may be required for continuous steering.

   Generally, in this method the reset period is unchanged, and the phase shifters are linked. This method might be used for large-angle beam steering. The change in the optical beam angle is given by

$$\Delta\theta \approx \frac{\lambda}{\Lambda}. \qquad (7.12)$$

7.2.1.2 Finest pointing angle

One major benefit of OPA nonmechanical beam steering is the ability to point very accurately. For many applications, this is very important. Modulo 2π beam steering using phased-array technology can enable very precise steering.[24] For an OPA-based beam-steering approach, very precise steering will not be difficult because a single electrode can be changed to make a small angle change. To consider the steering resolution of an OPA, start with an OPA that has 10 electrodes per a one-wavelength ramp (OPD of 1 λ). You might think that we can only change the steering angle by one electrode, so there can only be steps that are a 10% change in the angle. Consider a change in slope such that it would have a one-wavelength change in 10.2 electrodes. At the end of 10 electrodes of first ramp, the phase is (10/10.2) $\times$ λ. A negative phase change of one wavelength means that at the beginning of the next

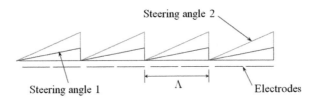

Figure 7.24 Variable-blaze, large-angle steering (reprinted from Ref. 34).

10-electrode ramp we start with a negative phase of –0.2/10.2 λ. At the end of this ramp the phase is 9.7/10.2 λ. At the beginning of the next 10-electrode ramp the phase is –0.4/10.2 λ, and at the end it is 9.6/10.2 λ. At beginning of the fourth 10-electrode ramp the phase is –0.6/10.2, and at the end it is 9.4/10.2. At beginning of the fifth ramp, the phase is –0.7/10.2, and at the end it is 9.2/10.2 λ.

Note that it would be possible to go to 11 electrodes instead of 10 electrodes for this delay. In that case, the end phase would be one wavelength. After the reset, the sixth ramp is just like the first—starting out at 0 phase. So, if 1 ramp out of 5 can change by 10%, we can make a device with an effective change of 10/5 = 2% instead of being limited to a 10% change in phase ramp as you might first think. Note also that there needs to be a phase swing of the device by not one wave, but by 1 + 0.7/10.2 = 1.074 waves, so a slightly larger OPD capability is required in the device. This will slightly increase fringing field issues. The same approach can be extrapolated to much smaller pointing angles that are 1/100th of a diffraction-limited spot size.

7.2.1.3 Largest steering angle for an OPA

The largest angle one can steer to is determined by the size of the smallest individually addressable phase element. If the full array of phase shifters is illuminated with a Gaussian beam, any individual phase shifter will have an approximately uniform irradiance distribution across it. For a circular aperture that is uniformly illuminated, the full-width beam divergence at the half-power point is

$$\theta \cong \frac{1.03\lambda}{d}, \qquad (7.13)$$

where θ is the beam divergence, λ is the wavelength of the electromagnetic radiation, and d is the width of the individual radiator. If phase can be locked among many individual radiators, the beam will become narrower in angle in proportion to the increase in the effective size of the radiator. If the full array is uniformly illuminated, it is possible to substitute $D = nd$ (where n is the number of individual radiators assembled) into Eq. (7.13) to determine how narrow the beam becomes. This definition is for full angular beam width at the half-power points. For Gaussian illumination of the full array, the effective size of the large aperture is reduced and the beam divergence increases. The allowed amount of clipping of the Gaussian beam by the aperture array determines how much the effective aperture size is reduced. By adjusting the phasing among the individual elements, the narrow beam can be steered under the envelope of the larger beam resulting from an individual radiator, as shown in Fig. 7.25 for uniform illumination.

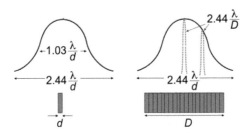

Figure 7.25 The beam-steering envelope and the angular size of the steered beam for a modulo 2π beam-steering array (reprinted from Ref. 34).

7.2.1.4 Liquid crystal OPAs

Birefringence of a nematic liquid crystal is produced by the rotation of molecules under applied voltage. Alignment layers create forces to make one orientation preferred in the absence of an electric field. Figure 7.26 shows a liquid crystal cell with voltage applied and without voltage. There are electrodes on the top and on the bottom of the cell. The liquid crystal molecules are oriented parallel to the electrodes with no applied field. The surface rubbings help align the molecules. When a field is applied, the molecules tilt, causing a change in the index of refraction of one of the polarizations. Light might enter from the bottom and leave from the top, somewhat steered. As seen in Fig. 7.26, the molecules are reoriented when an electric field (voltage) is applied between electrodes, therefore changing the refractive index as a function of position for polarized light (spatial phase modulation). Tens of thousands of electrodes may be used across a liquid crystal spatial light modulator device. A plot of the phase shift as a function of the applied voltage is shown in Fig. 7.27.

The speed of a liquid crystal steering device depends on how fast the liquid crystal molecules rotate from one orientation to another. The relaxation time for a liquid crystal cell to return to its no-voltage state once the electric field is removed is given by

$$\tau_d = \frac{t^2\gamma}{k\pi^2},\tag{7.14}$$

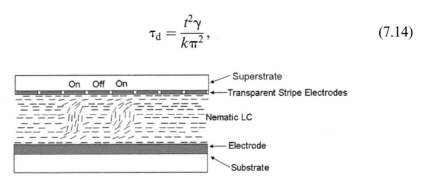

Figure 7.26 A side view of a nematic liquid crystal device showing the orientation of the liquid crystal molecules (adapted from Ref. 34).

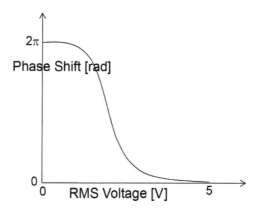

Figure 7.27 The phase shift versus applied voltage in a nematic liquid crystal device (reprinted from Ref. 27).

where τ_d is the time to return to no-voltage molecular orientation, t is the cell thickness, k is the effective elastic constant, and γ is the viscosity. Cutting the cell thickness in half reduces the switching time by a factor of four. The turn-off relaxation time is usually much slower than the time it takes to turn on the phase shift. Turn-on time is given by

$$\tau_r = \frac{\tau_d}{\left[\left(\frac{V}{V_t}\right)^2 - 1\right]}, \tag{7.15}$$

where V is the applied voltage, and V_t is the threshold voltage. Driving the turn on with higher voltage speeds it up by approximately the square of the ratio of the increased voltage. At high drive voltages, the turn-on time becomes much faster than the turn-off time. To speed up the turn-off response, scientists have developed dual-frequency liquid crystals (DFLCs), which drive the molecules both on and off. With DFLCs, the dielectric properties of the material change sign as a function of the frequency of the applied AC voltage. For example, at 1000-Hz drive frequency, the liquid crystals are driven one way, and at a 50,000-Hz drive frequency, the molecules are driven the other way. This increase in speed is very desirable, but using DFLCs increases the complexity of the drive circuitry. Also, DFLCs need to be temperature stabilized much more precisely than standard nematic liquid crystals.

Another method to speed up liquid crystal beam steering is by mixing polymers into the liquid crystals. This can increase the steering speed by about a factor of 300.[25] A polymer network liquid crystal increases steering speed by effectively making the liquid crystal layer thinner. The distance between the polymer material is made very small, increasing the switching speed. Alignment layers or stress/pressure can help to align the polymer networks, reducing scattering. Also, the refractive index of the polymer can be matched to the index of the liquid crystal, which reduces scattering. The biggest issue with scatter is seen when the beam is steered, as the beam is steered based on changing the index of refraction of the liquid crystal material. When the index of refraction of the liquid crystal material is changed, the index of the polymer material is not changed. This can increase the scattering.

Another approach to effectively decrease the thickness of the nematic liquid crystal material and increase the switching speed is to use stressed liquid crystals.[26] These are also polymer network liquid crystals, but they align the polymers by using stress rather than by an alignment layer. Stressed liquid crystal layers can be made thicker without increasing steering time compared to polymer network liquid crystals that are aligned based on an alignment layer.

7.2.1.5 Liquid crystal fringing field effect on steering efficiency

There is a spatial fly-back region when doing modulo 2π resets that are electrically addressed. This results from the fact that the electric field across the liquid crystal cell cannot change sharply at the reset, even if the voltage on the electrodes changes sharply in a spatial direction. During the fly-back portion of the phase profile, the beam is deflected mostly in the wrong direction. Steering efficiency due to fringing-field–based limitations, which result in a fly-back region, is given by[27]

$$\eta = \left(1 - \frac{\Lambda_F}{\Lambda}\right)^2,\qquad(7.16)$$

where η is efficiency, Λ_F is the width of the fly-back region, and Λ is the width between resets.

Figure 7.28 demonstrates how the electric field changes when voltage steps are applied in an attempt to impose modulo 2π beam steering. Fringing fields expand outward to each side of small electrodes. As a general rule, the narrowest width of an electric field region above an electrode is about the thickness of the liquid crystal layer between the electrode and the ground

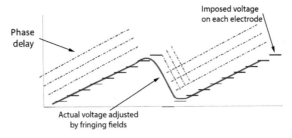

Figure 7.28 The electric field created inside of parallel electrodes when using step voltages intended to create a modulo 2π phase shift (adapted from Ref. 34).

plane. Since the thickness of the liquid crystal layer is often larger than the spacing between the electrodes, fringing fields can have a significant effect. The cell has to be approximately thick enough to obtain one wavelength (2π phase) of OPD. With a birefringence of 0.3, the cell has to be about 3.3 times one wavelength in thickness.

The 0.3 birefringence is a good value for current nematic liquid crystals. However, it will be difficult to achieve polymer network liquid crystals (or EO crystals) with a birefringence this large. With a birefringence of 0.3, efficiency versus angle drops off very fast. If birefringence were smaller, efficiency would drop even more quickly. To obtain high efficiency, the steering angles need to be limited when using modulo 2π beam steering with birefringence of 0.3 or less. Using 0.25-deg angular beam steering, Fig. 7.29 shows about 97% steering efficiency for transmissive beam steering. For 0.9-deg steering, the chart shows about 90% efficiency. Reflective beam steering is more efficient because the cell needs to be only half as thick to produce a certain OPO. This is due to the fact that light passes though the cell on its way into the cell, and, after reflection, it traverses the back surface of the cell on its way out of the cell. The reflective mode generated twice the OPD as a transmissive mode device of the same thickness.

7.2.1.6 Quantization-caused reduction in steering efficiency

A second contribution to steering efficiency for modulo 2π liquid crystal beam steerers is from the discrete nature of the phase steps. Equation (7.10), which is repeated here,

$$\eta = \left(\frac{\sin(m\pi/q)}{m\pi/q} \right)^2,$$

calculates this efficiency term. According to this equation, there is very little quantization loss with eight or more steps. At eight steps, there is ∼95%

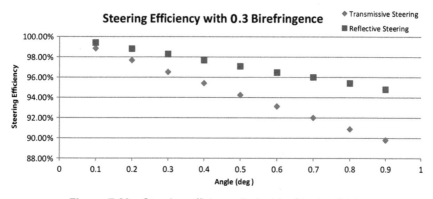

Figure 7.29 Steering efficiency limited by fringing fields.

Table 7.4 Steering efficiency due to quantization of the addressing steps.

Number of steps	Efficiency
4	81.1%
6	91.2%
7	93.5%
8	95.0%
10	96.8%
12	97.7%
14	98.3%
16	98.7%
18	99.0%
20	99.2%

efficiency. Twenty steps are needed to reach 99% efficiency. Table 7.4 shows the steering efficiency due to quantization of the steps.

In a liquid crystal beam steering device, quantification loss is not as high as either the equation or the table indicates. Fringing fields smooth out the steps, so a given discrete phase profile becomes blurred by fringing fields to look more like a linear ramp. Therefore, this loss is overestimated by Eq. (7.10).

7.2.1.7 Steerable electro-evanescent optical refraction

A steerable electro-evanescent optical refractor (SEEOR) uses liquid crystals as an active cladding layer in a waveguide architecture where light is confined to a high-index core and the evanescent field extends into the variable-index liquid crystal cladding. This allows very large (about 2 mm) optical path delays, so for small apertures, it eliminates the need for a 2π reset. Because of the long path, this approach can be a true time-delay OPA. There is therefore no fringing field issue because there are no resets. Also, because the liquid crystal layer is thin, these devices can be relatively fast, under 500 μs in response time. Inplane steering is accomplished by changing voltage on one or more prisms filled with liquid crystals, as shown in Fig. 7.30.[28] Out-of-plane beam steering is based on the waveguide coupler designed by R. Ulrich at Bell Labs in 1971 and is shown in Fig. 7.31.[29] In any waveguide, if the cladding is too thin, light will leak out of the guided mode. In a planar slab waveguide, Snell's law gives the propagation angle of the escaping light.

Since it is possible to tune the effective index of the waveguide, it is also possible to tune the angle of the escaping light. This waveguide-based liquid crystal beam steering can rapidly steer in one direction over wide angles, e.g., 40 deg. In the grating out-coupled dimension, the steering angle is more limited, e.g., to 15 deg in either direction. This is the second steering dimension. One main limitation to this technique is the size of the apertures,

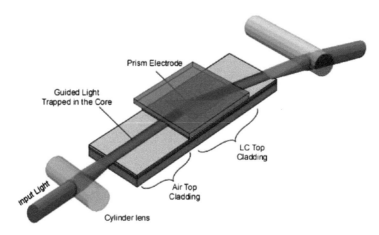

Figure 7.30 SEEOR beam-steering basic design (reprinted from Ref. 51).

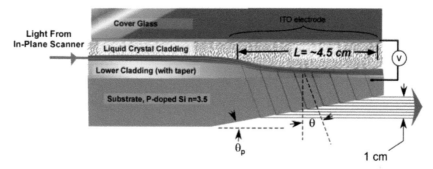

Figure 7.31 Out-coupling approach for the second dimension using SEEOR beam steering (reprinted from Ref. 51).

which is limited to slightly below 1 cm (or less) on a side. Currently, the loss using this technique is fairly high, on the order of 50%.[30]

7.2.1.8 Ferroelectric-SmC*–based beam steering

Ferroelectric liquid crystals are very fast, on the order of microseconds. Ideally, they could be considered either in a format similar to nematic liquid crystal OPAs, or in the polarization birefringent gratings, which is discussed in Section 7.2.6.1. The polarization-birefringent–grating, step-steering approach requires a variable wave plate for which ferroelectric liquid crystals can be a highly suitable material. Ferroelectrics are often used in the Pancharatnam geometry[31] to provide a phase change rather than an OPD change. As mentioned, ferroelectric liquid crystals are very promising for their speed. The variable-waveplate application for use with polarization birefringent gratings is currently in development and is close to becoming available. Using ferroelectric liquid crystals for OPAs is also under research at this time.

7.2.2 Chip-scale optical phased arrays

The physics of OPAs is the same whether you use a chip or birefringent liquid crystals to create the OPD. The major challenges with chip-scale OPAs are the large number of individual T/R modules required and their very small size at this time. One chip is 64×64 pixels with a pitch of 9 μm $\times$ 9 μm, making the array $\sim576 \times 576$ μm.[32] The individual antenna radiators inside a pixel are 3.0 μm in length and 2.7 μm in width. This OPA only transmits; it does not receive, creating a systems level difficulty. Also, to steer light being emitted from a larger aperture, the small size of the aperture means that significant magnification will be required, thus reducing the steering angle. Steering angle is reduced by magnification based on Eq. (7.1). A second issue with this unit is the 7-μm pitch, which limits the steering angle even for the very small aperture [see Eq. (7.13)]. Another chip-scale OPA uses random element pitch to reduce sidelobes but is also a small chip. When phased-array individual elements are larger than a wavelength, spurious sidelobes become an issue.

Another way to consider this is that the largest FWHM angle we can steer to without significant sidelobes is limited by the size of the individual phase shifters according to Eq. (7.13). If we want higher efficiency than 50% (half-maximum), we need to restrict our steering angle to maybe one-fourth of that value. If we use 3 μm as the size of the radiator and 1.5 μm as the wavelength, we can steer to about $1/7^{th}$ rad, or about 7 deg. Also, the chip in Ref. 32 is only about 576 μm, a very small aperture.

Let us assume that we would like to steer a 10-cm beam. This implies a magnification of about a factor of 174, reducing the 7-deg steering angle to 0.04 deg. However, over time, the steering arrays will become larger, and the individual steering elements will became smaller. As mentioned, these chips only transmit; they do not receive. Nanofabricated phased arrays could become an interesting option for steering to small angles by using magnification after steering.

There is current ongoing work with chip-scale OPAs. Most of the research emphasizes chip-scale OPAs steering in only one dimension, and then steering in the other dimension using wavelength tuning. This is very reasonable, except for the wide tuning band required to steer over a large angle. DARPA's Modular Optical Aperture Building Blocks (MOABB) program is pursuing devices up to 10 cm in diameter, but control of the number of required device connections is achieved by using wavelength tuning in one dimension.[33]

7.2.3 Electrowetting beam steering for LiDAR

Electrowetting is an option for steering optical beams but is currently not a very active beam-steering area of research. The approach uses small boxes that contain low-index water and high-index oil. Each sidewall of the boxes

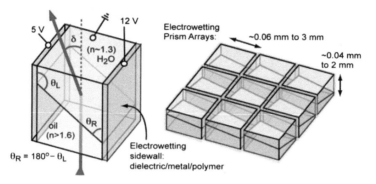

Figure 7.32 Electrowetting beam-steering approach (reprinted from Ref. 34).

has an electrode that modulates the liquid surface wetting contact angle. Figure 7.32 shows this approach.

A prism geometry is created in each box by applying sufficient voltage to any two opposing sidewalls to create a straight but tilted interface between the water and the oil. Electrowetting can easily change the apex angles of the prism by ±45 deg to allow transmissive beam steering out to ±45 deg. Compression and transmission ratios for the device can reduce the net beam-steering efficiency to ~50% at the maximum steering angles. Electrowetting is a fixed-period beam-steering approach. Steering occurs by changing the order of the blaze. If an additional phase layer is utilized, where the additional 0-to-2π–phase layer is the size of each box, continuous steering can be achieved. Without the additional phase layer, this approach will only steer to discrete angles. The size of these elements can be up to a few millimeters. Electrowetting can be used in either reflective or transmissive modes. Switching speed should be on the order of 1 ms for the smallest prisms (tens of microns) and tens to hundreds of milliseconds for millimeter-sized prisms. This size dependence is due to the fact that electrowetting is an electromechanical effect requiring movement of mass (i.e., the liquids). Optical absorption is practically a nonissue in the visible and near-IR regimes. However, for wavelengths >2 μm, it is well understood that the water inside the prism will be a strong absorber. To remedy this, either the water must be replaced with a different liquid or the prisms must be very small (tens of microns high) to minimize optical absorption.

7.2.4 Using electronically written lenslets for lenslet-based beam steering

Creating a moving microlens array electronically allows very precise positioning of the array since the positioning is based on electrode spacing and applied voltages. For simplicity, we assume that the microlens elements are plano-convex, as illustrated in Fig. 7.33. It is possible to calculate *d*, which

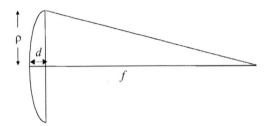

Figure 7.33 The OPD profile of a lens (adapted from Ref. 34).

is the lens' maximum thickness required to create the OPD using the standard sag formula:

$$OPD = (n-1)d = \frac{\rho^2}{2f_1},\qquad(7.17)$$

where n is the refractive index of the microlens, and f_1 is the focal length of the lens. The sag is considered to be the difference between a flat surface and the curved surface.[35] If $\Delta x = \rho$, the radius of the microlens, then the lenslet array will steer to the maximum allowed angle. Using $\Delta x = \rho$, it is possible to see that

$$OPD = \frac{\rho}{2}\tan(\theta_{max}).\qquad(7.18)$$

The effective resets would be at the spacing of the lenslet arrays. There is no physical reset however, so there will be no fringing field issues to address. There is only a change in slope of the OPD at the inflection point between lenslets, but no discontinuity. For small angles, it is possible to set sine and tangent equal to the angle in radians. An OPD four times thinner than a sawtooth phase profile is required to steer to the maximum angle in a lenslet array. A factor of 3.5 reduction in OPD using this approach is a good estimate. The best method of doing fine-angle beam steering between orders is by adding up to a one-wavelength phase delay behind each lenslet in one of the lenslet arrays. The additional phase delay can align the unfolded phases between lenslets. This additional phase delay can make small changes in the location of the orders of the fixed-period beam steerer, allowing us to steer to the desired angles. The additional phase delay can be included in one of the lenslet arrays, or can be a separate, fourth layer, depending on which is easier to implement.

7.2.5 Beam steering using EO effects

EO-crystal–based beam steering, or shaping lenses, has the benefits of being fast and robust, and having no moving parts. The big issue with EO crystals compared to liquid crystals is obtaining a high-index change while applying reasonable voltages that are less than the crystal damage threshold. A high-index change can then create a large OPD, which in turn can steer a beam. The angle–aperture product of an OPD-based nonmechanical beam-steering device depends on the OPD that can be generated. Consider a linear EO effect used in Pockels cells, or a quadratic Kerr effect used in generating an OPD in an EO crystal.

The most common material used when an EO effect is needed to change the index of refraction in an EO crystal is $LiNbO_3$. $LiNbO_3$ is used for modulators, Q switches, and sometimes for beam steering. $LiNbO_3$, GaAs, SBN, and barium titanate are all linear EO crystals. SBN and barium titanate can have large EO coefficients. For certain reasons, such as having a high piezoelectric effect and/or having a highly nonlinear EO tensor, crystals like PMN-PT, PLZT, and KTN are preferred to be used in their cubic phase region; therefore, those crystals are categorized as Kerr-effect EO crystals. KTN has the strongest Kerr effect, while PMN-PT has the second strongest. Since PMN-PT is a relaxor ferroelectric, the slope of its dielectric curve near its Curie temperature is much lower than that for KTN. Therefore, PMN-PT is easier to work with than KTN. Although all crystals exhibit some level of a nonlinear EO effect, we will consider only PMN-PT, PLZT, and KTN as Kerr-effect EO crystals because of their highly nonlinear EO-effect tensors. In Kerr-effect EO crystals, the index change is proportional to the square of the electric field. Although all crystals may be grown as single crystals or multiple crystals (ceramic), PMN-PT, with its high optical quality, is more available commercially in ceramic form as well as being available in single-crystal form. Either form can be used for beam steering.

For most of the crystals, we will consider either electrode patterns to grade the electric field or pulling patterns to grade the material, or both to create a prism-like OPD, as discussed for liquid crystals. For KTN we will discuss charge injection into the crystal. Charge injection can be considered for other materials with a high dielectric constant as well, but we will not address its use with these materials. Charge-injection–based KTN crystals are available as a commercial beam-steering device. For the charge injection case, since the index varies along the thickness of the crystal, we can create a virtual lens that moves up and down rather than creating a prism-like phase profile. Once the displaced beam hits a lens after the KTN crystal, beam steering occurs because the beam is displaced from the center of the lens that follows the KTN crystal. This is the same steering approach as was described in more detail for lenslet-based beam steering.

7.2.5.1 Prism-shaped-OPD–based beam steering

In what we refer to as bulk beam steering, we have light passing through an EO crystal and being steered by creating an OPD. For bulk beam steering in crystals, we do not need resets. Under current practice, one dimension is

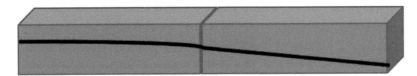

Figure 7.34 Beam deflection in a first crystal followed by the beam traveling through a second crystal. A half-wave-plate (not shown) is between the two crystals.

steered first in one crystal, and then the other dimension is steered in a second crystal. Figure 7.34 shows bulk beam deflection in the first crystal and the beam traveling though the second crystal, where it will be steered in a second dimension.

When the beam exits the first crystal, it is already deflected to a specific down angle. That beam direction is maintained in the second crystal. Both crystals have the same index of refraction so the beam angular deflection at the exit of the first crystal is the same as in the second one. A difficulty with this geometry is that the deflection angle must be kept small, or the beam will hit the side wall of the second crystal. The deflection of the beam inside of a crystal, resulting in the possibility of hitting the side wall, is often called beam walk off. If we have a 20-mm–long crystal to steer each dimension, then by the time the beam reaches the end of the second crystal, we should have an effective walk-off length of about 30 mm. In the first 20-mm–long crystal, we should have an approximated effective walk-off length of 10 mm because the steering occurs gradually over the length of the crystal such that, initially, the beam is not steered to the final steering angle within the crystal. The angle we obtain in the first crystal, however, continues into the second crystal. The steering has already occurred when light hits the crystal that steers the second dimension. In the other dimension, we should have about a 30-mm walk-off length because there is a second 20-mm crystal after the crystal that does the initial steering. In the dimension steered by the first crystal, the second crystal will have to be wider to avoid the beam hitting the side wall due to beam walk off.

The objective for beam steering using an EO crystal is to provide the maximum beam-steering angle–aperture product while minimizing the switching time and loss, assuming that a certain power is supplied to switch the crystal. We also would like to limit the required interaction length of the EO crystals. This not only makes the beam steerer smaller in the interaction length direction, but it also limits blockage of the beam due to the walk off. Defining the average power needed for switching the crystal requires specifying the number of times per second the angle changes, as well as the magnitude of the changes in angle, for each beam steerer. Each time the angle changes, a certain power from the power supply will be required for adding or subtracting charge on the capacitor that is the EO crystal. For EO crystals, steering time is often the time it takes to increase or decrease voltage across the crystal to the target value. This is the case if the steering time is not limited by the crystal parameters. Once voltage is at the target value on the correct electrodes, light will steer to the proper angle if the crystal response is fast enough. Because the steering angle is proportional to voltage, or voltage squared, we need precise voltage control if we wish to point very accurately. We also need to investigate what happens if we point to the same location when changing the voltage down to a value as when changing the voltage up to the same value. Steering to an angle may have some hysteresis.

We can think of an EO crystal as a capacitor. The capacitance of an EO crystal is given by

$$C = \frac{\varepsilon_{r_{ij}} \varepsilon_0 L w}{d},$$ (7.19)

where r_{ij} contains the dielectric constant ε_{ij}, which is a matrix. The dielectric constant can be different for different crystal orientations and for different crystals. We see that to keep the capacitance low, we would like a low dielectric constant; however, as will be explained, a high dielectric constant can allow steering to a larger angle so is desirable even if it means a larger capacitance. To steer to a given angle requires a certain voltage across a crystal. The specific charge needed to apply to the EO crystal (which is effectively a capacitor) to obtain a certain voltage is

$$C = \frac{q}{V}.$$ (7.20)

This requires a certain current over a period of time to place that charge on the crystal:

$$I = \frac{CV}{t},$$ (7.21)

where I is the current, V is the voltage, and t is the time required to charge the capacitor. If we only need to partially charge or discharge the capacitor, then either the time to reach the new angle is less, or the current required is less. Going from zero to a maximum angle requires a certain power, but going from zero to a smaller angle does not require as much voltage change on the crystal, or as much time to switch for a constant current limitation. Equation (7.21) could also be solved for time to switch:

$$t = \frac{CV^2}{P},$$ (7.22)

where P is the power used to charge the capacitor. We could also phrase this as current and power instead of voltage and power, but voltage across the crystal is what causes the optical beam to steer to a certain angle. This tells us how fast we can go from steering to an angle of zero to the maximum angle. We would like to minimize the power required from the power supply because that power will relate to the size of the power supply.

One way to steer to a certain angle using a prism-type OPD profile is to create an OPD on one side of the crystal, preferably with a linear prism profile in the OPD across the crystal. One side of the crystal will have a maximum OPD, and the other side will have zero OPD. This change in OPD across a crystal creates a tilt to the outgoing wavefront. The change in OPD is given by

$$OPD = \Delta nL, \tag{7.23}$$

where L is the length of the crystal, or the length of the crystal with a changed index of refraction, and Δn is the change in index of refraction. While we can develop a larger OPD by using a longer interaction length (up to the size of the largest available crystal), it is desirable to create a larger Δn instead of a larger interaction length as the method of developing a given OPD. As mentioned earlier, a larger interaction length not only makes the beam steerer larger, but may create more of a walk-off issue, where a portion of the beam could hit the side wall of the crystal unless the beam is kept small compared to the crystal size. The amount of OPD generated determines the angle–aperture product for beam steering. A wider aperture means steering to a smaller angle but does not change the angle–aperture product.

Because of the higher index of refraction, we will not need to steer to as large an angle inside the crystal to generate this angle upon leaving the crystal. We can use Snell's law at small angles to determine what angle we need to steer to:

$$\frac{\vartheta_0}{\vartheta_f} = \frac{n_f}{n_0}, \tag{7.24}$$

where n_0 is the refractive index in air, and n_f is the refractive index in the crystal. Solving for ϑ_f, we have

$$\vartheta_f \approx \frac{n_f}{n_0 \vartheta_0}. \tag{7.25}$$

The index of refraction in air is approximately 1. The approximate index of refraction for KTN is 2.29, for PMN-PT is 2.45–2.5, and for SBN is 2.35. Using these values reduces the required internal steering angle for prism-type steering. The index of refraction can vary with wavelength.

It is interesting that the OPD required to steer to a certain angle–aperture product remains the same for a given crystal regardless of the width of the crystal we start with. A wider crystal means a smaller steering angle. The angle–aperture product remains the same. If we magnify to a larger angle, the steering angle decreases, preserving the angle–aperture product even with the larger aperture.

We can generate a Δn that is proportional to the electric field or to the electric field squared, depending on whether the crystal exhibits a large linear EO effect (Pockels effect) or a large Kerr effect. It should be noted that all crystals, even cubic ones, exhibit some degree of Kerr effect; however, that effect is significant only in a few crystals like KTN and PMN-PT. Meanwhile, only certain crystals exhibit the linear EO effect (Pockels effect); in some of them, like PLZT, $BaTiO_3$, KTN, PMN-PT and SBN, that effect is considerable. In crystals like KTN and PMN-PT, which exhibit both large Kerr and Pockels effects, the Pockels effect is larger than the Kerr effect at a temperature below the crystal's Curie temperature; however, that linear effect as well as the piezoelectric effect will disappear above the Curie temperature where the crystal structure turns to cubic. Therefore, for the high-frequency applications that require being above the Curie temperature to eliminate the piezoelectric effect, the Kerr effect can still be advantageous. That is how KTN and PMN-PT are used in fast applications like microsecond beam steering for LiDARs.

As mentioned, KTN is a Kerr effect crystal, so it uses a quadratic function of the electric field to create a change in the index of refraction. When we say that KTN is a Kerr effect crystal, again, we mean that the most useful way to use KTN is in the Kerr effect situation rather than in the linear Pockels-effect situation because of the high piezoelectric effect and the highly nonlinear EO effect tensor or other reasons. This does not mean that it is impossible to use the linear Pockels effect in KTN, but that would be a non-optimum choice for high-frequency applications. This method of describing crystals as used on one mode goes for SBN and PMN-PT as well, even though other less-optimum modes are available. PMN-PT, which exhibits a high Kerr effect in its cubic phase, could also be used in a linear mode as well. The optimum situation for SBN crystal is when it is used in its ferroelectric region where it exhibits a high linear EO effect or Pockels effect. The EO crystals that are preferred to be used in the Pockels effect mode might be poled. That is because those crystals are used in their ferroelectric region and exhibit permanent dipole moments, which are better to be poled. In contrast, the crystals that are preferred to be used in the Kerr effect mode do not need to be poled because they are being used in their cubic phase and do not exhibit any permanent dipole moments.

As mentioned, an EO effect is a change in the refractive index of certain materials in response to an external electric field that varies slowly compared with the frequency of light. Applying a uniform electric field on the crystal exhibiting EO effects, the refractive index change is

$$\Delta\left(\frac{1}{n^2}\right)_{ij} = r_{ijk}E_k + s_{ijkl}E_kE_l + \pi_{ijkl}\sigma_{kl}, \qquad (7.26)$$

where, E_k and E_l are the applied electric field, r_{ijk} is the Pockels effect (linear EO effect) tensor, s_{ijkl} is the Kerr effect (quadratic EO effect) tensor, and π_{ijkl} is the piezo-optical tensor. The values of those tensors are available from the crystal supplier. Because of symmetry, r_{ijk} is reduced to a 6×3 matrix tensor r_{ij}, and s_{ijkl} is reduced to a 6×6 matrix tensor s_{ij}. Note that the stress field is related to the strain tensor T_{rs} according to Hooke's law:

$$\sigma_{kl} = c_{klrs}T_{rs},$$

and

$$\pi_{ijkl} = p_{ijrs}s_{rslk},$$

where p_{ijrs} and s_{rslk} are the elasto-optical tensor and compliances, respectively, and c_{klrs} is a fourth-order tensor (i.e., a linear map between second-order tensors) usually called the stiffness tensor or elasticity tensor. The stresses and strains of the material inside a continuous elastic material are connected by a linear relationship that is mathematically similar to Hooke's spring law and is often referred to by that name. For a crystal under an external field, although the direct photoelastic effect vanishes when no direct stress is applied to the crystal (free crystal), mechanical constraints can still exist due to the converse piezoelectric effect. That is because the electric field will cause a strain by the converse piezoelectric effect, and therefore the indicatrix will become modified by the photoelastic effect, which is called the secondary effect.

The strain produced by an electric field is calculated as $d_{kij}E_k$, where d_{kij} is the piezoelectric tensor. Therefore, the primary and secondary EO effects on a free crystal ($\sigma = 0$) can be calculated as

$$\Delta\left(\frac{1}{n^2}\right)_{ij} = r_{ijk}E_k + s_{ijkl}E_kE_l + \left(pd_{kij}\right)E_k. \qquad (7.27)$$

For a temperature below the Curie temperature, the Kerr effect is negligible compared to the Pockels effect. If the piezoelectric effect is also negligible, Eq. (7.27) reduces to

$$\Delta\left(\frac{1}{n^2}\right)_{ij} = r_{ijk}E_k. \qquad (7.28)$$

For example, if we have a 4-mm crystal such as PMN-PT and apply an electric field in the direction of the z axis of the crystal, considering that $\Delta(1/n^2) = -2\Delta n/n^3$, Eq. (7.28) simplifies to

$$\Delta n_e = \frac{1}{2} n_e^3 r_{33} E_z, \tag{7.29}$$

$$\Delta n_o = \frac{1}{2} n_o^3 r_{13} E_z, \tag{7.30}$$

where, n_e and n_o are the extraordinary and ordinary refractive indices, respectively. If the incident light is polarized perpendicular to the optical axis of the crystal, it sees the ordinary index, and if the light is not polarized perpendicular to the optical axis but propagates perpendicular to the optical axis, it sees the extraordinary refractive index.

If we suppose that n is the appropriate index and r_{ij} is the corresponding element of the Pockels effect tensor to the applied electric field of E_j, we can generally say that the refractive index change will be

$$\Delta n = \frac{1}{2} n^3 r_{ij} E_j. \tag{7.31}$$

Therefore, the optical path delay will be

$$OPD = L\Delta n = \frac{L}{2} n^3 r_{ij} E_j, \tag{7.32}$$

where L is the distance in units of meters, the electric field is in units of volts per meter, and r_{ij} has the units of meters per volt. E_j is voltage divided by crystal thickness. A key term in Eq. (7.32) is the EO tensor r_{ij}:

$$r_{ij} = 2g_{kl} P_j \varepsilon_{ij} \varepsilon_0, \tag{7.33}$$

where ε_{ij} and P_j are the relative permittivity and polarization, respectively, along the polar x_j axis, $\varepsilon_0 = 8.854 \times 10^{-12}$ F/m is the permittivity of vacuum, and g_{kl} is the electrostrictive constant of the para-electric phase, where we can substitute V/d for E. While a high dielectric constant does increase the required voltage and reduces the steering speed, it still is desirable to maximize the high dielectric constant in order to obtain the highest possible OPD.

As mentioned, for a temperature above the Curie temperature of the crystal, the crystal structure turns to cubic; hence, the linear EO effect as well as the piezoelectric effect will be zero. Therefore, only the Kerr effect is responsible for the index change. For the Kerr effect, the s tensor is reduced to a 6×6 matrix due to the symmetry.

For example, if we apply an electric field E_z in the z direction on an isotropic medium (with refractive index n), that medium will turn to a uniaxial crystal with the following indices:

$$n_e = n - \frac{1}{2} n^3 s_{11} E_z^2, \tag{7.34}$$

$$n_{\mathrm{o}} = n - \frac{1}{2}n^3 s_{12}E_z^2. \qquad (7.35)$$

The difference between n_{e} and n_{o} is used to make an OPD as follows:

$$\Delta n = n_{\mathrm{e}} - n_{\mathrm{o}} = \frac{1}{2}n^3(s_{12} - s_{11})E_z^2 = K\lambda E_z^2 = RE_z^2, \qquad (7.36)$$

where K is the Kerr effect constant of the crystal. Therefore, the OPD will be

$$OPD = L\Delta n = LK\lambda E_z^2 = LRE_z^2, \qquad (7.37)$$

where L is the distance in units of meters, the electric field has units of volts per meter, s_{ij} has units of square meter per square volt, K has units of meter per square volt, and R has units of square meter per square volt.

It should be noted that for the Kerr effect, the change in index is related to the dielectric constant squared. As a matter of fact, the high dielectric constant can make up the small–Kerr-effect tensor s_{ij}, but, again, it decreases the steering speed. In that case, the capacitance only increases linearly, but the OPD increases as the square of the dielectric constant. The main disadvantage of thicker crystals is that they require a higher voltage to produce the same electric field. A higher index of refraction is a huge enabler for EO-crystal–based beam steering. We see that Δn is proportional to n^3; however, in addition, the index change due to steering angle changes from the inside to the outside of the crystal adds another factor of n, so the ability to steer a beam is proportional to n^4. The way we switch steering angle is by changing the voltage. We see that angle–aperture steering is proportional to the fourth power of the refractive index, so it can have a lot of influence.

PMN-PT EO crystal (ceramic)–based beam deflection. As mentioned, PMN-PT is preferred to be used in Kerr effect mode because of its high piezoelectric effect and high–Kerr-effect tensor. The dielectric constant ε_{rij} for lithium-doped PMN-PT with varying temperature is shown in Fig. 7.35.[36] Lithium-doped KTN is easier to grow than KTN, but the doping slightly decreases the dielectric constant. A 2004 article by P. Kumar et al. discusses the dielectric constant of PMN-PT.[37] Considering the requested steering speed, we need to maintain a maximum ε_{rij} for PMN-PT. This may require operating the crystal at a higher temperature than room temperature, depending on the PMN-PT composition. Figure 7.35 also shows that in higher frequencies the dielectric constant is reduced, and therefore the refractive index change and deflection angle decrease.

We can substitute V/d for E in Eq. (7.28) to obtain Eq. (7.38) (the R value can be obtained from Ref. 38). Therefore, the refractive index change in the cubic phase of the crystal will be

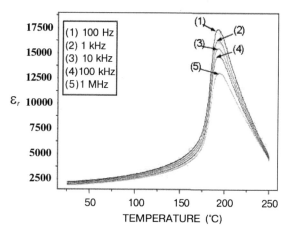

Figure 7.35 Dielectric constant versus temperature for PMN-PT (reprinted from Ref. 36 with permission; © 2004 Elsevier).

$$\Delta n = \pi n^3 R \left(\frac{V}{d}\right)^2. \tag{7.38}$$

This gives Δn in terms of a parameter R rather than in terms of the dielectric constant. It is important to note that d is measured in meters, not millimeters, even though thickness is most often given in millimeters in this book. We can calculate R by

$$R = -\frac{\pi}{3} g_{ij} \varepsilon_0^2 \varepsilon_{rij}^2 \tag{7.39}$$

to make Eqs. (7.38) and (7.39) consistent, while in the Kumar et al. article,[37] R was set to equal 13.3×10^{-16} m^2/V^2. The same article says that the highest R value as of 2004 was 27×10^{-16} m^2/V^2.[37] Boston Scientific Corp. states that their usual R value is about 7×10^{-16} m^2/V^2. If we use Eq. (7.39) and an R value of 7×10^{-16} m^2/V^2, we obtain the index change and steering angles shown in Table 7.5, which provides some large steering angles for PMN-PT. The difficulty is that we do not know the wavelength or temperature dependence of these measurements. Ideally, we will do some measurements on PMN-PT crystals. Note that the values in Table 7.5 are calculated using an R value of 7×10^{-16} m^2/V^2, but Boston scientific has demonstrated an R value of 23×10^{-16} m^2/V^2, and this is actually for a ceramic.

One interesting aspect of PMN-PT is its broadband transmission, as shown in Fig. 7.36.[39] PMN-PT has broader spectral coverage than some of the other possible EO crystal materials, such as KTN (see Fig. 7.37). Notice that KTN does not do as well in the mid-IR region.

Table 7.5 Steering angles for various voltages placed on PMN-PT.

Basic index	Electric field, V/m	Crystal width for 2400 V (mm)	Δn	OPD for 20 mm	Internal angle (mrad)	Walk-off amount (mm)	External angle (mrad)	External angle (deg)	Magnification	Effective angle for 15-cm aperture (deg)	Effective angle for 15-cm aperture (mrad)
2.5	900,000	2.67	0.032	0.64	238.56	2.39	596.41	34.17	37.50	0.91	15.90
2.5	850,000	2.82	0.028	0.57	200.97	2.01	502.43	28.79	37.50	0.77	13.40
2.5	800,000	3.00	0.025	0.50	167.55	1.68	418.88	24.00	37.50	0.64	11.17
2.5	750,000	3.20	0.022	0.44	138.06	1.38	345.15	19.78	37.50	0.53	9.20
2.5	700,000	3.43	0.019	0.38	112.25	1.12	280.62	16.08	37.50	0.43	7.48
2.5	650,000	3.69	0.017	0.33	89.87	0.90	224.68	12.87	37.50	0.34	5.99
2.5	600,000	4.00	0.014	0.28	70.69	0.71	176.71	10.13	37.50	0.27	4.71

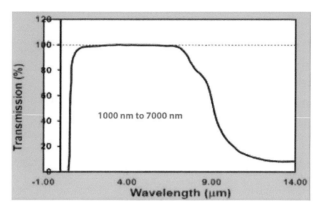

Figure 7.36 Transmission versus wavelength for PMN-PT (courtesy of Boston Scientific).

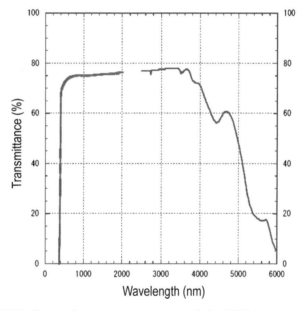

Figure 7.37 Transmission versus wavelength for KTN (courtesy of NTT).

Next, we can estimate refractive index versus wavelength, as given in Fig. 7.38. Refractive index does not changes substantially, but it does change. The equation for the fit to the blue curve is[40]

$$n = 2.4556 + 0.76322e^{-3.4527\lambda} + 35.53e^{-12.605\lambda}. \tag{7.40}$$

The absorption coefficient is given by

$$k = 0.036625 + 0.37695e^{-2.0477\lambda} + 7.1934e^{-11.624\lambda}. \tag{7.41}$$

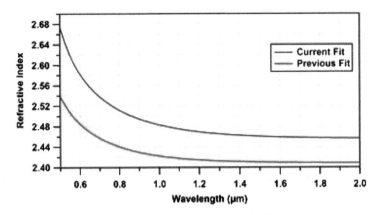

Figure 7.38 New fit to the refractive index of PMN-PT (red, bottom curve) along with the previous fit (blue, top curve) (courtesy of Boston Scientific).

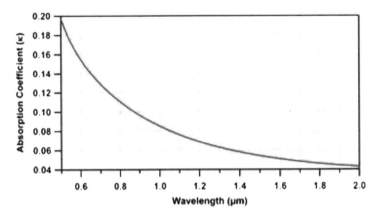

Figure 7.39 Absorption coefficient versus wavelength for PMN-PT. Note that the notation κ (Greek kappa) can also denote the absorption coefficient (courtesy of Boston Scientific).

This so-called absorption coefficient is not purely for absorption; it includes scattering. The actual absorption would be less. A plot of the absorption coefficient of PMN-PT versus wavelength is given in Fig. 7.39.

SBN EO-crystal–based beam deflection. SBN is a linear crystal, so we will be using the linear electric field, or Pockels, effect. Again, this does not mean that SBN only has a linear effect, but that its linear effect is the strongest effect. SBN has a Pockels coefficient between about 400 and 1400 pm/V. According to Ref. 7, the breakdown voltage for SBN is very high, 1300 kV/cm, so we should not have to worry about that issue.[41] Other data, however, indicate that we should restrict our voltages to 700 V/mm. The Altechna web site[42] provides various SBN crystal properties.

Table 7.6 shows the steering angles we can achieve using SBN. We can see from these values that SBN does not steer to as high an angle as PMN-PT. Also, SBN crystal availability is limited.

7.2.5.2 Space-charge-injection–mode KTN beam steering

The available commercial beam steerers using KTN make use of charge injection with titanium electrodes, so a single electrode can be placed on the top and bottom. This creates an effective lens that moves linearly with voltage to create angle steering. With KTN, using charge injection into the crystal, it is possible to make a virtual lens that moves up and down. This lens, when combined with a fixed lens, then steers the light. This is very similar to a moving lenslet. KTN is, however, a fickle material, and when using charge injection, its properties are difficult to control. There is a danger of damaging a crystal unless everything is exactly, correctly performed. Commercial beam-steering devices are available using KTN with charge injection.

If the density of electrons trapped in acceptors in KTN is far larger than that of free electrons in the conduction band, the field generated by trapped charge dominates KTN's EO effect. This is the situation for the commercial KTN beam steerer. Figure 7.40 shows charge density versus depth across two different 1-mm–thick crystals.[43] The voltages in the plot are calculated per millimeters of thickness.

When titanium electrodes are used, charge is injected more strongly into the KTN, resulting in a nonlinear change in electric field that results in a linear change in the index of refraction.[44] Using this approach, NTT has developed a commercial beam-steering product with 0.5-mm–thick KTN crystal, and with 0.7-mm–thick KTN crystal. Since these beam-steering devices are small, significant magnification would be required to use them for larger-aperture beams. The maximum charge density in a KTN crystal for uniform charge density is given approximately by

$$\rho_0 = \frac{\varepsilon V_{\text{DC}}}{d^2}, \tag{7.42}$$

where ρ_0 is the charge density, V_{DC} is the applied voltage, ε is the dialectic constant, and d is the thickness. Illuminating the crystal with 405-nm light to knock electrons out of trapped states into a conduction band, it is possible to clear the crystal of trapped charges. Moderate UV light breaks loose the electrons by pumping the trapped electrons into the conduction band. Table 7.7 shows transit times across the crystal, given an electron velocity of 13.2 m/s.

The lower charge density creates a lens with less power. We want to have the KTN as thin as possible to remain close to the already-tested operating parameters and to maximize the charge in the crystal; however, the thinner the

Table 7.6 SBN steering angles for various EO coefficients.

n_0	Pockels cell coefficient (pm/V)	E (v/m)	Δn	Interaction length (mm)	OPD (mm)	Internal deflection angle (mrad)	Walk-off amount (mm)	Voltage across a 3-mm cell	Angle in air (mrad)	Steering angle (deg)	Magnification	Angle for a 15 cm aperture (deg)	Angle for a 15 cm aperture (mrad)
2.35	440	8.00E+05	0.0023	20	0.046	15.33	0.15	2400	36	2.06	50.00	0.04	0.72
2.35	880	8.00E+05	0.0046	20	0.092	30.67	0.31	2400	72.1	4.13	50.00	0.08	1.442
2.35	1000	8.00E+05	0.0052	20	0.104	34.67	0.35	2400	81.5	4.67	50.00	0.09	1.63
2.35	1400	8.00E+05	0.0073	20	0.146	48.67	0.49	2400	114.4	6.55	50.00	0.13	2.288
2.35	2800	8.00E+05	0.0145	20	0.29	96.67	0.97	2400	227.2	13.02	50.00	0.26	4.544

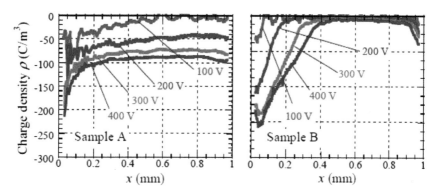

Figure 7.40 Spatial distributions of charge density for two samples (courtesy of NTT).

Table 7.7 Transit times across the crystal.

Distance (mm)	Time (ms)
1.5	0.114
2	0.152
2.5	0.189
3	0.227

crystal the larger the angle we have to relay through the telescope. We will have an electrode on the top and a ground plane electrode on the bottom.

Figure 7.41 is a diagram of a KTN crystal with uniform electrodes on top and bottom, and driven by a power supply. The charge injection method has frequency limitation. It cannot be applied for high and low frequency (from hertz to megahertz). The maximum and minimum frequencies are in the kilohertz region.

Figure 7.42 shows the dielectric constant versus temperature for a particular KTN composition. We see that the dielectric constant peaks at a very high value, and we also see that the peak is very sharp. This means that it is important to carefully control the temperature. A complicating factor is that charge injection can slightly change the Curie temperature. Allowing the temperature to cross the Curie temperature can crack the crystal. This can be an incentive to allow some margin in approaching the Curie temperature,

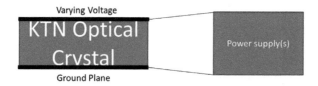

Figure 7.41 Power supply drive characteristics.

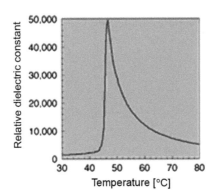

Figure 7.42   Dielectric constant versus temperature for KTN (courtesy of NTT).

although being as close as possible to the Curie temperature will yield the maximum beam-steering effect but will also reduce the steering speed. We need to maximize the dielectric constant with respect to the given steering speed.

If we can control the Curie temperature and the temperature of the crystal, we can use a dielectric constant ε_n of 30,000, which is the same as was assumed in Ref. 45. NTT often uses a dielectric constant of only 17,500 instead of 30,000, so we will use that value, too, for the following reason. The Curie temperature varies along the crystal due to a change in the crystal mix versus length, and it also varies with the amount of charge injected. Notice that to obtain ε_n of 30,000, we need to maintain the temperature of the KTN to a narrow range, close to the Curie temperature. This specific range depends on the exact crystal composition, and the specific ε_n versus temperature may vary along a crystal. Some KTN crystals do not have this large a value of ε_n. It is convenient to have a KTN material with a Curie temperature slightly above room temperature to control temperature by only heating the crystal rather than by both heating and cooling. There are crystal growth techniques that minimize the variation with length of the dielectric constant.

In the pre-charge mode, high voltage (e.g., ±300 V for each 1-mm thickness of KTN) is applied to KTN for 10 s to each polarity. This 20-s procedure is the pre-charge. After the pre-charge, the varying operating voltage is applied. Because of electron injection from both electrodes, the charge distribution is almost uniform throughout the KTN crystal for thin crystals, although it might be better modeled using a hyperbolic cosine function rather than a constant function. However, the trapped electrons slowly dissipate due to their own electric field, as shown in Fig. 7.43. Transmission versus wavelength for a 1-mm path through KTN is shown in Fig. 7.44 (which is the same as Fig. 7.37). KTN looks like a good choice for the near-IR region, but not for the MWIR wavelengths. For the MWIR, PPMN-PT would be preferable.

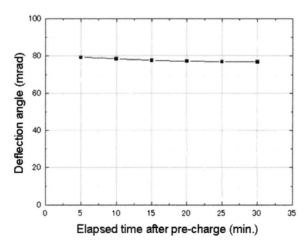

Figure 7.43 Lifetime of trapped charge (courtesy of NTT).

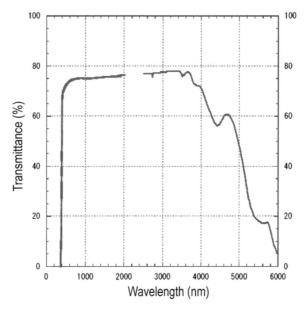

Figure 7.44 Transmission versus wavelength for KTN (courtesy of NTT).

If the density of electrons trapped in acceptors in KTN is far greater than that of free electrons in the conduction band, the field generated by trapped true charge dominates KTN's EO effect. With charge injection, the voltage forms a lens that moves back and forth across the crystal, as shown in Fig. 7.45.

For a Kerr effect crystal that is in the same crystal class as KTN, the refractive index change will be

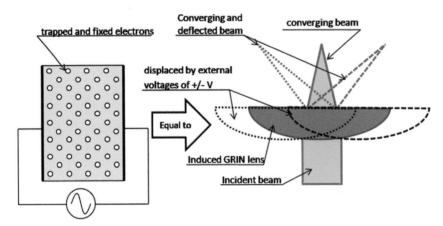

Figure 7.45 The effect of putting a voltage across a KTN crystal with trapped electrons (courtesy of NTT).

$$\Delta n = -\frac{1}{2}\varepsilon^2 n_0^3 g_{11} E^2. \tag{7.43}$$

We also have Gauss' law:

$$\nabla \cdot \mathbf{E} = \frac{\rho}{\varepsilon}. \tag{7.44}$$

For constant charge density ρ, the electric field only depends on the position normal to the electrodes. In that case, Eq. (7.44) can be simplified to

$$\frac{\delta E}{\delta x} = \frac{\rho_0}{\varepsilon}. \tag{7.45}$$

Integrating Eq. (7.45) with respect to x, we obtain

$$E = \frac{\rho_0 x}{\varepsilon} + a, \tag{7.46}$$

where a is a constant. We can then integrate to obtain voltage:

$$\int_{\frac{-d}{2}}^{\frac{d}{2}} E dx = V. \tag{7.47}$$

Integrating Eq. (7.47) and considering the case where $x = 0$, we have

$$a = \frac{V}{d}. \tag{7.48}$$

So Eq. (7.46) becomes

$$E = \frac{\rho_0 x}{\varepsilon} + \frac{V}{d}.$$ (7.49)

We can then substitute to obtain

$$\Delta n(x) = -\frac{1}{2} n_0^3 g_{11} \left(\rho_0 x + \frac{\varepsilon V}{d} \right)^2.$$ (7.50)

We can expand the squared term in Eq. (7.50) into three terms:

$$\Delta n(x) = -\frac{1}{2} n_0^3 g_{11} \left[\rho_0^2 x^2 + 2 \frac{\varepsilon V}{d} \rho_0 x + \left(\frac{\varepsilon V}{d} \right)^2 \right],$$ (7.51)

where $(\varepsilon V/d)^2$ is a constant term that changes index across the whole crystal. This is just a phase delay in the crystal and does not steer the beam. We will ignore that term from now on, reducing Eq. (7.51) to

$$\Delta n(x) = -\frac{1}{2} n_0^3 g_{11} \left[\rho_0^2 x^2 + 2 \frac{\varepsilon V}{d} \rho_0 x \right],$$ (7.52)

except for a constant index change resulting from the dropped term. Equation (7.52) assumes a constant charge density. We have a term that is linear with x, $2(\varepsilon V/d)\rho_0 x$, and one that is quadratic with x, $\rho_0^2 x^2$. The term of $\rho_0^2 x^2$ is called the virtual lens effect term.

The quadratic term provides a lens effect that is proportional to the charge density squared and is independent of voltage. We can see that the power of the lens is proportional to one over the thickness to the fourth power, so a thicker crystal could be an issue. The quadratic term in Eq. (7.52) is

$$\Delta n(x) = -\frac{1}{2} n_0^3 g_{11} (\rho_0 x)^2.$$ (7.53)

Equation (7.53) gives the index profile of the lens. The index of refraction of KTN is 2.29. The g_{11} factor can be estimated from the plot in Fig. 7.46. If we estimate a value of 1.05 at 1550 nm for the product graphed on the y axis of Fig. 7.46, and we have an n_0 of 2.29, then g_{11} will be 0.0774.

The charge distribution within the crystal creates a radial gradient in the index profile within the crystal. A radial gradient index lens has an index profile given by[46]

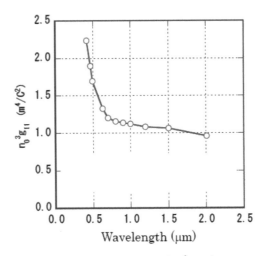

Figure 7.46 Wavelength dispersion of $n_0^3 g_{11}$ (courtesy of NTT).

$$n = n_0\left(1 - \frac{k}{2}x^2\right).$$ (7.54)

We see that we can set k as follows:

$$k = n_0^2 g_{11}\rho_0{}^2.$$ (7.55)

NTT has used a ρ_0 value of 51.2 coulombs (C) and $n_0{}^3 g_{11} = 1.1$. With these values and the value of n_0 given for KTN, we can calculate the value of k.

For a radial gradient lens created by a crystal of length t, the focal length is given by

$$f = \frac{1}{n_0\sqrt{k}\sin(t\sqrt{k})}.$$ (7.56)

We could substitute in for k, but Eq. (7.56) does not simplify that well anyway, so we can just calculate k and then substitute it in. For a 10-mm–long crystal, we calculate k to be 1259 and the focal length to be 34.7 mm.

For constant charge density across the crystal, we can solve for the x location where there is zero change in Δn versus change in x. This will occur at

$$x = \frac{\varepsilon V}{d\rho_0},$$ (7.57)

which can be simplified to

$$x = \frac{Vd}{V_{\text{DC}}}.$$ (7.58)

The position of x with the largest value is related to the voltage and the thickness, and the larger we make V_{DC} the better.

The other aspect of steering with this approach is that with an applied voltage, the centroid of the gradient lens formed in the crystal changes. This change in the location of the centroid moves the location of the focal point of the gradient lens, as illustrated in Fig. 7.47. When combined with a following optic, the displaced focal point produces a beam that is steered. The angle of steering is given by Eq. (7.7), which is repeated here:

$$\tan \vartheta = \frac{\Delta x}{f},$$

where Δx is the amount that the centroid of the gradient lens moves, and f is the focal length of the following lens. There is no beam deflection internal to the crystal and therefore no walk-off issue. This may allow a narrower crystal. The lens we write moves up and down, but we do not steer until we add another lens into the optics system. This means that we do not have to worry about beam walk off for charge-injection–based beam steering. There is, however, an issue of where to add the second lens.

We note that in Fig. 7.47 the location of the beam that is steered changes with the steering angle. This may cause issues when two steering stages must be combined to produce steering in two dimensions. The steered beam can be keep at a constant position on the output lens through the use of a field lens, as shown in Fig. 7.48.

An additional issue that must be addressed is that the steering produced by the first stage in a two-stage (2D) beam steerer may cause the location of the beam at the next steering stage to change. This can be compensated with an additional (fixed) optic, as shown in Fig. 7.49. The relay lens images the

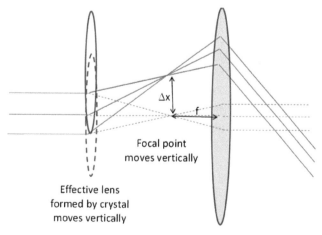

Figure 7.47 Simple decentered lens-based steering (drawing courtesy of Ed Watson).

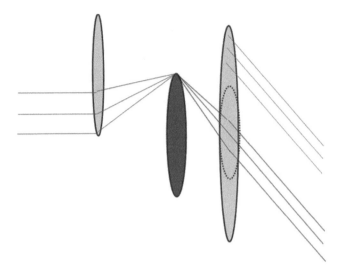

Figure 7.48 Lenslet-based steering using a field lens to enhance fill factor (drawing courtesy of Ed Watson).

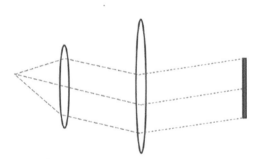

Figure 7.49 Use of a relay lens in lenslet-based steering (drawing by Ed Watson).

plane at the exit of the first steering stage onto the plane of the entrance of the next steering stage (entrance to the next crystal). The relayed image will have unit magnification if the lens is set up in a 2*f*-2*f* configuration. (In a 2*f*-2*f* configuration, when a lens with focal length *f* is placed a distance 2*f* from the image plane, then another image sensor is placed 2*f* beyond the lens, that lens will relay the first image to the second image with 1:1 magnification). The relay lens may present a design challenge because it needs to be of sufficient aperture to gather all of the steered light, but of short enough focal length such that the distance between the steering lens and the next steering stage does not get exceptionally large since the distance between the two is four times the focal length of the relay lens.

We now have a steering angle of

$$\tan \vartheta = \frac{Vd}{fV_{\text{DC}}}. \tag{7.59}$$

Charge injection beam steering is like a moving-lens, beam-steering system for which we need to steer both an outgoing beam and an incoming image.

7.2.5.3 Analysis of EO-crystal-film–based beam steering

The index change of a Kerr material is given by Eq. (7.43). Voltage over thickness gives the electric field. We do not care if the voltage is positive or negative since it is squared for a Kerr effect material. We will only consider Kerr effect materials in this thin crystal analysis. Our baseline will be to use PMN-PT. R is given by Eq. (7.39), repeated here for convenience:

$$R = -\frac{\pi}{3} g_{ij} \varepsilon_0^2 \varepsilon_{rij}^2,$$

where R equals 13.3×10^{-16} m^2/V^2 as a typical value. Kumar et al.[37] state that the highest R value to date is 27×10^{-16} m^2/V^2. The dielectric constant ε_{rij} for PMN-PT is shown in Fig. 7.35. Notice that to maintain a high ε_{rij} for PMN-PT requires operating the crystal at a high temperature. This could be an issue for some applications. ε_0 is 7.7754×10^{-12}.

We also need to consider the maximum steering efficiency that could be obtained based on fringing field effects. Assuming that fringing field effects are the main limitation, we then have diffractive steering efficiency given by Eq. (7.15), repeated here for convenience:

$$\eta = \left(1 - \frac{\Lambda_F}{\Lambda}\right)^2.$$

Equation (7.15) holds if we assume that the horizontal distance for a reset due to fringing fields is the thickness of the crystal layer. The distance between resets for small angles is

$$\Lambda = k \frac{\lambda}{\Delta \vartheta}, \tag{7.60}$$

where k is the number of wavelengths contained in a single reset. We can then obtain

$$\vartheta = \Delta n k (1 - \eta^2) \tag{7.61}$$

with the small angle approximation for transmissive beam steering, and twice that angle for reflective beam steering. The angle in Eq. (7.61) is in radians.

As the electric field increases, we can decrease the cell thickness and required voltage while increasing the electric field. Too much of an increase in the electric field may result in issues with the crystal damage threshold. The thinner we can make the active crystal layer the less degrading influence fringing fields will have on steering efficiency.

7.2.6 Phased-based nonmechanical beam steering

7.2.6.1 Polarization birefringent grating beam steering

A promising, wide-angle steering approach is based on polarized gratings using circularly polarized light. This approach directly creates a phase delay, which eliminates the need for a 2π fly-back region, increasing steering efficiency. This steering is based on the optical model of the quarter-wave-plate, half-wave-plate, quarter-wave-plate (QHQ) device of Pancharatnam, shown in Fig. 7.50.[34] Linear incident light becomes circularly polarized after the first quarter-wave-plate, which can be defined as E_{in} according to Jones calculus notation:

$$E_{in} = \begin{bmatrix} E_{x_{in}} \\ E_{y_{in}} \end{bmatrix} = \begin{bmatrix} E_{x_{in}} \\ i \cdot E_{x_{in}} \end{bmatrix}. \quad (7.62)$$

For convenience, we assume that it is right-hand circularly polarized light. $E_{x in}$ and $E_{y in}$ are vector components along the x axis and y axis, respectively. The transmitted light E_{out} is defined as a linear mapping of the incident light E_{in} by a Jones matrix that represents the half-wave-plate:

$$E_{out} = \begin{bmatrix} \cos\beta & -\sin\beta \\ \sin\beta & \cos\beta \end{bmatrix} \cdot \begin{bmatrix} 1 & 0 \\ 0 & e^{i\phi} \end{bmatrix} \cdot \begin{bmatrix} \cos\beta & \sin\beta \\ -\sin\beta & \cos\beta \end{bmatrix} \cdot \begin{bmatrix} E_{x_{in}} \\ i \cdot E_{x_{in}} \end{bmatrix}, \quad (7.63)$$

where β represents the angle between the slow axis of the half-wave-plate and the x axis, and ϕ denotes the phase retardation of the half-wave-plate, which equals π. The final relationship can be simplified as

$$E_{out} = \begin{bmatrix} E_{x_{in}} e^{i\cdot 2\beta} \\ -i \cdot E_{x_{in}} e^{i\cdot 2\beta} \end{bmatrix}. \quad (7.64)$$

In the last expression, the transmitted light is left-hand circularly polarized with a common phase factor of $e^{i\cdot 2\beta}$. The phase of the transmitted circularly polarized light can be accurately controlled by the azimuth angle β. If β varies horizontally from 0 to π, the spatial phase profile of transmitted light will vary horizontally from 0 to 2π. If we have a liquid crystal cell with an inplane director, the azimuth angle linearly rotating from 0 to π, and the total OPD across the cell agreeing with the half-wave retardation for the design

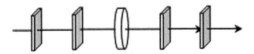

Figure 7.50 The structure of a Pancharatnam beam-steering device. The first and last elements are linear polarizers. The second and fourth elements are quarter-wave-plates, and the middle element is a half-wave-plate (reprinted from Ref. 34).

wavelength, then the final spatial phase profile of transmitted light will linearly change from 0 to 2π. Even though the optical thickness is only a half-wave, it is possible to create a constant large phase gradient over an aperture size that is limited only by manufacturing constraints.

Fixed liquid crystal polarized gratings (LCPGs) with nearly ideal diffraction efficiencies (>99.5%) have been experimentally demonstrated over a wide range of grating periods, wavelengths, and areas. Each fixed–polarization-grating stage can double the maximum steered angle in one dimension without major efficiency reductions, so very large steered angles are possible (at least to ±40-deg field of regard). This is in contrast to volume-holographic, wide-angle steering that requires fine-angle steering both before and after the holographic stack of glass. The structure at the heart of these devices is a polarization grating implemented using nematic liquid crystals. The nematic director is a continuous, inplane, bend–splay pattern established using a UV polarization hologram exposing photo-alignment materials. When voltage is applied, the director orients out of plane, effectively erasing the grating. Diffraction occurs according to

$$\eta_{m=0} = \cos^2\left(\frac{\pi \Lambda nd}{\lambda}\right), \tag{7.65}$$

where η_m is the diffraction efficiency of the mth order, λ is the wavelength of incident light, and $S'_3 = S_3/S_0$ is the normalized Stokes parameter corresponding to the ellipticity of incident light. Note that only these three orders are possible, and that when the retardation of the liquid crystal layer is on the order plus or minus 1, as show here,

$$\eta_{m=\pm 1} = \left[\frac{-\mp S'_3}{2}\right] \sin\left(\frac{\pi \Delta nd}{\lambda}\right), \tag{7.66}$$

then 100% of the incident light can be directed out of the zeroth order. Note further that when the input polarization is circular, all light can be directed into a single first order, with the handedness ($S'_3 = \pm 1$) selecting the diffraction order. Figure 7.51 shows how the molecules rotate without needing any resets as the distance across the cell increases.[34] The molecules rotate around and around, just as the phase goes around and around. No

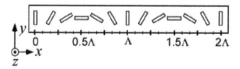

Figure 7.51 Rotation of phase in a Pancharatnam beam-steering device with no resets required (reprinted from Ref. 34).

resets are needed because this is phase-based instead of OPD-based beam steering.

Multiple-stage LCPGs. Multiple stages of polarization birefringent gratings can be cascaded to implement digital beam steering, as shown in Fig. 7.52.[34] The number of stages within the beam-steering stack limits the overall efficiency, which is strongly affected by four factors: steering efficiency, scatter, absorption, and Fresnel loss per layer. Total steering loss from all four factors combined is the fraction of received light not steered into the intended order. Fresnel loss is reflection due to the refractive index difference between layers, primarily between air and the first and last layer. Absorption is mostly due to the transparent conductor layers if indium tin oxide is used. Newer transparent conductors can limit this loss. Scattering is due to imperfections in the material.

With careful engineering, this approach combined with narrow–field-of-regard liquid crystal steering should be able to achieve >90% efficiency over a very wide FOV, with incremental beam steering into ≥1-deg angle increments. Each layer doubles the number of steering angles. Speed is limited by the ability to switch polarization. A polarization rotation stage is between each layer. Room-temperature nematic liquid crystals will switch at ~10 ms. Polymer network liquid crystals can speed this up but may suffer from additional scattering loss. Ferroelectric liquid crystals can also speed this up, hopefully without the additional scattering loss. The difficulty with ferroelectric liquid crystals is obtaining 45 deg of rotation in a given liquid crystal layer. With ferroelectric liquid crystal, the polarization rotation switching speed can be reduced to 10–20 µs. At this time, two ferroelectric liquid crystal layers are required to obtain enough phase shift, but that should change over time. Ferroelectric dynamic half-wave-plates are not quite ready for significant commercial use, but will be before long.

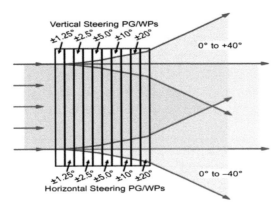

Figure 7.52 An example of a stack of polarization birefringent gratings (PG/WPs is polarization-grating wave plates) (reprinted from Ref. 34).

Holographic gratings for beam steering. Volume (thick) holograms offer the potential to implement large-angle steering with high efficiency. Since they are gratings, they are also a phased-based steering device, but are not based on the classic paper by Pancharatnam.[31]

Through the use of multiple holograms, multiple discrete steering angles can be addressed. With volume holograms, the number of steering states only increases linearly, not geometrically, such as with polarization birefringent gratings. Holographic gratings are very selective in angle. A slight change in input angle means that the light either goes straight through the grating without any interaction, or it is deflected. Because of this high angular selectivity, we can chain angle slightly as a method of addressing each holographic grating. With angle addressing, a small-angle beam steerer selects which hologram is to be addressed, resulting in the desired large angle of deflection. Then another small-angle continuous beam steerer is required to go between the large angles addressed by each hologram. Figure 7.53 shows an adaptive optics layer followed by elements for azimuth and elevation small-angle steering followed by a stack of volume holograms. Lastly, we need another set of azimuth and elevation beam steerers.

Advances in volume holographic gratings in photo-thermal refractive glass make volume holograms an interesting large-angle, step-steering approach. At one time it was a preferred approach, but it has mostly been abandoned because of the development of polarization birefringent gratings. Each glass holographic grating can have >99% efficiency. When two holograms are written in a single piece of glass, the efficiency can still be over 97%. Losses due to scattering and absorption can be less than 0.5% for each holographic grating. These gratings are very capable of handling high power. Currently, the largest-diameter holographic glass available is about 5 cm in diameter. The biggest disadvantage of volume holograms for

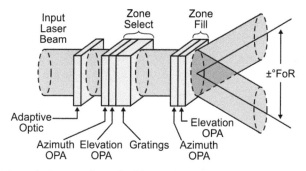

Figure 7.53 Volume-hologram–based wide-angle beam steering. The first element is an adaptive optics layer that can compensate for radial beam distortion, the second and third elements perform fine-angle steering to aim the beam to the correct holographic element, and the last two device layers steer the beam within the area covered by each holographic element (adapted from Ref. 34).

wide-angle step steering is that we need another hologram for each new set of wide angles to be addressed. Many layers of holographic glass can be placed back to back with low loss. For example, we could use eight holograms in each direction: azimuth and elevation. If each hologram steers to an angle separated by 5 deg from the adjacent angle, we have a total field of regard of 40 deg broken up into 8 zones of 5 deg each. Volume holograms only respond to light input at the proper angle. Other light passes through them with very little loss. This means that they have a small acceptance angle and are not suitable for imaging, except over a very small angle.

7.3 Some Optical Design Considerations for LiDAR

7.3.1 Geometrical optics

Geometrical optics is a very important area for LiDAR. There are, of course, whole books on the subject. One of the important areas of research addresses long–focal-length optics to provide high resolution at long range for object identification. Often LiDAR is cued by another sensor, and the LiDAR's task is to identify something. To achieve this, we need as much resolution as possible. Also, it is likely that the acquisition sensor is a thermal imager at longer wavelengths than the LiDAR, so its diffraction-limited resolution will not be as good as a LiDAR at say ~1.5-μm wavelength. A key equation is

$$\text{focal spot diameter} = 2(f/\#)\lambda, \tag{7.67}$$

where the factor of 2 will capture most of the energy in the spot. $f/\#$ is the focal length over the aperture diameter:

$$f/\# = f/D. \tag{7.68}$$

It is clear that the wavelength and the f-number, not the size of the optical aperture, are important.

Figure 7.54 shows the imaging section of an optical system. While for a small $f/\#$ the focal spot can be about one wavelength, we often need to use a large $f/\#$ in order to have detectors large enough to sample the incoming field

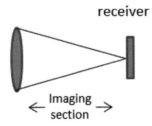

Figure 7.54 Imaging section of an optical system.

as accurately as we would like. Often the DAS times the focal length is set approximately equal to the spot diameter, although there could be interest in oversampling the smallest spot diameter. Another term often used is the numerical aperture, which is given by

$$NA = \frac{1}{2(f/\#)}.$$ (7.69)

The focusing ability might also be important if we decide to use stops to block radiation outside of a certain region. An intermediate focus might be used on receive with solid material around it to block any radiation much larger than the spot size.

7.3.2 Adaptive optics systems

If the receive optics for a sensor is larger than the Fried parameter r_0 of the atmosphere,[47] an adaptive optics mirror can be put in the receive path to compensate for phase distortions. If the transmit optics are larger than the r_0 of the atmosphere, again, adaptive optics can be used to compensate and thereby form a beam with less beam divergence. There are two main parts to any adaptive optics system. One part is the actual device that imposes the phase shift. This can be a mirror with many actuators behind it, or it can be a liquid crystal device or some other device to modulate phase. Adaptive optics mirrors may use electro-strictive (lead magnesium niobate) or PMN actuators, piezoelectric-driven actuators, or other approaches. The second part is the control system that decides which phase shifts to impose. To determine which phases need to be imposed, the incoming signal is first measured. While multiple measurement instruments may be used, the most popular is the Shack–Hartman sensor, shown in Fig. 7.55. A second method of determining the required phase shift is to use a metric to judge the quality of the compensation. One such approach is called stochastic parallel gradient

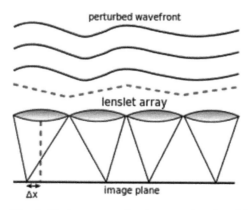

Figure 7.55 Shack–Hartman wavefront sensor (reprinted from Ref. 48).

descent (SPGD), where increments toward an ideal metric are chosen along a gradient.

An issue for adaptive optics systems can be having a point source to use as a guide for finding the optimum phase corrections. Sometimes an artificial point source is created, called a guide star.[49]

7.3.3 Adaptive optical elements

All of the phase modification approaches discussed in the previous section can be used to create a parabolic shape or another shape, as well as to try to create a prism phase ramp. Parabolic shapes can use modulo 2π phase shifting or can use true time delay phase shifting. Section 7.2.4 discusses this capability for the case of writing lenslets. This is, however, only one application of imaging optical systems with writeable lenses that can be changed on demand. The focal length of a LiDAR can certainly be changed. This will be very helpful for adapting to viewing different ranges. Adaptable optical elements based on phase changes will certainly be an important new area of research.

Problems and Solutions

7-1. Calculate the allowed steering angle if we have an OPA with a 3-μm pitch and a 1.5-μm wavelength, assuming that we require 90% steering efficiency. For this calculation, we can assume a Gaussian envelope to the steering efficiency.

7-1 Solution: The full-width, half-maximum width of the steering envelope is approximately 0.5 rad, or 1.5 μm / 3 μm. For simplicity, we can assume a Gaussian distribution in one dimension. Let $f = e^{-\vartheta^2/2a^2}$. For our half-maximum case, the steering angle is 0.25 rad and f is 0.5. We can use those values to calculate the value of a: 0.212. We have a steering angle of 0.0973 rad or 5.57 deg.

7-2. If we use Risley prisms to steer to ±20 deg, what steering angle should one of the prisms have?

7-2 Solution: Each prism should steer to 10 deg. Then when both are tilted, we have a 20-deg deflection.

7-3. A nematic liquid crystal cell has a beam steering speed of 5 ms.

   a) If the width of the cell is cut in half, what is the steering time?
   b) If the voltage is increased by a factor of ten over the threshold voltage, how do the turn-on and turn-off times change?

c) If we use DFLCs instead of nematics, and the DFLCs have the same steering time with voltage matching the DFLC, what is the steering speed if we increase the voltage by a factor of ten?

7-3 Solution:

a) 1.25 ms.
b) The turn-off time remains the same. The turn-on time will decrease by a factor of 99, to about 50.5 μs.
c) If we use a DFLC, then both turn-on and turn-off times decrease to about 50.5 μs.

7-4. If we have a 5-mm–wide crystal, and we steer to ±15 deg using that crystal, then we expand the beam to a 15-cm–diameter aperture maintaining the same beam fill factor for the aperture, what angle will the beam be deflected on exiting the 15-cm–diameter aperture?

7-4 Solution. The angle–aperture product remains the same. For a 15-cm–diameter aperture instead of the 0.5-cm–diameter aperture, we need to magnify the beam 30 times. This decreases the steering angle by a factor of 30, so the steering angle will be 0.5 deg out of a 15-cm–diameter aperture.

7-5. Using PMN-PT with $R = 8 \times 10^{-16}$, an index of refraction of 2.45, a cell thickness of 5 mm, and 5000 V across the cell, what is the index of refraction change? The cell length is 20 mm. What is the steering angle?

7-5 Solution: We use Eq. (7.38): $\Delta n = \pi n^3 R(V/d)^2$. The refractive index change is 0.037. The steering angle is 148 mrad, or 8.47 deg inside of the crystal, but is 2.45 times larger once outside the crystal, or 20.75 deg.

7-6. If the wavelength is λ and the reset period is L, find the blaze term for each saw tooth shown in Fig. 7.21.

7-6 Solution:

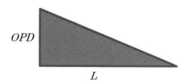

The light at the right side will not take an OPD, but the light at the left side will have an OPD. Therefore, for $x = 0 \rightarrow \phi = 0$, and

for $x = L \rightarrow \phi = kOPD \rightarrow \phi = kOPDx/L = (2\pi/\lambda)OPDx/L$.

7-7. Fraunhofer diffraction theory predicts a far-field pattern for grating illuminated by a plane wave. The grating transmission function is given as

$$T(x,y) = \left[b\left(\frac{x}{d}, \frac{y}{d}\right) \otimes \frac{1}{d^2}\text{comb}\left(\frac{x}{d}, \frac{y}{d}\right) \right] a(x,y),$$

where $b(x,y)$ is the blaze profile, $a(x,y)$ is the aperture shape, and comb(x,y) is the sampling function. The far-field (ff) pattern will be

$$U(x_{ob}, y_{ob}) \propto \left[d^2 B\left(\frac{dx_{ob}}{\lambda r}, \frac{dy_{ob}}{\lambda r}\right) \text{comb}\left(\frac{dx_{ob}}{\lambda r}, \frac{dy_{ob}}{\lambda r}\right) \right] \otimes A\left(\frac{x_{ob}}{\lambda r}, \frac{y_{ob}}{\lambda r}\right),$$

where comb defines the location of the grating modes, B determines the amount of energy in each mode, and A determines the beam profile at each mode. For the blazed term calculated in Problem 7-6, the center of the efficiency pattern is wavelength independent:

$$B = \sin c\left[\frac{dx_{ff}}{\lambda z} - \frac{OPD}{\lambda} \right],$$

where $x_{ff} = \frac{zOPD}{d}$.

a) Show that the zeros-of-efficiency term is wavelength dependent.
b) Plot the normalized amplitude in the far field. What is observed in the plot?

7-7 Solution:

a)
$$x_{o,ff} = \frac{n\lambda z}{d} + \frac{zOPD}{d} \rightarrow$$

the zeros-of-efficiency term is wavelength dependent.

b)

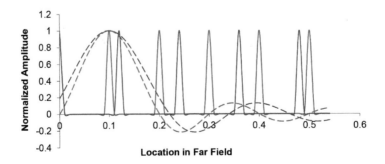

If the phase across a reset period depends on wavelength, then only for the design wavelength will the zeros-of-blaze term lie on grating orders.

7-8. Repeat Problem 7-7 for the case of wavelength-independent blaze ($\phi = 2\pi x/L$).

7-8 Solution:

(a) In this case, the center of the efficiency pattern will be wavelength dependent. We will have

$$B = \operatorname{sinc} c\left(\frac{d x_{ff}}{\lambda z} - 1\right),$$

where

$$x_{ff} = \frac{z\lambda}{d},$$

Therefore, the zeros-of-efficiency term is wavelength dependent:

$$x_{o,ff} = \frac{n\lambda z}{d} + \frac{z\lambda}{d}.$$

If the phase across a reset period does *not* depend on wavelength, then the zeros-of-blaze term always lies on grating orders. We still have dispersion in the primary order.

(b)

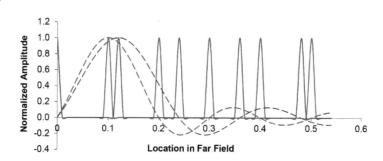

7-9 Plot the minimum thickness of thin film crystal with respect to applied voltage ($\lambda = 1.55$, $n = 2.5$, and $R = 8.2 \times 10^{-16}$ and 1.2×10^{-15})

(a) for reflective steering and
(b) for transmissive steering.

7-9 Solution:

(a)

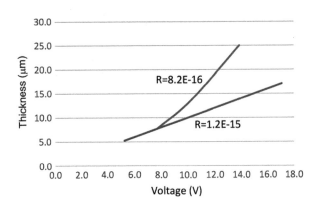

(b)

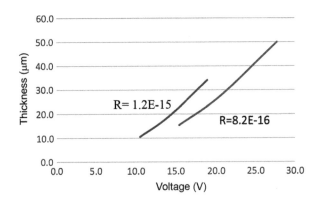

7-10. Using data given in Problem 7-9, plot the steering angle versus applied voltage considering 90% and 95% steering efficiency.

7-10 Solution:

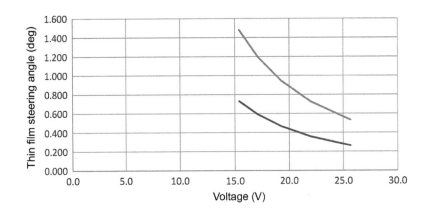

7-11 If $\lambda = 1.55$, and with a steering efficiency of 95%, calculate the reset period, the allowed fly-back distance, and the electrode pitch for 8 steps per ramp for

(a) Steering angle = 5 deg
(b) Steering angle = 30 deg

7-11 Solution:

a) Reset period = 17.76 μm, allowed fly-back distance = 2.83 μm, electrode pitch for 8 steps per ramp = 2.22 μm.
b) Reset period = 2.96 μm, allowed fly-back distance = 0.47 μm, electrode pitch for 8 steps per ramp = 0.37 μm.

7-12 For a reflective thin film, plot the reset distance versus the angle of deflection ($\lambda = 1.55$).

7-12 Solution:

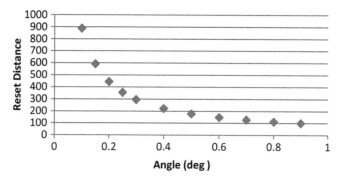

7-13. For the reflective thin film in Problem 7-12, plot the beam-steering efficiency versus angle of deflection in the following cases: birefringence = 0.1, 0.131, and 0.2.

7-13 Solution:

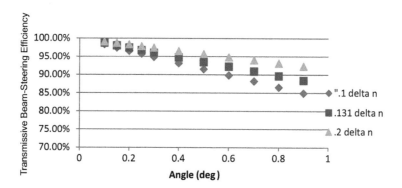

7-14. Plot the refractive index and absorption coefficient for visible-spectrum light for PMN-PT crystal.

7-14 Solution:

$$n = 2.4556 + 0.76322e^{-3.4527\lambda} + 35.53e^{-12.605\lambda}.$$

$$k = 0.036625 + 0.37695e^{-2.0477\lambda} + 7.1934e^{-11.624\lambda}.$$

Visible light falls approximately from 450 nm to 750 nm.

```
l=1e-3*(450:0.01:750);
n=2.4556+0.76322*exp(-3.4527*l)+35.53*exp(-12.605*l);
k=0.036625+0.37695*exp(-2.0477*l)+7.1934*exp(-11.624*l);
figure
plot(l*1e3,n)
xlabel('wavelength [nm]')
ylabel('Index of Refraction')
figure
plot(l*1e3,k)
xlabel('wavelength [nm]')
ylabel('Absorption')
```

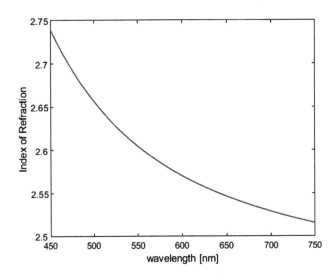

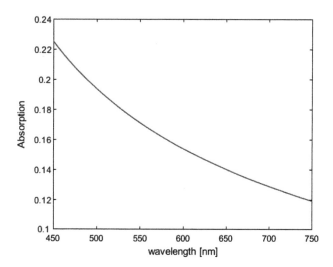

7-15. Plot the refractive index and the absorption coefficient for the spectrum from 750 nm to 2.0 μm for PMN-PT crystal. What is the absorbance at 1550 nm, 1064 nm, and 750 nm?

7-15 Solution:

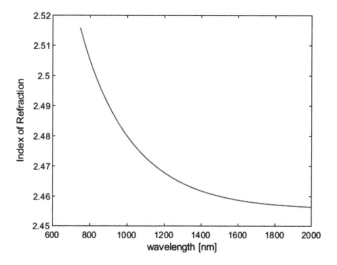

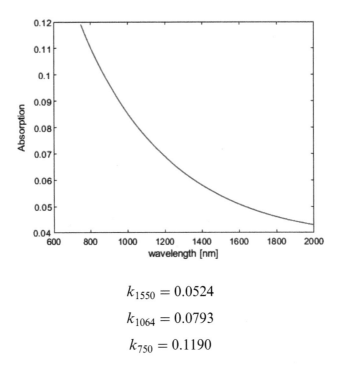

$$k_{1550} = 0.0524$$

$$k_{1064} = 0.0793$$

$$k_{750} = 0.1190$$

7-16. Consider an imaging lens with a numerical aperture of 0.20 and an aperture diameter of 5 cm. Estimate the focal length of the lens.

7-16 Solution:

$$NA \approx \frac{1}{2(f/\#)},$$

$$(f/\#) = \frac{f}{D},$$

$$NA \approx \frac{D}{2f} \rightarrow f \approx \frac{D}{2NA} = 125\,\text{mm}.$$

Notes and References

1. National Academy of Sciences, *Laser Radar: Progress and Opportunities in Active Electro-Optical Sensing*, P. F. McManamon, Chair (Committee on Review of Advancements in Active Electro-Optical Systems to Avoid Technological Surprise Adverse to U.S. National Security), Study under Contract HHM402-10-D-0036-DO#10, National Academies Press, Washington, D.C., p. 231 (2014).

2. Princetel, Inc. tutorial on fiber optics rotary joints: http://www.princetel.com/tutorial_forj.asp.

3. Churchill Navigation website: http://www.churchillnavigation.com.

4. Personal communication from Jeanette Lurier, Raytheon, July 31, 2013.

5. National Academy of Sciences, *Laser Radar: Progress and Opportunities in Active Electro-Optical Sensing*, P. F. McManamon, Chair (Committee on Review of Advancements in Active Electro-Optical Systems to Avoid Technological Surprise Adverse to U.S. National Security), Study under Contract HHM402-10-D-0036-DO#10, National Academies Press, Washington, D.C., p. 232 (2014).

6. Wikipedia page on galvanometers: http://en.wikipedia.org/wiki/Galvanometer.

7. Cambridge Technology website: http://www.cambridgetechnology.com/technology.

8. Private discussion with Denny Adams from Lockheed Martin in Nov. 2018.

9. Encyclopedia Britannica entry on Lissajous figure: https://www.britannica.com/topic/Lissajous-figure.

10. C. Schwarze (Optra Inc.), "A new look at Risley prisms," *Photonics Media*, June (2006).

11. C. Oh, J. Kim, J. Muth, S. Serati, and M. J. Escuti, "High-throughput, continuous beam steering using rotating polarization gratings," *IEEE Photonics Technology Letters* **22**(4), 220–202 (2010).

12. National Academy of Sciences, *Laser Radar: Progress and Opportunities in Active Electro-Optical Sensing*, P. F. McManamon, Chair (Committee on Review of Advancements in Active Electro-Optical Systems to Avoid Technological Surprise Adverse to U.S. National Security), Study under Contract HHM402-10-D-0036-DO#10, National Academies Press, Washington, D.C., p. 233 (2014).

13. Industrial Laser Solutions Editors, "OPTRA introduces Risley prism pairs for fast steering laser beams to laser machining market," PennWell Corp., Tusla, Oklahoma (2011).

14. D. J. Adams, A. F. Lindsay, B. L. Stamper, C. T. Knorr, A. P. Seibel, and M. R. Crano, "Risley integrated steering module," U.S. Patent No. 7, 797, 792, March 12, 2011.

15. G. F. Marshall, "Risley prism scan patterns," *Proc. SPIE* **3787**, 47–86 (1999) [doi: 10.1117/12.351658].

16. P. C. D. Hobbs, *Building Electro-Optical Systems: Making It All Work*, Second Edition, John Wiley & Sons, Hoboken (2009).

17. L. Wu, S. Dooley, E. A. Watons, P. F. McManamon, and H. Xie, A tip-tilt-piston micromirror array for optical phased array applications, *Journal of Micromechanical Systems* **19**(6), 1450–1461 (2010).

18. B. Borovic, A. Q. Liu, D. Popa, H. Cai, and F. L. Lewis, "Open-loop versus closed-loop control of MEMS devices: choices and issues," *J. Micromech. Microeng.* **15**(10), 1917–1924 (2005).

19. Email from Huikai Xie dated Aug. 20, 2018.

20. Discussions with Huikai Xie from August 2018.
21. E. A. Watson, "Analysis of beam steering with decentered microlens arrays," *Opt. Eng.* **32**(11), 2665–2670 [doi: 10.1117/12.148100].
22. P. J. M. Suni, J. Bowers, L. Coldren, S. J. B. Yoo, and J. Colosimo, "Coherent LiDAR-On-A-Chip," Presented at IEEE National Aerospace & Electronics Conference (NAECON) 2017.
23. Wikipedia page on Fresnel lens: https://en.wikipedia.org/wiki/Fresnel_lens.
24. B. R. Hatcher, "Granularity of beam positions in digital phased arrays," *Proc. IEEE* **56**(11), 1795–1800 (1967).
25. Y.-H. Fan, Y.-H. Lin, H. Ren, S. Gauza, and S.-T. Wu, "Fast-response and scattering-free polymer network liquid crystals for infrared light modulators," *Applied Physics Letters* **84**(8), 1233–1235 (2004).
26. J. L. West, G. Zhang, and A. Glushchenko, "55.1: stressed liquid crystals for electrically controlled fast shift of phase retardation," *Society for Information Display International Symposium Digest of Technical Papers* **34**(1), 1469–1471 (2003).
27. P. F. McManamon, T. A. Dorschner, D. L. Corkum, L. J. Friedman, D. S. Hobbs, M. Holz, S. Liberman, H. Q. Nguyen, D. P. Resler, R. C. Sharp, and E. A. Watson, "Optical phased array technology," *Proc. IEEE* **84**(2), 267–297 (1996).
28. National Academy of Sciences, *Laser Radar: Progress and Opportunities in Active Electro-Optical Sensing*, P. F. McManamon, Chair (Committee on Review of Advancements in Active Electro-Optical Systems to Avoid Technological Surprise Adverse to U.S. National Security), Study under Contract HHM402-10-D-0036-DO#10, National Academies Press, Washington, D.C., p. 237 (2014).
29. ibid.
30. The company making these devices has been purchased by Analog Devices, which has an interest in auto LiDAR applications. This will likely mean that additional money will be invested in the engineering to bring down this loss and to make these steering devices inexpensive. It is likely that after probable investment, this will be an attractive beam-steering technique when small apertures are applicable.
31. S. Pancharatnam, "Achromatic combinations of birefringent plates: Part II: An achromatic quarter-wave plate," *Proc. Indian Academy of Sciences* Vol. **41**, No. 4, Sec. A, pp. 137–144 (1955).
32. J. Sun, E. Timurdogan, A. Yaacobi, E. S. Hosseini, and M. R. Watts, "Large-scale nanophotonic phased array," *Nature* **493**, 195–199 (2013).
33. G. Keeler, "Opportunities for LIDAR and free-space optical communication using micro-scale photonics technologies," presented at the OSA conference on Application of Lasers for Sensing and Free Space Communication, 25–28 June 2018, Orlando, Florida.

34. P. F. McManamon, P. J. Bos, M. J. Escuti, et al., "A review of phased array steering for narrow-band electrooptical systems," *Proc. IEEE* **97**(6), pp. 1078–1096 (2009).

35. Wikipedia page on lens sag: https://en.wikipedia.org/wiki/Lens_sag.

36. Personal communication from Shogo Yagi (NTT Corp.), August 27, 2018.

37. P. Kumar, S. Sharma, O. P. Thakur, C. Prakash, and T. C. Goel, "Dielectric, piezoelectric and pyroelectric properties of PMN-PT (67:32) system," *Ceramics International* **30**, 575–579 (2004).

38. H. Jiang, Y. K. Zou, Q. Chen, K. K. Li, R. Zhang, Y. Wang, H. Ming, and Z. Zheng, "Transparent electro-optical ceramics and devices," *Proc. SPIE* **5644**, 380–394 (2004) [doi: 10.1117/12.582105].

39. Personal communications from Xiaopei (Aileen) Chen at Boston Applied Technologies, Inc. in 2017 and in August 2018.

40. Personal communications from Xiaopei (Aileen) Chen to Michael Finlan on August 8, 2014, then to the author from Michael Finlan multiple times, the latest being August 27, 2018.

41. T. Liu, G. Chen, J. Song, and C. Yuan, "Crystallization kinetics and temperature dependence of energy storage properties of niobate glass-ceramics," *Transactions of Nonferrous Metals Society of China* **24**(3), 729–735 (2014).

42. Altechna website: http://www.altechna.com/.

43. T. Imai, J. Miyazu, and J. Kobayashi, "Charge distributions in $KTa_{1-x}Nb_xO_3$ optical beam deflectors formed by voltage application," *Optics Express* **22**(12), 14114 (2014).

44. K. Nakamura, "Optical beam scanner using Kerr effect and space-charge-controlled electrical conduction in $KTa_{1-x}Nb_xO_3$ crystal," *NTT Technical Review* **5**(9), 1–8 (2007).

45. S. Yagi, "KTN crystals open up new possibilities and applications," *NTT Technical Review* **7**(12), 1–5 (2009).

46. M. Riedl, *Optical Design Fundamentals for Infrared Systems*, Second Edition, SPIE Press, Bellingham, Washington, p. 95 (2001) [doi: 10.1117/3.412729.ch5.]

47. The Fried parameter is a measure of the quality of optical transmission through the atmosphere due to random inhomogeneities in its refractive index. It is defined as the diameter of a circular area over which the root-mean-square aberration due to passage through the atmosphere equals 1 rad. As such, imaging from telescopes with apertures much smaller than r_0 is less affected by atmospheric inhomogeneities than by diffraction due to the telescope's small aperture. However, the imaging resolution of telescopes with apertures much larger than r_0 (thus, including all professional telescopes) will be limited by the turbulent atmosphere, preventing the instrument from approaching the diffraction limit.

48. Wikipedia page on adaptive optics: http://en.wikipedia.org/wiki/Adaptive_optics.

49. Wikipedia page on laser guide star: http://en.wikipedia.org/wiki/Laser_guide_star.

50. P. McManamon, *Field Guide to Lidar*, SPIE Press, Bellingham, Washington (2015) [doi: 10.1117/3.2186106].

51. S. R. Davis, G. Farca, S. D. Rommel, S. Johnson, and M. H. Anderson, "Liquid crystal waveguides: new devices enabled by >1000 waves of optical phase control," *Proc. SPIE* **7618**, 76180E (2010) [doi: 10.1117/12.851788].

Chapter 8
LiDAR Processing

8.1 Introduction

A LiDAR receiver gathers data from each detector. This data can then be converted to a 2D image, a 3D point cloud image, object velocity, or some other information that a LiDAR can generate by processing the received data, depending on the particular LiDAR system. One of the objectives of this chapter is to discuss how to go from detections to images for various LiDAR types. Another level of processing (automatic object recognition) uses that image data to automatically detect, recognize, or identify an object, but this will not be part of the discussion. What will be considered is the human interface and getting data into a format (often an image) that can be absorbed by a human. A second objective of the chapter is to discuss sensor metrics for evaluating sensor performance. There are well-established 2D sensor metrics, but for 3D sensors and other dimensions available from LiDAR, no effective metrics have been established. Some metrics are evolving and can be discussed.

8.2 Generating LiDAR Images/Information

8.2.1 Range measurement processing

A LiDAR measures range by taking the convolution of the "pulse" emitted by the LiDAR with the range profile of the target. This creates a measured range profile of the target corrupted by the fact that the LiDAR pulse is not a delta function. The word pulse is placed in quotation marks because it may in fact be a pulse, but there are other methods of marking the range for a LiDAR, such as linear frequency modulation (LFM) or pseudo-random codes. The range to the target is a measured range profile. A simple rectangular pulse can be assumed with a certain pulse width, or a Gaussian pulse shape or more-complex pulse shapes can be used. A linear frequency chirp, or a pseudo-random–coded waveform, will have an equivalent pulse shape as a method of determining range. A convolution is given by

$$f * g = \int_{-\infty}^{\infty} f(\tau)g(t - \tau)d\tau. \tag{8.1}$$

This can, of course, be calculated digitally. The MATLAB® command 'conv(u,v)', where u and v are the two range profiles in our case, performs a digital convolution. Equation (8.1) can be made discrete for a digital convolution, becoming

$$w(k) = \sum_{j} u(j)v(k - j + 1). \tag{8.2}$$

Another processing issue with respect to range measurement accuracy is the LiDAR clock frequency. If the receiver is digitally sampled, this sampling rate will limit the range resolution. Let's say that the digital clock frequency is 0.25 ns. This will work well for sampling a 3-ns pulse but would not work if we had to sample a 0.5-ns pulse. Also, if this frequency drifts, it would cause an error because we would assume in our processing that the signal is sampled every 0.25 ns. Any deviation in the clock frequency would cause an error. The matched filter of an object is the autocorrelation of the target with an impulse response, so if we had a delta function waveform and we did ideal processing, we would have a matched filter.

8.2.2 Range resolution of LiDAR

A LiDAR measures range by timing the travel time of a laser pulse to the target and to return:

$$\Delta R = \frac{c}{2B}. \tag{8.3}$$

The factor of two is included because the laser pulse has to go to the target and return. For calculating LiDAR range resolution, it is easy to think of the speed of light as 1 ft per ns or 0.3 m per ns. This means that a 1-ns pulse will have about 6 in. or 0.15 m of resolution. Often LiDAR pulses are on the order of nanoseconds. If the LiDAR receiver has sufficient bandwidth to follow the return signal, the measured return signal for a pulsed LiDAR will be the convolution of the laser pulse and the target.

If the LiDAR pulse was a delta function and we had sufficient SNR, the range profile of the target would be captured exactly if we capture a set of range returns. Figure 8.1 shows how the return signal is spread by the target range profile. Delta functions do not spread the return, but more extended targets do spread it. Not all LiDARs use a pulsed waveform. We need some type of time tag to determine how long it takes the laser pulse to go to the

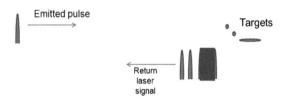

Figure 8.1 Returns from targets of various range depths (reprinted from Ref. 5).

target and return. This time tag can be provided by frequency chirping the LiDAR signal. It can also be provided by pseudo-random changes in frequency or in phase. Range resolution is proportional to 1 over the signal bandwidth, regardless of how the bandwidth is controlled. To use LFM or pseudo-random–code modulation, we need to use heterodyne detection, or else we will not see the frequency/phase change. Direct detection LiDARs cannot use LFM or pseudo-random–coded waveforms because these LiDARs do not measure phase.

8.2.2.1 Nyquist sampling a range profile

The Nyquist sampling theorem states that a signal can be reconstructed from a set of samples if the sampling rate is at least twice its maximum frequency. We need to meet this criteria to reconstruct the range profile. To accurately reconstruct the shape of the return range profile, it is beneficial to sample more frequently than twice during the highest frequency. If we do not sample often enough, we can have aliasing, in which a high frequency is aliased down to an incorrect lower frequency. We can look at the sampling either in Fourier space or in real space. In Fourier space, we convert the return range profile pulse into a series of sine waves representing each frequency. You can see from Fig. 8.2 how the same data could look like more than one sine-wave frequency. If you do not sample a sine wave often enough, you end up thinking it is a lower frequency than it really is.

We could consider a Gaussian-shaped return pulse. Figure 8.3 shows a Gaussian wave [blue curve (1)], a Gaussian wave sampled once every unit on the x scale [green curve (2)], and a Gaussian wave sampled once every other

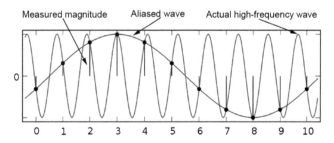

Figure 8.2 Sampling a sine wave (adapted from Ref. 5).

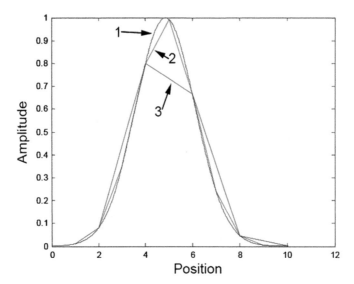

Figure 8.3   Sampling a Gaussian-shaped pulse.

unit on the x scale [red curve (3)]. We can see that when sampling is reduced, the accuracy of replicating a Gaussian wave goes down.

8.2.2.2 Unambiguous range

If the range to an object is sufficiently long such that more than one laser pulse is traveling to and from the target at a given time, there could be an ambiguity in range. For a fixed pulse repetition rate, the unambiguous range is given by

$$R_{\text{unambig}} = \frac{c\tau}{2},$$

(8.4)

where c is the speed of light, and τ is the time between pulses. This limits how far away an object can be without ambiguity. When an object is farther than the unambiguous range, it is ambiguous which pulse we are receiving because more than one pulse is in the air at the same time. One way to extend the unambiguous range is to change the time between pulses. Then, until the pattern of changes starts to repeat, we can still tell exactly which laser pulse we are receiving. Figure 8.4 shows a pattern in which a small increment is added to each interval between pulses.

Figure 8.4   Varying the pulse repetition frequency to expand the unambiguous range.

As long as the period keeps changing, we can determine which pulse most recently traveled from the sensor to the target and back again, allowing us to measure range without any ambiguity. There will still be a range ambiguity at some range, but it will depend on when the pulse pattern repeats itself. We can have as many pulses in the air at one time as the number of different time intervals between pulses. If we had an unambiguous range of some distance (e.g., 40 km) with a fixed pulse repetition frequency (PRF), and we made a minor change in the time between pulses for ten pulses in a row, we would end up with an unambiguous range of approximately 400 km.

Unambiguous range is only relevant if you care about the exact range of an object. For tracking or imaging an object, we may not need to resolve range ambiguity. Range ambiguity is only a concern when we want to determine absolute range.

8.2.2.3 Threshold, leading edge, and peak detectors

Some detectors such as GMAPDs use the entire pulse to sample range (because in that case there is some probability of an avalanche any time during the pulse). Other detectors can use the leading edge of a pulse to determine range, or can use peak (or threshold) detection. Of course, leading-edge or threshold detection means that you are using only a portion of the pulse energy for range measurement, but it also means more range accuracy for a given pulse width. In theory, the pulse width does not matter if you are only using the leading edge. Only using a portion of the pulse requires more laser power.

Another issue with leading-edge detection is that it can be biased by higher versus lower SNR, as can be seen in Fig. 8.5, in which the object is at the same

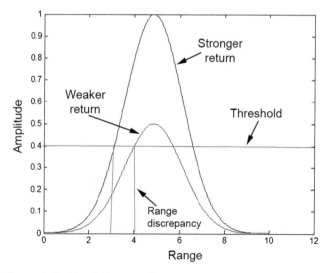

Figure 8.5 The influence of amplitude on threshold detection.

range, but a higher SNR shows it as closer because the threshold for detection is reached at a closer range. If, however, we can measure range as a ratio of pulse amplitudes, always measuring at a percentage level such as 40% or the 50%, we can eliminate this bias, but we then have to measure both the peak and the threshold level. You can see from Fig. 8.5 that the 50% value from peak is the same for both signal amplitudes. This method is sometimes called constant fraction detection. Peak detection is somewhat unreliable because any noise causes fluctuations in the peak value.

8.2.2.4 Range accuracy, range precision, and range resolution

One area that causes controversy involves the discussion of range resolution, range precision, and range accuracy. Often people do not handle these terms carefully, resulting in unnecessary arguments. Range resolution measures the ability to split two objects. It quantifies the ability to detect two objects separated in range along the same line of sight, as shown in Fig. 8.6. If, when we sum the two range profile intensities, there is a dip between them, we have two separate returns. The bandwidth B of the signal determines the range resolution. The resolution is limited to $c/(2 \times B)$.

Range precision quantifies the relative uncertainty of a range measurement to an object. High SNR allows us to measure range precision to a small fraction of the range resolution. The better the SNR the higher the ratio of range precision to range resolution.

Range accuracy then depends on range precision as well as systematic errors (e.g., clock rate, drift, timing offsets, etc.). It quantifies the degree to which a range measurement yields the absolute or true value of range. To

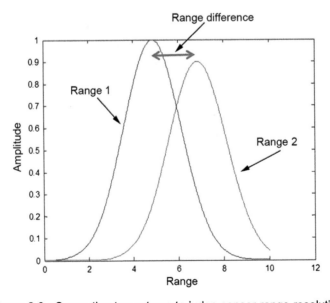

Figure 8.6 Separating two returns to judge sensor range resolution.

make use of range resolution, we need to be able to make multiple range measurements in order to see the two peaks, or at least a dip between the peaks. For a LMAPD, this necessitates a ROIC that can capture multiple range measurements. For a GMAPD, this means making multiple range measurements on multiple pulses.

8.2.3 Angle LiDAR processing

8.2.3.1 Point spread function

The point spread function (PSF) is a spatial impulse function of the system. The PSF, along with the detector angular subtense (DAS), allows the LiDAR to measure angle/angle resolution. The DAS samples the angular PSF. Diffraction effects of the LiDAR receiver, blurring effects of the atmosphere, and any optical aberrations combine to produce the PSF. In most cases, we will place the FPA in the image plane to sample the PSF. Alternatively, using a direct-detection LiDAR, the pupil plane field can be captured with a coherent LiDAR or can be estimated by guessing at the phase and using a hill-climbing metric. For a coherent LiDAR, the FPA can be in the pupil plane or the image plane. Once the field is known, we can convert from image plane to pupil plane and back again. Then the size of the pupil plane sample along with the atmosphere will determine the PSF. Perfect, nonaberrated imaging optics can be modeled as a phase screen with a phase profile of

$$t_{\text{lens}}(w_{\text{az}}, w_{\text{el}}) = e^{\frac{-j\pi(w_{\text{az}}^2 + w_{\text{el}}^2)}{\lambda f}}, \tag{8.5}$$

where λ is wavelength, f is focal length, and the two w terms are the distance from the center of the lens, limited of course by the radius of the lens. This phase function will bring the received pupil-plane function to an image at a distance f from the lens. The smallest spot a lens can make depends on the diffraction limit of the lens. The diameter of the Airy disk, which is the PSF produced by a circular lens, can be defined as

$$D_{\text{Airy}} = 2.44 \cdot \frac{\lambda f}{D}, \tag{8.6}$$

where D is the lens diameter. The Airy disk is the diameter out to the zeros of the diffraction-limited spot at the focus for a circular aperture. The focal length over D is the $f/\#$ of the lens. Usually the DAS is set at approximately the angular extent of D_{Airy}/f. For a scanned LiDAR, we could have an illuminated region smaller than the DAS. In that case, the angular resolution would be limited by the illuminated spot size in angle, although the optics and detector will have their own PSFs that also need to be considered.

8.2.3.2 Microscanning of LiDAR images for improved sampling

Sometimes we would like to sample at a higher sampling density than provided by the DAS. We can take an image with the FPA in one location, then move the image on the FPA by a fraction of a DAS (often one-half of the DAS) to make a new image, as shown in Fig. 8.7.[1,2] This process increases the sampling in angle/angle space. It can be used with passive sensors, but also with LiDARs. Of course, a method to impose the small motion is needed. An ideal method of imposing the small motions is failure to stabilize the LiDAR well, combined with an inertial measurement unit (see Section 8.2.5.2) that measures and reports the location and orientation of the LiDAR. This is random microscanning, unlike microscanning imposed by uniform motion. Microscanning can interact with the various beam-steering approaches. We can chose to step the whole image on the FPA from one location to another to increase sampling, or we can impose micro-motion, which can increase sampling just as much, but with much less required angular motion. Another aspect to consider when microscanning is the size of the focal spot. Resolution can be limited by the DAS or by the size of the focal spot. With micro-scanning, we are increasing sampling but not resolution, so both the DAS and the diameter of the focal spot are larger than the amount of microscan micro-motion. This limits the focal lengths that can be considered in LiDAR design.

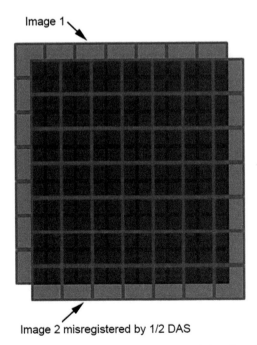

Image 1

Image 2 misregistered by 1/2 DAS

Figure 8.7 Microscanning uses two captured images slightly misregistered, resulting in a higher sampling (reprinted from Ref. 5).

8.2.3.3 Multiple-subaperture spatial heterodyne processing

In spatial heterodyne processing, we measure the field across a FPA in either the image plane or the pupil plane. If we measure the field in the pupil plane and Fourier transform it, we have a central autocorrelation area that is bright. This is the autocorrelation of the LO summed with the autocorrelation of the signal, which is usually of much lower intensity than the LO:

$$I = |u_t + u_{LO}|^2 = |u_t|^2 + |u_{LO}|^2 + u_t u_{LO}^* + u_t^* u_{LO}, \tag{8.7}$$

where u_t is the signal returned from the target, and u_{LO} is the LO signal. The first two terms in Eq. (8.7) describing this beat signal are seen in the center of Fig. 8.8, but the LO intensity dominates. The remaining two terms are spatially separated and are complex conjugates of each other. The complex terms provide field information as well as intensity. We then have to gather multiple "snaps," defined as single sample images, and transform each snap to the image plane. We do not know the motion of each subaperture, although an inertial system might provide an indication of motion. We dither the location one subaperture snap at a time and look at an image quality metric such as image sharpness to optimize the location of the LiDAR on each snap. If we have many subapertures, moving only one subaperture does not change the sum metric much. For each snap, we determine the optimum location for the best image quality by summing the dithered subapertures. Once the subaperture location is determined in each snap, we sum the snaps for each subaperture and add them to determine the speckle-averaged image. Figure 8.9 shows the capture of multiple snaps, which can then be combined to obtain the desired intensity image.

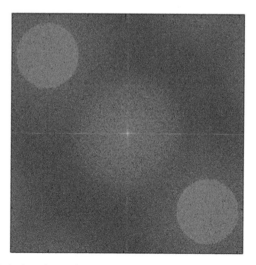

Figure 8.8 Spatial heterodyne image measured in the image plane and Fourier transformed into the pupil plane (reprinted from Ref. 3).

Add fields in each snap.   Sum intensities of
                                             high-res. snap.

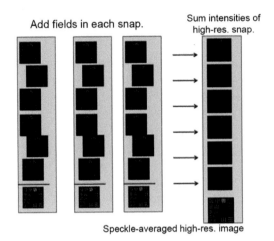

Speckle-averaged high-res. image

Figure 8.9   Processing spatial heterodyne images (reprinted from Ref. 5).

8.2.4 Gathering information from a temporally coherent LiDAR

One of the interesting aspects of temporally coherent LiDAR is that the LiDAR needs sufficient samples in range. This is not an issue for GMAPD cameras because they have only one sample per pulse, but it can be an issue for LMAPD cameras. One benefit of LMAPDs is that they can capture a full scene on a single pulse. Another is that the full 3D image can be read out simply. The disadvantage is that we need a complex readout circuit, which can limit the readout rates. Each pixel needs significant electronics under it. Ideally, we can have a full-waveform sample in each detector element, but often a LMAPD camera only has the ability to capture one or a limited number of samples in each angle/angle pixel. The Voxtel VX 806 camera captures only three samples in each detector. The VX 819 camera provides 24 samples in range that can be separated by 2 ns to 16 ns per sample. It has a 128 × 128 pixel format, a 100-MHz bandwidth, a 94-Hz frame rate, a jitter of <100 ps, and a 50-μm pitch. If we window the VX 819 camera down to a 32 × 32 pixel format, it has a frame rate of 990 Hz, and for a 16 × 16 pixel format, it has a frame rate of 2075 Hz. We can see from Fig. 8.10 that if all we want to do is sample a single sine wave, 24 samples is not bad, as long as the samples are appropriately separated.

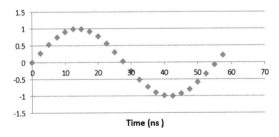

Figure 8.10   24 samples @ 2.5 ns with 18-MHz beat frequency.

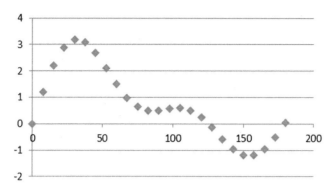

Figure 8.11 Sum of 4 sine waves: 11-, 7-, 5-, and 3-MHz beat frequency.

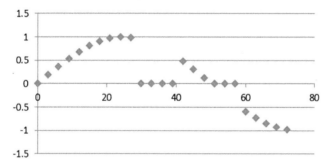

Figure 8.12 10-ns chip length, 3-ns samples, and 10-MHz beat frequency.

Figure 8.11 shows the sampling of a single beat frequency, as well as the sampling of a more complex waveform: the sum of four frequencies. We can see that, as the waveform grows more complex, 24 samples does not provide good insight into the underlying waveform.

Figure 8.12 shows an attempt to detect pseudo-random codes using only 24 samples. In this case, we use a chip length of 10 ns and 3-ns sampling. With 24 samples and 3-ns sampling, we only have a 72-ns period over which to take samples.

We can see that 24 samples is not sufficient. An ideal number of samples for a LMAPD performing temporally coherent detection would be at least 50 with a high frame rate of at least 10 KHz. There could also be a continuous sample rate, which could be as high as 1–10 MHz.

8.2.4.1 Velocity resolution of LiDAR

When a LiDAR pulse returns, its frequency will have shifted based on the Doppler shift of the return light:

$$\Delta f = \frac{2V}{\lambda}, \tag{8.8}$$

which indicates that the higher the velocity to or away from the sensor the higher the frequency shift. The factor of 2 is included because of the two-way path of the light: to the object and returned from the object. In astronomy we have a Doppler shift in the color of stars, but this is radiation traveling only one way, not two ways, so it would not have the factor of two. In astronomy a star receding from the sensor is shifted to longer wavelengths in what is called a red shift. A star coming toward the sensor would be shifted to shorter wavelengths, which is called a blue shift. Obviously, if people can see a red or blue shift, it is possible to measure the color and determine Doppler shift.

Direct-detection LiDARs can measure Doppler shift with wavelength filters. If, however, we want very accurate measurement of the Doppler frequency shift, we need a coherent LiDAR, which measures phase of the returned field using heterodyne detection. The frequency shift is the rate of change of a phase shift, so Doppler shift also changes frequency. To measure the Doppler shift, we need to measure for a sufficiently long time period. In order see that frequency, we need to measure for a time period equal to one over the frequency shift we want to measure. To sample it more effectively, we need to measure for a time period that is twice that long. This sets up the conflict between range resolution and velocity resolution discussed earlier. Because optical carrier frequencies are so high, the Doppler shift is large enough to be measured, even for very small velocities on the order of microns or tens of microns. Figure 8.13 conceptually shows an optical beam bouncing off of a moving object. The return beam will have a slightly different wavelength due to the Doppler shift. The beam bouncing off of the objects explains the factor of 2 in Eq. (8.8).

8.2.4.2 Processing laser vibrometry data

One of the applications of measuring the Doppler shift using coherent LiDAR is laser vibrometery, where the Doppler shift in frequency oscillates as a surface going back and forth. Laser vibrometry generates a signal in frequency space. High Doppler shift implies high velocity between the object and sensor. The point of processing a laser vibrometer signal is to (1) highlight specific frequencies contained in the return signal, (2) convert them to

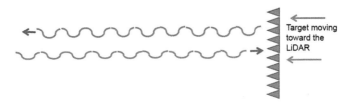

Figure 8.13 Change in frequency due to Doppler shift (reprinted from Ref. 5).

velocities and, in turn, (3) convert these to vibration spectra. To measure a certain frequency takes time. To measure the Doppler shift will require at least one over twice the Doppler shift:

$$\Delta t = \frac{2}{\Delta v}, \tag{8.9}$$

where Δv is the Doppler shift in frequency. If we want to measure a Doppler shift of 2 Hz, we need to stare at a given location for a full second. If we want to measure low-frequency Doppler shifts, meaning micro-Doppler velocities, we need to stare at a given location for a while. The spectrogram processing approach, however, starts to break down when the Doppler frequencies become comparable to vibrational frequencies. Figure 8.14 shows a diagram of LiDAR measuring the Doppler shift of a signal bounced off of a target surface.

The LiDAR engineer will need to determine Doppler shift versus time at each given pixel. With vibration, that Doppler shift will vary. We can measure the Doppler shift and make a plot of that frequency versus time. A Fourier transform of that frequency versus time will provide the vibration spectrum at that point.

Gathering vibration data sequentially is not the same as gathering them at the same time. For a 3D image all you have to do is account for LiDAR motion, but in the case of measuring vibration, the object is moving with time. Ideally, we would like a flash image of the entire modal structure of an object, e.g., a bridge, taken at an instant in time to determine the modal structure of that object. At that instant, we can see which portions of the object are moving toward or away from the LiDAR and with what velocity. It will be difficult, if not impossible, to reconstruct the modal structure a piece at a time because portions of the bridge change at different times.

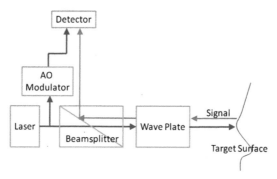

Figure 8.14 Diagram of a LiDAR measuring the Doppler shift of a signal bounced off of a target surface. (adapted from Ref. 5).

8.2.5 General LiDAR Processing

8.2.5.1 Definitions of various LiDAR processing steps for geospatial images

The definitions below are used by the National Geospatial Agency (NGA) in the U.S. and by many organizations, but are not necessarily used by the entire LiDAR community.

Level 0 (L0) – Raw data and metadata: L0 data are the raw data that are in the form in which they were stored when collected from the mapping platform.

Level 1 (L1) – Unfiltered 3D point cloud: L1 data comprise a 3D point data (point cloud) representation of the objects measured by the LIDAR mapping system. This point cloud is the result of applying algorithms to project the L0 data into 3D space.

Level 2 (L2) – Noise-filtered 3D point cloud: Noisy, spurious data have been filtered from the dataset, intensity values have been determined for each 3D point, and relative registration has been performed. The impetus behind separating L1 from L2 is due to the nature of Geiger-mode LIDAR (GML) data. Derivation of L1 GML data produces very noisy point clouds, which require specialized processing (coincidence processing) to remove the noise.

Level 3 (L3) – Georegistered 3D point cloud: L3 datasets differ from L2 in that the data have been registered to a known geodetic datum.

Level 4 (L4) – Derived products: L4 datasets represent LIDAR-derived products to be disseminated to standard users. These products could include digital elevation models (DEMs), viewsheds, or other products created in a standard format and using a standard set of tools.

Level 5 (L5) – Intel products: L5 datasets are a type of specialized products for users in the intelligence community. These datasets may require specialized tools and knowledge to generate.

8.2.5.2 Inertial measurement units

To make a LiDAR image we often have to stitch together sections of the image gathered at various times. An inertial measurement unit (IMU) is an electronic device that measures and reports on a LiDAR's velocity and orientation, using a combination of accelerometers and gyroscopes. It is possible that the LiDAR will use an IMU on the vehicle carrying the LiDAR, but it is more likely that the LiDAR will have its own IMU to precisely identify changes in position or velocity of the LiDAR. Unless the vehicle is perfectly rigid, an IMU in one location will not record exactly the same motion as an IMU in another location. A flash LiDAR image that is gathered using a LMAPD may generate a complete image with a single frame, but all of the other types of LiDARs require stitching together multiple samples taken at different times. An IMU is critical to this function. A flash image generated using a GMAPD is not truly a single image. As many as, e.g., 50 images may be taken of the same target to create one flash image. Motions between samples must be compensated before coincidence processing can occur. With a laser repetition rate of 40 KHz, we need to stich frames over as much as 1 ms. A scanning imager, of course, collects data at various times, and motion must be compensated. Synthetic-aperture LiDAR requires motion compensation. An IMU works by detecting the rate of acceleration using one or more accelerometers and detects changes in rotational attributes such as pitch, roll, and yaw using one or more gyroscopes. IMUs are used for navigation but need to be reset periodically. They are critical for stitching together LiDAR images when portions of the image are gathered at different times. In order to stich images together, relative motion may be sufficient. If absolute motion is required, IMU data may need to be supplemented with GPS or other navigation information.

8.2.5.3 Data product types

In direct-detection systems, two types of information are typically recovered. The first is the range from the sensor to the target on a pixel-by-pixel basis, which is often called a 3D point cloud image. Here, the range information is gathered (often through some form of timing circuitry) as a function of position on the receiver focal plane. Hence, the contrast information that is provided from pixel to pixel is a variation in range. The second type of information that can be gathered is reflectance, e.g., irradiance, often called grayscale. The contrast from pixel to pixel in this case is derived by quantifying the energy deposited on each pixel, which is related to the reflectivity of the surface illuminated by the laser. Grayscale is discussed in the next section. We are interested in determining the number of photodetections required for each of three types of data products: geometry only (i.e., just a point cloud), geometry plus reflectivity measured with a resolution of $N_{bits} = 3$ bits, and geometry plus reflectivity measured with a resolution of $N_{bits} = 8$ bits. Once we have the number of photodetections for each sensing modality, we can use that information and a standard link budget approach to calculate total energy required for each modality. Figure 6.3 shows grayscale over a range from 5% to 15% reflectivity.

8.2.5.4 Grayscale calculations

The approach presented here for active grayscale measurement using laser illuminator photons is as follows. We divide the distribution into a defined number of reflectivity levels (gray levels). We assume that object reflectivities range between a minimum of $\rho_{min} = 0.05$ and a maximum of $\rho_{max} = 0.15$, as illustrated in Fig. 6.3. Then the LiDAR system is required to discern a reflectivity bin size of

$$\epsilon = (\rho_{max} - \rho_{min})/2^{Nbits}/\rho_{max} = 0.0104 \qquad (8.10)$$

for the 8-bit case, or 0.0833 for the 3-bit case (i.e., the reflectivity intervals are 8 times wider). We must be able to distinguish one gray level from another, even in the presence of noise in the LiDAR receiver. Our LiDAR measurements are done with enough SNR such that there is a 80% probability of assigning the target reflectivity to the correct bin ($P_c = 0.8$). All measurement modalities are subject to shot noise arising from the fact that the quantization of the received light obeys Poisson statistics. Other sources of instrument noise and distortion will add to this minimum noise level. An example of the reflectance bins and the effects of shot noise is indicated in Fig. 8.15 for the simple case of $N_{bits} = 3$ and $P_c = 80\%$. The eight grayscale bins are indicated by the vertical, gray, dotted lines. The colors represent 8 different mean numbers of events, from 242.6 (dark blue) to 727.9 (dark red). These different mean numbers of returns could indicate the number of photons received from different reflectivity targets. The solid colored lines indicate the cumulative distribution function (CDF) of the results of 2000 random trials. The dotted colored lines indicate the Poisson distribution function indicative of a shot-noise–limited LiDAR for each mean number of events. The shot noise is widest at the highest reflectivity, so it is this limit that sets the minimum required number of received photons.

While this chapter presents only the various conditions that might exemplify an advantage to one detection mode or another, it is interesting to see the effect that various levels of grayscale have on imagery. This can be seen in Fig. 8.16 for grayscale ranging from 1 to 6 bits.

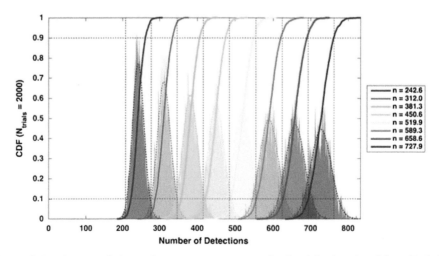

Figure 8.15   Impact of shot noise on measurement of reflectivity (reprinted from Ref. 4).

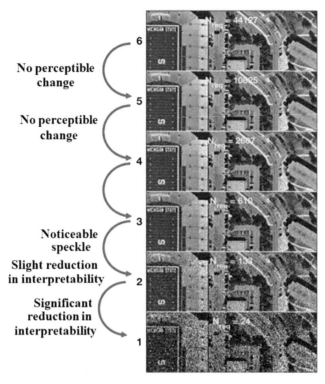

Figure 8.16 The effect of grayscale on a particular image (reprinted from Ref. 4).

8.2.5.5 Fourier transforms

Computationally, the difference between the pupil plane and the image plane is just a Fourier transform. Alternatively, in physical space, a lens serves the same function as a Fourier transform in digital space. Once the field is known in either domain, the other domain is also known. The aperture receives the pupil plane field. We can use a lens to focus that field into the imaging plane. Once in the imaging plane, we can either capture the full field using heterodyne techniques, or we can capture only intensity, as we do in direct detection. If, however, we capture the full field in the pupil plane, we do not have to use a lens to go to the image plane. We can use a Fourier transform. This is what is known as lensless imaging. The Fourier transform of a pupil plane field is given by

$$F\{E(x,y)\} = \iint E(x,y)e^{[-i2\pi(f_x x + f_y y)]}dxdy. \tag{8.11}$$

We need to transform the electric field into an image-plane spatial-frequency–based complex term that contains the far-field image information. The translation from the far field to the pupil plane is a Fourier transform, which is replicated by the lens or a Fourier transform of the image plane field, bringing us back to an image of the object being viewed in the far field. Of course, today people often perform a Fourier transform digitally, using the fast Fourier transform (FFT) and MATLAB commands, examples of which are included here:

- Y = fft2(X) returns a 2D digital FFT of a 2D variable *x*.
- Y = fft2(X,m,n) pads the area around *x* with zeros to make the Fourier transform larger.
- Y = fft(X) returns an array that is just as large as the original array. To provide more fidelity in frequency space, sometimes the returned array needs to be larger than the original array. Another option that is often used to provide more fidelity involves changing the zero location.
- Y = fftshift moves the zero-frequency components to the center of the array. It swaps the locations of the terms in a 2D array.
- Y = fftshift(fft2(X,m,n)) can be used if you want to both pad the array and shift the center.

8.2.5.6 Developing 3D maps from LiDAR

There are multiple ways of obtaining angle/angle information for a LiDAR. We can have scanning LiDAR or flash LiDAR; either type can use GMAPDs or LMAPDs. All LiDARs obtain range by sending out a waveform and timing how long it takes the waveform to return. The waveform can be a series of pulses or some other waveform, but time of flight can be measured and the speed of light is used to determine range to the object. This range measurement occurs in each angle/angle pixel, creating 3D pixels, or voxels. Figure 8.17 is an example of a 3D map created in this way.

Scanning LiDARs require rapid movement in angle. For example, assume that we want an image that is 480 × 640 pixels in angle space, and we have 8

Figure 8.17   A 3D map of San Francisco (reprinted from Ref. 5).

individual detectors. We could have 80 rows of 8 pixels times 840 pixels in each row. At 30-Hz frame rate, we need to move $60 \times 640 \times 30 = 1,152,000$ times each second and have a laser with that repetition rate. For this case it will take $1/30^{th}$ of a second to make an image, so we have to know how the LiDAR and target moved over that $1/30^{th}$ of a second.

Alternatively, we could make the same image with a 32×128 pixel GMAPD camera with an array of 15 snapshots high providing the 480 pixels, by 5 snapshots wide providing the 640 pixels. Assuming 50 pulses per snapshot, we would need a laser repetition rate of $50 \times 5 \times 15 \times 30 = 112,500$. We would need to stabilize only to a fraction of the 32×128 pixel array, and stitching together the frames would occur in a little over 1 ms. We then need to stitch the 128×32 pixel section together. It will take $1/30^{th}$ of a second to make that image.

A third option is to take the same 32×128 pixel array and make the image with a LMAPD. The laser would need to pulse at a rate of only $5 \times 15 \times 30 = 2250$ Hz. Of course, if the LMAPD array was 480×640 pixels, the laser would only need to pulse at 30 Hz.

8.2.6 Target classification using LiDAR

3D shape is an ideal characteristic to use for object identification because it usually does not change significantly. It is, of course, true that someone can pile brush on top of a vehicle to disguise its shape; however, the basic shape of an object is much less variable than the grayscale of 2D image illuminated with variable light sources, which might even illuminate only a portion of an object.

3D point cloud images are rendered in terms of voxels (which define a point in 3D space), rather than pixels (which define a point in 2D space). Figure 8.18 shows a 3D image of a car. It is usual for a LiDAR to have a range dimension with equally high or higher resolution than the angle/angle dimensions because

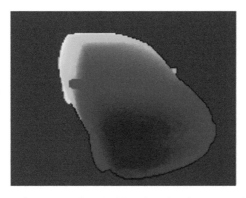

Figure 8.18 3D image of a car rendered with color showing range (reprinted from Ref. 5).

the range resolution depends on the LiDAR bandwidth, not on the diffraction limit. At long distances, it is therefore easy to have excellent range resolution with a LiDAR. This is in contrast to a stereo 3D image, where the range resolution will not be as good as the angle/angle resolution.

One useful benefit of a 3D point cloud image is that it can be viewed from multiple angles. While data from angles other than the angle viewed by the LiDAR might be missing, often there are sufficient data to provide excellent 3D views from many different locations. Not only does this make it much easier to hand off an image from one sensor to another, it also improves the ability to recognize an object. An object is often more recognizable from one particular aspect than from another aspect.

Problems and Solutions

8-1. Calculate the Fourier transform of the following functions:

a) $\operatorname{rect}\left(\dfrac{x - x_0}{a}, \dfrac{y - y_0}{b}\right)$

b) $\operatorname{gaus}\left(\dfrac{x}{a}, \dfrac{y}{b}\right)$

c) $\operatorname{comb}(x,y)$

d) $\operatorname{tri}\left(\dfrac{x - x_0}{a}, \dfrac{y - y_0}{b}\right)$

e) $\cos(2\pi\xi_0 x)$

8-1 Solution:

a) $a b \operatorname{sinc}(a\xi, b\eta) e^{-i2\pi(x_0\xi + y_0\eta)}$

b) $a b \operatorname{gaus}(a\xi, b\eta)$

c) $a b \operatorname{comb}(\xi, \eta)$

d) $a b \operatorname{sinc}^2(a\xi, b\eta) e^{-i2\pi(x_0\xi + y_0\eta)}$

e) $0.5[\delta(\xi - \xi_0) + \delta(\xi + \xi_0)]$

8-2. Find the convolution $\operatorname{rect}(x) * \operatorname{rect}(x - a)$ and sketch the result.

8.2 Solution:

$\operatorname{tri}(x - a)$

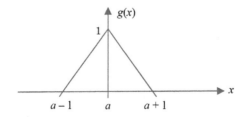

8-3. In Problem 8-2, what is the width of the resulting convolution? What result would you expect for rect(x) ∗ rect(x) ∗ rect($x - a$) in terms of width, location, and sharpness of features?

8-3 Solution:
The width of the convolution is determined by the sum of widths of the two functions being convolved. In this case, it is one function being convolved with itself, so the convolution is double the width of the function ($1 + 1 = 2$). Sharpness keeps reducing; if we continue, we will end up with a Gaussian function.

8-4. Consider a camera with a detector array having the following characteristics: square detectors with a detector pitch (center-to-center spacing) of 25 µm and a detector width (each individual element) of 20 µm. There are 128×128 detectors in the array. Assume that a light intensity given by $E(x,y)$ is incident on the detector array.

    a) What are the Nyquist rate and Nyquist frequency associated with this detector array?
    b) Write an expression that describes the response function of an individual detector in the array.

8-4 Solution:
    a) Sampling frequency $= 1/\text{pitch} = 1/\Delta x = 1/25$ µm $= 40$ mm^{-1}.
        Nyquist rate $=$ sampling frequency.
        Bandwidth $= 2 \times$ max frequency of image $\rightarrow$ Nyquist frequency $=$
        $0.5BW = 0.5 \times 40 = 20$ mm^{-1}.
    b) Detector is square $\rightarrow$ rect is good to describe the detector.
        $\rightarrow \text{rect}\left(\frac{x}{20\,\mu\text{m}}, \frac{y}{20\,\mu\text{m}}\right) = \text{rect}\left(\frac{x}{0.02\,\text{mm}}, \frac{y}{0.02\,\text{mm}}\right)$.

8-5. Using the response function in Problem 8-4, write an expression for the discrete, sampled intensity produced by the detector array.

8-5 Solution:

$$g_s(x,y) = \left[\text{rect}\left(\frac{x}{dx}, \frac{y}{dy}\right) * E(x,y)\right] \frac{1}{\Delta x \Delta y} \text{comb}\left(\frac{x}{\Delta x}, \frac{y}{\Delta y}\right) \frac{1}{(84 \times 25)^2}$$
$$\times \text{rect}\left(\frac{x - 12.5}{84 \times 25}, \frac{y - 12.5}{84 \times 25}\right).$$

8-6. Fourier transform the expression in Problem 8-4 to derive the spatial frequency description of the sampled intensity.

8-6 Solution:

$$G_s(f_x, f_y) = dxdy \Big[R(dxf_x, dyf_y) \cdot \tilde{E}(f_x, f_y) \Big] \times \Big[\text{comb}(\Delta xf_x, \Delta yf_y)$$

$$\times \text{sinc}(128 \times 25f_x,\ 128 \times 25f_y) e^{-i2(12.5f_x + 12.5f_y)} \Big].$$

8-7. In Problem 8-4, assume that $E(x,y) = 1 + \cos(2\pi f_{x0}x)$, where $f_{x0} = 10$ mm^{-1}. Is this intensity pattern aliased by the detector array? If so, what spatial frequency does the intensity alias to?

8-7 Solution:

$f_{x0} = 10$ mm^{-1}.

$$\tilde{f}\{E(x,y)\} = \delta(f_x) + 0.5\Big[\delta(f_x - f_{x_0}) + \delta(f_x + f_{x_0})\Big].$$

$$\tilde{f}\left\{ \text{rect}\left(\frac{x}{0.02}, \frac{y}{0.02}\right)\right\} \approx \text{sinc}(0.02f_x,\ 0.02f_y).$$

In this array for sampling we have

$\Delta xf_x = n$ and $n = 1 \rightarrow f_x = n/\Delta x \rightarrow f_x = 1/0.025 = 40$ mm^{-1}

$\rightarrow 40/2 = 20$ mm^{-1} is the maximum frequency allowed.

$2f_{x0} = 2 \times 10 = 20$ and $20 < 40 \rightarrow$ no aliasing occurs.

8-8. In Problem 8-4, assume that $E(x,y) = 1 + \cos(2\pi f_{x0}x)$, where $f_{x0} = 30$ mm^{-1}. Is this intensity pattern aliased by the detector array? If so, what spatial frequency does the intensity alias to?

8-8 Solution:

$f_{x0} = 30$ mm^{-1}. Similarly, we have

$2f_{x0} = 2 \times 30 = 80$ and $80 > 40 \rightarrow$ aliasing occurs.

$f_a = f_s - f_{x0} = 40 - 30 = 10$.

8-9. In Problem 8-4, assume that $E(x,y) = 1 + \cos(2\pi f_{x0}x)$, where $f_{x0} = 50$ mm^{-1}. Is this intensity pattern aliased by the detector array? If so, what spatial frequency does the intensity alias to?

8-9 Solution:

$f_{x0} = 50$ mm^{-1}. We have

$\text{sinc}(0.02f_x) = \text{sinc}(0.02 \times 50) = \text{sinc}(1) = 0$.

Therefore,

$R(dxf_x, dyf_y) \cdot \tilde{E}(f_y, f_y) = 0 \rightarrow$ no aliasing.

8-10. Consider the detector array shown below, which is drawn to scale.

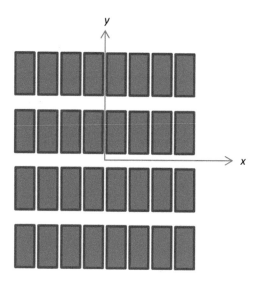

a) In which dimension (x or y) is the Nyquist frequency the largest? Why?

b) In which dimension (x or y) is the blur due to the detector the largest? Why?

c) What is the transfer function (TF) for this array? (Assume that the detector has width a and height b.)

d) Assume that the intensity of the light on the detector array is given by $f(x,y) = 2 + 1.5 \cos(2\pi x/3a)$. What is the modulation of this intensity?

e) For the input intensity of part (d) and using the results of part (c), find the modulation of the output.

8-10 Solution:

a) Sampling rate = 1/sampling period = 1/pitch.
The pitch in the x direction is smaller than in the y direction. Therefore, the Nyquist rate in the x direction is higher.

b) Blur depends on size $\rightarrow r(x,y) \times E(x,y)$. The amount of blur produced is determined by the size of the individual detector. Because Δy is larger in the y direction, blur is larger in y direction.

c) TF = Fourier transform of PSF $\rightarrow$ rect($x/a,y/b$) $\rightarrow$ TF = absinc(af_x, bf_y).

d) $F(x,y) = 2 + 1.5\cos(2\pi x/3a)$ $\rightarrow$ (max $-$ min)/(max $+$ min) = ¾.

e) $f_x = \pm 1/3a$ $\rightarrow$ sinc($a \times 1/3a$) = sinc(1/3) = 0.8.
$\rightarrow$ Output modulation = input modulation $\times$ MTF = 0.75 $\times$ 0.82 = 0.82.

8-11. Assume that the intensity of the light on the detector array is given by $f(x,y) = 2 + 1.5 \cos(\pi x/3a)$.

a) Plot the intensity over *x* ranging from 0 to *a*.

b) Assume that you can sample, at best, the MTF down to 0.1. What is the resolution in cycles/millimeters?

8-11 Solution:

(a)

```
x = 0.001:0.001:1;
I = 2+1.5*cos(pi*x);
plot(x,(I.^2-min(I.^2))/(max(I.^2)-min(I.^2)),x,
ones(1,1000)*0.1)
```

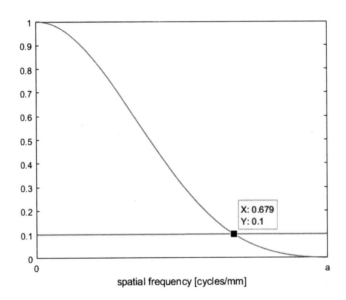

spatial frequency [cycles/mm]

(b) The best spatial resolution is 0.679*a*.

8-12.

(a) Roughly speaking, what would be the convolution of a triangle function with a Gaussian?

(b) Roughly speaking, what would the convolution of a rectangular function with a Gaussian produce?

8-12 Solution:

(a) Because the triangle function would sharpen the Gaussian, it would trend the Gaussian curve towards a Lorentz curve.

(b) A rectangular function convolved with a Gaussian would produce a super- Gaussian, roughly speaking, by widening the top of the curve while trending the edges flatter.

8-13. Convolve a Gaussian with itself. That is,

$$C(x) = \exp\left[-(x - \mu)^2/2w^2\right] \times \exp\left[-(x - \mu)^2/2w^2\right].$$

Plot the result for $6w < x < 6w$, $\mu = 0$, as well as the original Gaussian.

8-13 Solution:

$$C(x) = \int_{-\infty}^{\infty} \exp\left[-(t - \mu)^2/2w^2\right] \times \exp\left[-(x - \mu - t)^2/2w^2\right] dt$$

$$= \sqrt{\frac{\pi}{2}} e^{-\frac{1}{2}(x - 2\mu)^2/2w^2}.$$

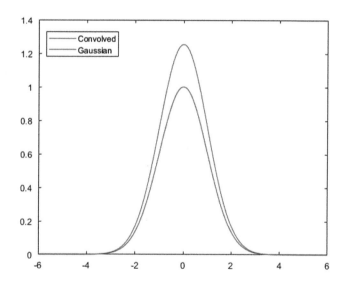

8-14. Fourier transform the result obtained from Problem 8-13. Use widths of 1 mm, 1.25 mm, and 0.75 mm for the Gaussian.

8-14 Solution:
Since the result from Problem 8-13 is still a Gaussian, just use the Fourier transform table.

$$\sqrt{\frac{\pi}{2}} F[\text{gaus}(x)] = \sqrt{\frac{\pi}{2}} \pi w^2 \exp[-w^2(k_x^2)/2].$$

Specifying that $w = 500$ μm, k_x will range from -60 mm^{-1} to 60 mm^{-1}.

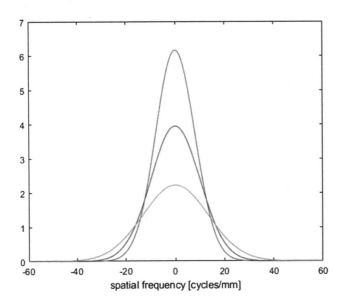

References

1. R. C. Hardie, K. J. Barnard, J. G. Bognar, E. E. Armstrong, and E. A. Watson, "High-resolution image reconstruction from a sequence of rotated and translated frames and its application to an infrared imaging system," *Optical Engineering* **37**(1), 247–260 (1998) [doi: 10.1117/1.601623].
2. E. A. Watson, R. A. Muse, and F. P. Blommel, "Aliasing and blurring in microscanned imagery," *Proc. SPIE* **1689**, 242–250 (1992) [doi: 10.1117/12.137955].
3. J. R. Kraczek, "Noniterative Multi-aperture and Multi-illuminator Phasing for High-Resolution Coherent Imaging," Ph.D. thesis, University of Dayton (2017).
4. P. F. McManamon, P. S. Banks, J. D. Beck, D. G. Fried, A. S. Huntington, and E. A. Watson, "Comparison of flash lidar detector options," *Optical Engineering* **66**(3), 031223 (2017) [doi: 10.1117/1.OE.56.3.031223].
5. P. McManamon, *Field Guide to Lidar*, SPIE Press, Bellingham, Washington (2015) [doi: 10.1117/3.2186106].

Chapter 9
Figures of Merit, Testing, and Calibration for LiDAR

9.1 Introduction

We need to be able to quantitatively measure the performance of a LiDAR. This will be more difficult than measuring the performance of a traditional 2D sensor because we have more dimensions to measure. The most common LiDAR is a 3D LiDAR, which involves measuring range response as well as angle/angle response. This chapter first discusses LiDAR figures of merit, which are the characteristics we need to measure to determine LiDAR performance (the discussion may not describe how to measure all of the figures of merit). Simple direct-detection LiDAR will not have as many relevant figures of merit as a coherent LiDAR. Next, the chapter discusses LiDAR testing, focusing primarily on 3D direct-detection LiDAR. Finally, to obtain optimum performance, we need to calibrate the LiDAR. The chapter concludes by providing methods to remove intensity and range measurement nonuniformities. Again, this discussion on calibration will focus on 3D direct-detection LiDAR.

9.2 LiDAR Characterization and Figures of Merit

Here it might be useful to remind the reader about the differences between resolution, accuracy, and precision. We define resolution as the ability to detect two objects some in dimension. Resolution could be in angle, range, or another dimension. For range resolution, this is limited to Eq. (8.3), repeated here:

$$\Delta R = \frac{c}{2B},$$

where c is the speed of light, and B is bandwidth.

We define precision as the uncertainty of a measurement relative to a reference location that dimension. We define resolution as the ability to separate two objects in that dimension. To separate them we will need a dip, or at least a flattening, in the response between the two objects. We define accuracy as the degree to which a measurement yields an absolute position in space. For range this depends on range precision as well as systematic errors (e.g., clock rate, drift, timing offsets, etc.). We can focus on measuring resolution or precision for most LiDARs. Accuracy involves using other combined sensors, such as a global positioning system (GPS) or an inertial measurement unit (IMU). The most fundamental characteristic to measure is resolution. Precision depends on both resolution and SNR. Therefore, we will focus on measuring resolution in each dimension.

The resolution and precision as defined above depend on the point spread function (PSF) of a sensor. The PSF can be thought of as the blur in an image that represents an unresolved object. The PSF is the spread in response in that dimension based on sensor characteristics. The image of an object in a particular dimension is the convolution of the object and the PSF, which for a point object is just the PSF. When the detected light is coherent, image formation is more complicated. The complex field can have negative values and can interfere with another complex field.

The range response of a target is the convolution of the target response and the sensor response. If we have a point target in range, the range response will be due to only the LiDAR. For a 3D LiDAR, we need the PSF and the range resolution in three dimensions rather than in just two dimensions. Range PSF is likely to be significantly different from the angle PSFs because range resolution is dependent on LiDAR bandwidth. On the other hand, angle/angle resolution is dependent on the diffraction limit of the sensor, and the sampling in angle/angle is dominated by the DAS and the range to the LiDAR. There are coherent LiDARs that measure velocity. We could have a PSF in velocity, which will characterize the ability of the LiDAR to separate the velocity of an object from the velocity of another object that has almost the same velocity. Those are the main dimensions considered by a LiDAR, although we have also talked about a spectral dimension and speckle considerations. Characterizing the effect of speckle is another potential complication.

Each imaging mode has multiple figures of merit that describe the LiDAR performance. Some of these figures of merit are only relevant to certain types of LiDAR.

9.2.1 Ideal point response mainlobe width

The PSF describes the response of an imaging system to a point source or point object. This is essentially the same as the ideal point response (IPR). The resolution, or the minimum resolvable target separation, is defined in range or cross-range as the mainlobe width of the IPR. The IPR mainlobe width is defined as the distance between the full-width, half-maximum (FWHM) points of the 1D cross-section of the IPR intensity. This is equivalent to the points that are 3 dB below the peak power and is consistent with our usage of beam width as the FWHM value of the beam width. As mentioned before, this is the way the radar community defines beam width, but not all optical folks define it this way.

9.2.2 Integrated sidelobe ratio

Although resolution is usually the first metric considered in an imaging system, the integrated sidelobe ratio (ISLR) can be very important, especially for a coherent LiDAR. The ISLR quantifies the fraction of the waveform energy compressed into the main lobe relative to that present in the side lobes. In radar a main lobe can be described as the fraction of energy in the main grating mode. In single-aperture LiDAR, the main lobe is the energy in the lobes of the PSF. However, in multi-aperture LiDAR, the main lobe can take on a meaning similar to its definition in radar technology. Energy within the side lobe contributes to image clutter and reduces contrast within the image. Ideally, all of the signal energy would be in the main lobe; however, side lobes are always present. A low-ISLR image may indicate an image with extremely low quality even though it may have excellent resolution. The ISLR is calculated as the ratio of energy within the main lobe out to the first null to the energy outside the main lobe.

9.2.3 Peak sidelobe ratio

The peak sidelobe ratio (PSLR) is also an important metric, but unlike the ISLR, which focuses on the average sidelobe level, the peak sidelobe ratio specifically identifies the strongest side lobes. The peak side lobes may be significantly greater than the average sidelobe level. High peak side lobes may not significantly decrease the overall contrast, but their appearance can generate secondary targets or "ghost" images, leading to degradations that range from loss of interpretability to resolution loss.

9.2.4 Spurious sidelobe ratio

The spurious sidelobe ratio (SSLR) is similar to the PSLR but is important when the PSLR is calculated for a restricted region of interest. The SSLR provides a measure of the energy in clutter sources outside the region of interest that could induce artifacts in the region of interest.

9.2.5 Noise-equivalent vibration velocity

The noise-equivalent vibration velocity (NEVV) is the square root of the signal spectral amplitude that equals the measurement noise floor. The noise floor is determined by system parameters such as measurement contrast-to-noise ratio (CNR), master oscillator linewidth, and shot noise level, as well as external parameters such as atmospheric turbulence environment, platform residual motion, and pointing system angular jitter. NEVV is measured in units of $\mu m/(s\sqrt{Hz})$.

9.2.6 Abiguity velocity

The ambiguity velocity V_{amb} is the maximum velocity that can be sensed uniquely and is referenced to the wavelength of the source. It is the maximum velocity that will produce a wavelength-equivalent displacement over the shortest measurement time, usually $1/PRF$, where PRF is the pulse repetition frequency. For macro-Doppler imaging, large V_{amb} values are required since targets may be moving at many meters per second. High PRFs make V_{amb} values larger, but make micro-Doppler measurements less sensitive.

9.2.7 Unambiguous range

Unambiguous range is given by

$$R_{unambig} = \frac{c}{2PRF},\tag{9.1}$$

unless a multipulse waveform is used to extend the unambiguous range. Note that the only difference between Eqs. (9.1) and (8.4) is that the latter uses the time between pulses, and the former uses the PRF, which is one of the time periods between pulses.

9.3 LiDAR Testing

9.3.1 Angle/angle/range resolution testing

A traditional 2D sensor might be characterized using the standard U.S. Air Force 1951 test target, shown in Fig. 9.1. This test target has both azimuth and elevation features that start large and become increasingly smaller. The test of a sensor using this target is how small the test target features can be when viewed from a given distance while still having the same average viewer able to resolve a rectangle of a certain size. In order to expand the Air Force 1951 test target to 3D features, we will need a target distributed in range as well as in angle/angle space. One difficulty with this is that the range resolution may not be similar in size to the angle/angle resolution. Therefore, the extent of range separation may be different from the extent of angle/angle separation.

An alternative target for characterizing 2D sensors is shown in Fig. 9.2. This target has the advantage of providing a slant edge and a continuously expanding set of lines, both of which can be useful in characterizing an optical sensor. The ISO 12233 target, a portion of which is shown in Fig. 9.2, was specifically designed to be used with several different MTF methods, including the slant edge method. We used the ISO 12233 target because the slant edge MTF method requires a thick edge that is tilted to get a MTF

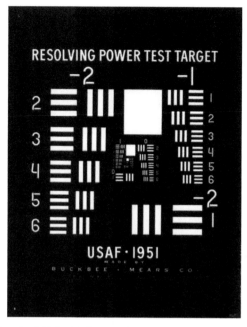

Figure 9.1 Standard U.S. Air Force 1951 test target (reprinted from Ref. 1).

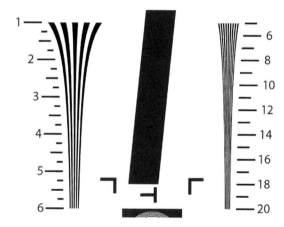

Figure 9.2 Potential 2D sensor target for sensor characterization (reprinted from Ref. 2).

estimate. We integrated a MATLAB® routine (called 'sfrmat3' created by Dave Burns[3]) into our simulations to make the MTF estimate. A slant edge provides many more points to determine the PSF. This target will be more difficult to expand into the third dimension.

As an alternative to specifically shaped targets, LiDAR engineers can use point targets to characterize a sensor and measure the MTF. Obviously, for a 3D LiDAR we need to extend this to three dimensions. When discussing target cross-section we mentioned the use of a gold ball reflector. This can be used as a point reflector in three dimensions. Point reflectors may be useful for LiDAR characterization. We can separate the points in angle/angle space, in range, and even in velocity for a coherent LiDAR. Point targets are not often used because of low signal from a true "point." Edges are more common (such as the targets in Fig. 9.2) because edges produce a line spread function from which point spread functions can be estimated.

An example of a less-complex 3D target is shown in Fig. 9.3. This target has some similarities to the 2D target shown in Fig. 9.1 but is 3D. It was used in a flight test of a 3D LiDAR.

Another type of 3D test target is shown in Fig. 9.4, which is from Jason Stafford's Ph.D. dissertation "Range Compressed Holographic Aperture LiDAR" (University of Dayton).[4] This target also provides a 3D target that can be used to characterize a 3D sensor in all three dimensions. In angle/angle space we see a large number of triangles, providing a variety of sizes as the triangles become larger. The triangles are also separated in range, providing a third dimension.

Other LiDARs can measure additional target features and require a different set of targets and a different set of metrics. For example, a laser vibrometer can measure target velocity and can infer target acceleration. Metrics for this type of LiDAR are discussed next.

9.3.2 Velocity measurement

Measurement of velocity will mean measurement of moving objects. People have been known to use a spinning disk from a drill with a sander on it to provide a velocity target. A spinning disk will provide a range of velocities, with higher velocities toward the outer portion of the disk. It does provide known velocity at any particular radius of the disk. The biggest benefit of a spinning object is that the object does not have to translate to provide

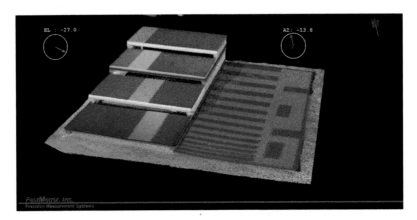

Figure 9.3 An example of a 3D test target (courtesy of Gary Kamerman of FastMetrix, Inc.).

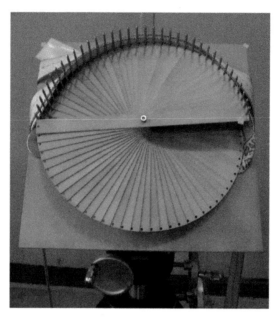

Figure 9.4 Another 3D LiDAR target possibility (reprinted from Ref. 4).

significant velocity. Linear velocity requires object motion, which can complicate testing. In general, velocity measurement is more pertinent to coherent LiDAR, which is capable of measuring Doppler shift. Velocity can, however, be measured by successive measurements of range using a direct-detection LiDAR. One thing to always keep in mind with respect to Doppler LiDAR is that the Doppler shift is only directly to and from the LiDAR; it is only along that dimension.

9.3.3 Measuring range walk

To measure range walk we need a target at a fixed distance and we want to vary the laser power. As the power increases, the LiDAR may measure a shorter range. This issue was discussed in Chapter 8. If the LiDAR uses a simple threshold detection scheme, the result is a more-powerful laser-crossing threshold at a shorter range. Range walk can be characterized in this manner; however, one needs to be careful to specify an origin when doing so. The plot in Fig. 9.5 specifies the origin of the range walk as relative to zero intensity and decreasing quadratically in range until saturation. The physical characteristics of range walk are provided by Fig. 9.6, which has already been presented as Fig. 8.5.

Timing jitter can provide some uncertainty to the measurement, as well. The pulse will be delayed due to a decreased amplitude and threshold condition, which causes the range walk effect by producing the effect of a longer range for a shorter amplitude pulse.

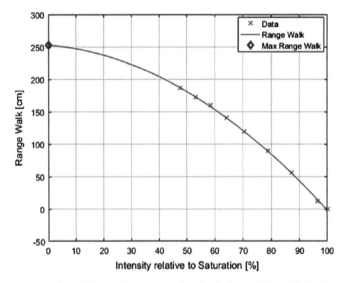

Figure 9.5 Example range-walk plot (adapted from Ref. 5).

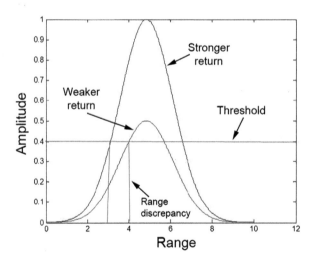

Figure 9.6 Plot showing the physical characteristics of range walk.

9.4 LiDAR Calibration

For LiDARs with multiple detectors in a camera, we need to calibrate nonuniformity in both intensity and range. Fixed-pattern noise (FPN) is prevalent in LiDAR cameras, as well as other cameras. This type of noise can cause a variation in intensity measurements or, for LiDARs, can cause a variation in range measurements. Fixed-pattern noise can be caused by amplification gain, pixel response, dark current, and column amplification offset. It is defined as the difference of response in a given pixel compared to the spatially averaged mean of the image, whether for intensity or range.

We can do dark frame subtraction to eliminate dark current nonuniformity. This is one of the types of nonuniformity mentioned above. Any time we subtract a frame average, we want to average over many samples, possibly as many as hundreds. For dark frame subtraction, we prevent any light from entering the camera and measure camera response. While the camera response should be zero, there is dark current, and each detector can have a different amount of dark current. A dark nonuniformity measurement is a good way to start nonuniformity calibration. Reference 5 discusses dark nonuniformity correction and range walk.

In order to judge pixel response nonuniformity and gain nonuniformity, we need to use a laser source. The best way to judge this nonuniformity is typically to illuminate a Lambertian scattering wall that has uniform reflectance. The wall is at a fixed distance, so we can measure range nonuniformity as well as intensity nonuniformity. The wall needs to be far enough away that the LiDAR can respond at that range.

Fixed-pattern noise is prevalent in IR cameras and can be caused by variation in pixel-to-pixel amplification gain, pixel response, dark current, and column amplification offset.[5] It is often caused by electronic feedback in the system. Here, we focus on an experiment to eliminate one of the sources for nonuniformity: dark-current–based nonuniformity. The camera used in this experiment was a p–i–n architecture camera; therefore, amplification and gain were not of concern. Pixel-response nonuniformity—a method of removing the nonuniformity caused by the illuminated pixel response of the camera—has been addressed, and the results are published in Ref. 6.

Dark nonuniformity correction is one part of a multistep process to correct the FPN in a camera. The dark nonuniformity correction removes an estimation of the dark-current–induced FPN, while pixel response nonuniformity correction removes an estimation of the FPN induced by the pixel response to illumination. These two components of nonuniformity correction are required for correcting this camera without gain and, in particular, for future experiments using a camera with linear gain.

Fixed-pattern noise can be thought of as being related to the difference between the spatially varying mean and the spatially averaged mean of the image:

$$\sigma_{\text{total FPN}} = \left[\frac{1}{XY - 1} \sum_{n=1}^{N} \sum_{y=1}^{Y} \sum_{x=1}^{X} (\bar{p}(x,y) - \bar{p}) \right]^{1/2}, \qquad (9.2)$$

where $\bar{p}$ and $\bar{p}(x,y)$ are given, respectively, in the following equations:

$$\bar{p} = \frac{1}{NXY} \sum_{n=1}^{N} \sum_{y=1}^{Y} \sum_{x=1}^{X} p_n(x,y), \qquad (9.3)$$

$$\bar{p}(x,y) = \frac{1}{N} \sum_{n=1}^{N} p_n(x,y), \qquad (9.4)$$

where X, Y, and N are the number of rows, columns, and frames, respectively, in the image.

Dark frame subtraction is a method of nonuniformity correction where the average over many (perhaps hundreds) dark frames is subtracted from an illuminated frame as a bias. The averaging process reduces temporal noise by a factor $1/\sqrt{N}$, so the higher number of frames taken produces a more pristine dark frame. However, since the fall off is as $1/\sqrt{N}$ in large limit, this will trend towards 0 with little variation. At larger values of N that are greater than a few thousand, the computational processes required to decrease the temporal noise become far greater than the benefits produced.

The dark frame can be thought of as subtracting out the FPN associated with the no-illumination case and multiplying a bias β to an estimate of the underlying temporal noise:

$$p_{offset} = \beta \sigma_{temporal}. \qquad (9.5)$$

The bias is typically a value of about $6\sigma_{temporal}$. By offsetting the value by a small bias, the offset prevents unsigned integer rollover and very large values (greater than response) from occurring. Finally, by taking the captured frame, subtracting the FPN, and adding this bias, a correction is applied to the uncorrected image $p(x, y)$:

$$p'(x,y) = p(x,y) - \sigma_{total\ FPN} + p_{offset}. \qquad (9.6)$$

It should be noted that gain is not accounted for in this process due to the camera having a p–i–n diode architecture, and that therefore unity gain is also not accounted for because the experiments done were at dark level.

Range walk is the variation of the range as the input intensity is varied. Characterization of the camera includes characterizing the range walk as well as the camera intensity and range gain. Characterizing the camera intensity uses the same experimental setup and the same dataset as characterizing the range walk because the range walk relates the corresponding, varied intensity data to the passively varied range data. For this camera, the camera range gain should be linear in response. The camera intensity gain should be linear in response until saturation.

9.4.1 Dark nonuniformity correction

The intensity return or LiDAR calibration is given in raw data number units. An 8-bit return will provide $2^8 = 256$ possible values per pixel. The conversion from this raw digital value to photon count is the electronic gain G of the array multiplied by the difference between the intensity return and the minimum of that return at dark level:

$$a = G(A - A_0), \tag{9.7}$$

where a is the return amplitude in photon count, A is the return amplitude in data numbers, and A_0 is the dark offset, estimated as the minimum value of the average dark frame.

The corrected range return r is displayed in meters and is calculated according to the following conversion:

$$r = d_0 + \frac{(T - T_0)}{T_s} \tau_{\text{gate}}, \tag{9.8}$$

where τ_{gate} is the range gate in meters (in this experiment, it was 300 m); d_0 is some offset determined by the jitter of the laser used in the setup (a high-jitter laser relative to r/c will have an appreciable deadzone); T_0 is the minimum value of the average noise frame, i.e., an offset bias; T is the uncorrected range return; and T_s is the the saturation value for range (this is often about one-half of the full bit length of the pixel), where $T_s = T_{s,0} - T_0$ for the corrected range.

Figure 9.7 shows images of a uniform object before and after correction, and Fig. 9.8 shows the relative intensity profiles of the images in Fig. 9.7.

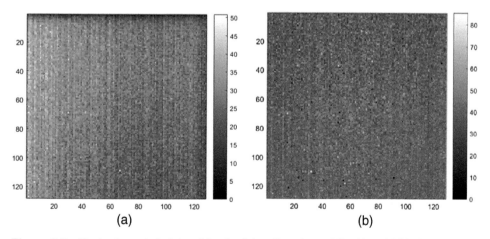

Figure 9.7 First return at dark level for the intensity return of the Voxtel VX-806 3D flash LiDAR camera (a) before and (b) after correction; units are in meters (adapted from Ref. 5).

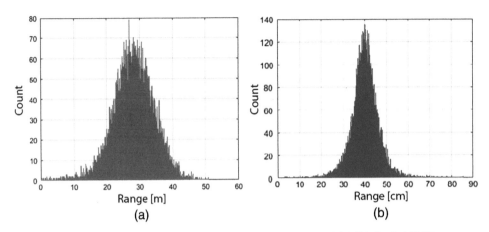

Figure 9.8 Histograms of the dark return of the Voxtel VX-806 3D flash LiDAR camera (a) before correction and (b) after correction (adapted from Ref. 5).

A distribution and an image are typically normalized by the input. For the purposes of seeking a best-case threshold for programmatic filtering (using nearest neighbors) of high-amplitude, somewhat spatially static temporal noise caused by being near threshold, a standard-deviation–normalized histogram and image are used. These are produced by taking the non-normalized image minus the mean value of the non-normalized image, and dividing this by the standard deviation of the non-normalized image:

$$p_{\text{norm}} = \frac{p - \bar{p}}{\sigma(p)}. \tag{9.9}$$

This is then used to generate a normalized distribution and create a normalized image. In conjunction with one another, an optimal threshold in units of standard deviation from the mean is used, typically in the positive direction only, but allowable in both directions, to generate thresholds for the nearest-neighbor filter algorithm. Nearest-neighbor correction is a method by which the nearest neighbor of a pixel replaces its value if that pixel is marked as defective and in need of replacing. This method is required to replace several defective pixels beyond a specified threshold in the intensity return:

$$x(i,j) = \frac{x(i \pm 1, j \pm 1) + x(i, j \pm 1) + x(i \pm 1, j)}{n}, \tag{9.10}$$

where n is the number of pixels taken from the neighborhood of the pixel being corrected; typically, $n = 8$, but it could be 5 or 3 if the pixel to be detected is an edge or corner pixel, respectively. An edge pixel is along the edge of the image, and a corner pixel is in the corner. A threshold is set in standard deviations from the norm to filter for the nearest-neighbor algorithm. This can be done automatically using MATLAB's 'histfit' function and some clever programming.

9.4.2 Results of correction

In our experiment, analysis of the corrections requires capturing raw image data using MATLAB image acquisition, as well as use of proprietary software for control of the camera. Nonuniformity correction tables are loaded as TXT files into the proprietary software's camera pipeline using a microSD card on one of the stack boards for the camera. Five hundred dark frames are captured (with corrections enabled) independently for range return and for intensity return at dark level. These dark frames are then averaged to generate a distribution plot and a sample correction image for the range and intensity returns. Finally, the noise and mean values of the distribution are generated (Table 9.1) from the average of the 500 corrected dark frames.

The preliminary data for the range return include noise levels significantly higher than the background (5.6087 m) at a mean of 38.9148 m, ranging from roughly 9.0 m to roughly 62.0 m. The nonuniformity correction for the range return reduces noise levels to those comparable to temporal noise (7.97 cm) at a dark level of 9.0 m minimum and a mean of 9.4406 m. This correction reduces the dark level from approximately 38 m to approximately 9 m, or the effective dark level for the range. This correction also reduces range noise by a factor of 70.4. The plot in Fig. 9.9 shows a narrow distribution of corrected range returns.

Corrections on the intensity return (Fig. 9.10) are more difficult to analyze, in large part due to the high-amplitude, low–temporal-mobility noise that was filtered out of the dark, near-threshold input frames and therefore is still present in the most-ideal test case of being in full dark for the intensity return (Fig. 9.11). Thus, the intensity return requires either analyzing the results from the entire return (including the near-threshold misfires of the camera) or statistically excluding those results, using the same statistical bounds as were used in the nearest-neighbor correction (Figs. 9.12 and 9.13).

The dark intensity return is measured in W/m^2 radiant intensity, converted from raw digital format to photon count. The preliminary data for the

Table 9.1 Noise and mean values for range before and after nonuniformity correction for the Voxtel VX-806 3D flash LiDAR camera.

Frame	Noise	Mean Value
Range after correction	5.9376 cm	40.0185 cm
Range before correction	5.9227 m	28.1833 m

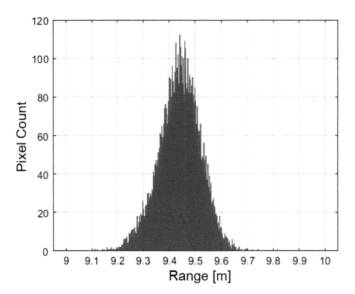

Figure 9.9 Noise distribution of the corrected range returns (adapted from Ref. 5).

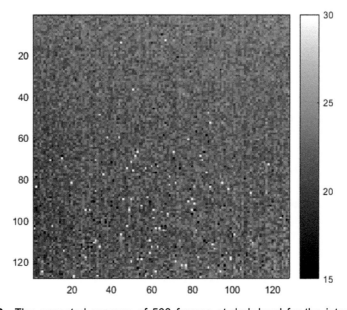

Figure 9.10 The corrected average of 500 frames at dark level for the intensity return. Units are in W/m² (adapted from Ref. 5).

intensity return includes noise levels significantly higher than the background: $28.6277 \ \text{W/m}^2$ at a mean of $235.0232 \ \text{W/m}^2$. The nonuniformity correction for the intensity return reduces noise levels to those comparable to temporal

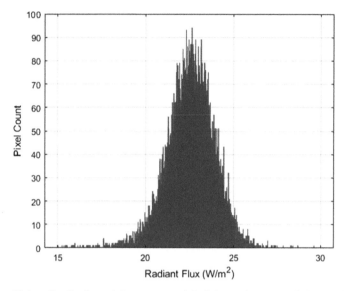

Figure 9.11 Noise distribution of the corrected dark intensity return (adapted from Ref. 5).

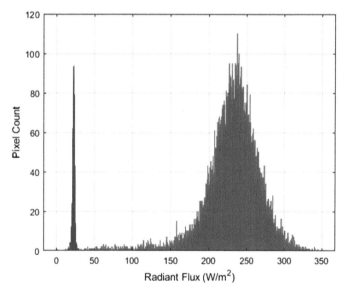

Figure 9.12 Noise distributions of the corrected (blue) and uncorrected (red) intensity returns (adapted from Ref. 5).

noise: 1.2992 W/m² at dark level with a mean of 22.5528 W/m². This correction reduces the dark level from approximately 235 W/m² to approximately 23 W/m² for the effective dark level of the intensity return. This correction also reduced intensity noise by a factor of 22.03.

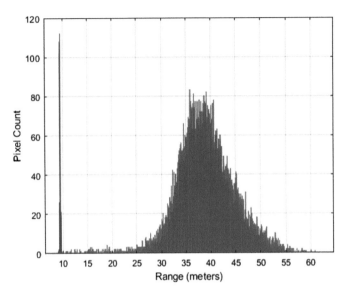

Figure 9.13 Noise distributions of the corrected (blue) and uncorrected (red) range returns (adapted from Ref. 5).

Problems and Solutions

9-1. For a flight altitude of 300 m, the object is 10 m high directly below the sensor. What's the laser pulse travel time?

9-1 Solution:
1.93 μs

9-2. What is the maximum unambiguous range (pulse laser) for the following non-overlapping pulse rates?
Pulse rates: 10 kHz, 71 kHz, 100 kHz, and 167 kHz

9-2 Solution:

PRF (Hz)	Unambiguous range (m)
10,000	15,000
71,000	2113
100,000	1500
167,000	898

9-3. We have two aircraft as our targets. Show the range resolution for a pulse width of 1 μs for a well-designed LiDAR system. What does the system "see" if the spacing between two aircraft is too small, or if it is large enough?

9-3 Solution:

$$B = \frac{1}{\Delta t} = \frac{1}{1\,\mu s} = 1\,\text{MHz},$$

$$\Delta R = \frac{c}{2B} = \frac{3 \times 10^8\,\text{m/s}}{2(1\,\text{MHz})} = 150\,\text{m}.$$

As seen above, the LiDAR sees only one target if the spacing between the two aircraft is small. If that spacing is large enough, the system will see both targets.

9-4. What is the range resolution for bandwidths of 100 MHz, 200 MHz, 300 MHz, 400 MHz, and 500 MHz.

9-4 Solution:

Bandwidth (MHz)	Resolution (m)
100	1.5
200	0.75
300	0.5
400	0.375
500	0.3

References

1. Wikipedia page on 1951 USAF resolution test chart: https://en.wikipedia.org/wiki/1951_USAF_resolution_test_chart.
2. ISO 12233 Test Chart: https://www.graphics.cornell.edu/~westin/misc/reschart.html.
3. P. D. Burns, "sfrmat3: SFR evaluation for digital cameras and scanners" http://losburns.com/imaging/software/SFRedge/sfrmat3_post/index.html.
4. J. W. Stafford, "Range Compressed Holographic Aperture Ladar," Ph.D. thesis, University of Dayton (2016).
5. A. D. Reinhardt, D. Miller, A. Lee, C. Bradley, and P. F. McManamon, "Dark non-uniformity correction and characterization of a 3D flash lidar camera," *Proc. SPIE* **10636**, 1063608 (2018) [doi: 10.1117/12.2302818].
6. C. P. Bradley, S. S. Mukherjee, A. D. Reinhardt, P. F. McManamon, A. O. Lee, and V. Dhulla, "3D imaging with 128 × 128 eye-safe InGaAs p-i-n lidar camera," *Proc. SPIE* **11005**, 1100510 (2019) [doi: 10.1117/12.2521981].

Chapter 10
LiDAR Performance Metrics

10.1 Image Quality Metrics

Most early image metrics have been derived for 2D imaging. Change detection and video are currently becoming more popular as well. 3D imaging is a revolutionary sensing technology on the horizon, and other sensing technologies such as laser vibration detection are developing. Eventually, we will need to develop image quality metrics applicable to these new sensing modalities, including 3D sensing. 3D LiDAR is a very likely method for developing a 3D point cloud image. Passive stereo sensors can also develop such an image, and 3D image metrics can be used for both methods of developing a 3D image. A stereoscopic, passive 3D image will likely have lower resolution in range than in angle/angle, with range resolution depending on angle resolution and the size of the stereo angle used, whereas time-of-flight LiDAR has range resolution that is independent of angle/angle resolution.

The goal of using a LiDAR is often to detect, recognize, or identify an object. To achieve the desired information, the object to be recognized will strongly influence the LiDAR parameters. Recognizing a person's face will require a small field of view, but will require resolution on the scale of centimeters. A military tank, on the other hand, is much larger, and its features are not as fine, so a larger field of view is required, but resolution can be relaxed to about 15 cm for recognition or identification. We do not need as fine a resolution to recognize a tank as we need to recognize a face. The so-called Johnson criteria,[1] developed by Johnny Johnson, recognizes this target-dependent aspect of sensor-based detection and will be further discussed. The Johnson criteria was the first set of sensing metrics the author is aware of. The National Imagery Interpretability Rating Scales (NIIRS) was developed later. The NIIRS is based on image interpretability for a larger variety of tasks as a method to evaluate the relative value of various image sources. The tasks are listed in Tables 10.1, 10.2, and 10.3 according to sensor-specific modality.

For the Johnson criteria, we start with an object to be detected, recognized, or identified. Johnny Johnson also considered orientation as an information objective. Detection highlights the area of interest where the object is located. We use the term recognition to identify the class of object, such as truck or tank. We use the term identification to provide the type of tank or truck, such as a T-72 or an M1 tank. Sometimes there is a desire to go even farther and tag a specific vehicle, such as a specific car someone is driving. To do this, unique features of that vehicle must be identified, such as dents in certain locations on the vehicle. Tagging a specific vehicle can require high resolution, but at times can be of strong interest. Once we have considered the characteristics of the object to be viewed, we have to consider a LiDAR to measure those parameters of the object. This LiDAR generates data about the object, which is processed into information.

The LiDAR views the object through a medium, such as air or water. The medium will impose certain constraints, which could include loss of contrast or resolution. The LiDAR has limitations in the following parameters: its field of view, its resolution, its detector sampling (both the detector angular subtense and any motion that allows a dither of the portions sampled), its optical aberrations, its grayscale dynamic range in each pixel or voxel, its noise characteristics, possibly its range resolution and number of stored samples in range, possibly its ability to measure polarization properties of light reflected from the object, possibly its ability to measure the spectrum of reflected light, its ability to measure motion of the object being viewed, possibility its ability to illuminate the object with light having some structure or pattern in angle/angle space, and possibly in other specific LiDAR parameters that influence the ability of the LiDAR to capture information truly representative of the object on which we would like to achieve some information-based judgment.

In addition to measuring certain parameters, we may need to locate the object in the world. The LiDAR can measure relative object parameters between portions of the object, or from the LiDAR to the object, but unless we know where the LiDAR is in the world, and in what direction it is pointing, we cannot locate the object in the world. For this reason, navigation information such as GPS or an inertial navigation system may be required.

10.1.1 Object parameters

The size of an object influences the required LiDAR field of view. Ideally, the LiDAR will see the whole target. Then, for various objectives, you will need a certain number of samples across the target. That is the idea of Johnson's criteria,[2] which is for 2D imaging. The minimum required resolution according to Johnson's criteria is expressed in terms of line pairs of image resolution across a target for several tasks. These amounts of resolution provide a 50% probability of an observer discriminating an object to the specified level.

Table 10.1 provides data from Johnny Johnson's original article on night vision goggles.[1] The table shows that the Johnson criteria are applied to minimum dimension, and it shows the variability of his data with respect to various target types.

Table 10.1 Line pairs required to achieve certain tasks, as recorded in developing the Johnson criteria (from Ref. 1).

Broadside View	Detection	Orientation	Recognition	Identification
Truck	0.90	1.25	4.5	8.0
M-48 Tank	0.75	1.20	3.5	7.0
Stalin Tank	0.75	1.20	3.3	6.0
Centurion Tank	0.75	1.20	3.5	6.0
Half-Track	1.00	1.50	4.0	5.0
Jeep	1.20	1.50	4.5	5.5
Command Car	1.20	1.50	4.3	5.5
Soldier (standing)	1.50	1.80	3.8	8.0
105 Howitzer	1.00	1.50	4.8	6.0
Average Line Pairs:	1.00 ± 0.25	1.40 ± 0.35	4.0 ± 0.8	6.4 ± 1.5
	Resolution in line pairs per minimum dimension			

In the Johnson criteria there is no variation in the required number of line pairs, or pixels (two pixels per line pair), depending on the object viewed, even though Johnson's actual data show some variation based on the object being viewed. In reality, an object with more detail might require more line pairs; also, if an object is viewed against a complex background, it may require more line pairs compared to when viewed against a plain background. These situations are not considered in the Johnson criteria. One advantage we will have in considering 3D LiDARs is that much of the background complexity can be removed by discriminating in range. A range-gated 2D LiDAR can also eliminate background clutter, possibly allowing easier discrimination, even though it is still only a 2D sensor. Johnson's criteria also does not indicate the significant influence of contrast on the required number of line pairs. The contrast is a function of both the inherent contrast at the object (the reduction in contrast due to the medium through which the object is being viewed) and the sensor's effect on contrast. One effect on contrast is that, if the object does not completely fill a pixel, the intensity viewed will be an average across the objects viewed within the pixel. The shape of an object could be a factor. A long, thin object might not require the same number of pixels across its smallest dimension as an object that is more balanced with respect to the two dimensions being viewed. Human visibility in the U.S. is defined as the range at which contrast is reduced to 2% of its value. Johnny Johnson's goal was a simple rule of thumb; his criteria worked much of the time, and he succeeded for traditional 2D imagery. When Johnny Johnson did this work, he emphasized night vision goggles. This work has stood up well since the 1950s for most 2D objects.

More recently, people have used pixels on target instead of line pairs. Each line pair is equivalent to two pixels. A further development of the same idea suggested by Johnny Johnson is to consider which spatial frequencies are contained in the target. If the target is a white board that is 3 m across in one dimension, the board is still only a point as viewed from the LiDAR if resolution is decreased from 100 m to 50 m, then to 20 m, and finally to 10 m.

In all cases, the object is smaller than the resolution. Also, if it is a uniform white board (low–spatial-frequency content), then, except for exact placement of the board, it will not matter if resolution is decreased from 2 m to 1 m to 1 cm, or to 1 mm. At 1-mm resolution, you will have many pixels that look exactly the same. The ability to detect, recognize, or identify remains the same in those cases except for SNR considerations. If the object were 5 m across but had significant internal structure (high–spatial-frequency content), then increasing the resolution could make a difference. One of the things we can do to assist the process of developing metrics is to characterize the available spatial frequencies in each target set of interest (as defined by the tasks described in the two tables in Section 10.3). If we have a very high-resolution image of an object, we can Fourier transform it to determine the object's spatial frequency content. We show this method next for images of a tank. This could be part of a future comprehensive metric evaluation. The smallest scale in Fourier transform space will be

$$df_x = \frac{1}{Nd_x},$$ (10.1)

where d_x is the pixel size, and N is the number of pixels in a given dimension. One can see that the smallest increment in spatial frequency space is governed by the number of samples taken in that dimension. If a tank is 8 m across and the full number of samples is used across that width, then that full width sets the smallest spatial frequency. The pixel pitch and number of pixels (in conjunction with the focal length of the system) then sets the spatial frequency limits, or field of view. With very fine resolution, you can go to high spatial frequencies at the expense of field of view.

We can take an image and Fourier transform it. Figure 10.1(a) shows a T-72 tank image taken from the Internet.[3] Figure 10.1(b) shows the magnitude of the image's Fourier transform in two dimensions.

The MATLAB[®] code used in the Fourier transform is provided in Appendix 10-1. The point of these images is that the spatial frequency content of an image determines which spatial frequencies are of interest in characterizing that object. Once you have sampled all of the spatial frequencies of interest, higher resolution does not provide a benefit. Figure 10.2(b) shows the Fourier transform of the Air Force 1951 bar target.[4] Looking at this 2D Fourier transform, you can see that there is more higher-frequency content in the Air Force 1951 target than in the tank.

An interesting question is whether the same criteria apply to range resolution in the third dimension as they do in the first two dimensions. If we have an object with a lot of structure in range, the object should show higher spatial frequencies in range. We could consider spatial frequencies in 1D, 2D, or 3D imaging.

Figure 10.1 A photograph of a T-72 tank and (b) the Fourier transfer of the image [part (a) reprinted from Ref. 3].

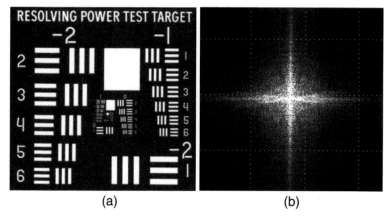

Figure 10.2 (a) Air Force 1951 test target and (b) the 2D Fourier transform to show its spatial frequencies [part (a) reprinted from Ref. 4].

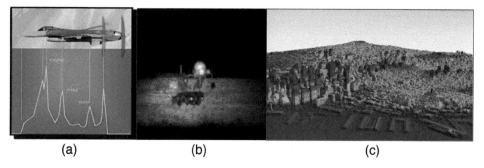

Figure 10.3 (a) 1D image (graph) below a 2D image (aircraft) shown to indicate where the 1D range scatter comes from; (b) 2D image, and (c) 3D image [part (a) provided by Matthew Dierking, Air Force Research Lab; part (b) reprinted from Ref. 27, and part (c) reprinted from Ref. 5].

Figure 10.3 shows a 1D image, a 2D image, and a 3D image. Adding a second dimension instead of having only a range profile obviously makes it easier for humans to identify objects. 2D imagery is the human baseline because our eyes are 2D imagers. We see angle/angle images. We can also see the grayscale and color that add to that 2D image. Gated LiDAR 2D imaging can be range gated to reduce clutter. Adding a third dimension provides additional information, so it seems likely that any additional information will make detection, recognition, identification, or other tasks easier to accomplish. Another aspect to the third dimension is that value can be gained when viewing an object from a diverse set of angles. An object you can see through will provide some 3D range information from a single viewing angle, but solid objects will only provide complete range relief information as viewed from the angle from which the image was taken. A 3D range relief image is shown in Fig. 10.3(c). We can 3D map the closest portion of the object at a given location from a single viewing location. To view the back of a solid object, we will need angular diversity. For 2D imaging, the viewing angle changes the perspective of the object, but changing the viewing angle does not increase information as rapidly as with a 3D imaging LiDAR. Another significant benefit from 3D images is the ability to hand off an image to a different viewing position. If we have a 3D image of an object, it is easier to adjust the viewing angle for handoff considerations. A 2D image can be generated as viewed from multiple angles.

10.1.1.1 Additional object characteristics

Grayscale is a parameter that is limited by the variation in reflectivity of the object to be sensed. Higher–spatial-frequency 2D images have a more rapid change in grayscale. If an entire object is a single value of reflectivity, grayscale will be constant if illumination is constant. If there is a wide variation in reflectivity, a LiDAR should have significant grayscale dynamic range to accurately sense that object. Illumination can also influence grayscale. Some LiDARs have patterned illumination.

Some objects are all one color. Other objects contain a large variety of colors. This is another criterion that can determine the required LiDAR parameters to accurately sense an object. Multiple wavelengths will be required to see the "color" of an object.

Motion can assist in detecting, recognizing, or identifying an object, but of course, the limitation on using motion will depend on how much the object moves, how long the motion is observed, and how unique that motion is compared to the object's background. Change detection is a slower version of motion. It can be used with 3D imaging as well as with 2D imaging. For example, if someone is building a tunnel, you might be able to watch a pile of dirt next to the tunnel entrance increase in size, and then estimate the size of the tunnel based on the 3D size of the dirt pile. Change detection can be a powerful image discriminant; however, to date, the author is not aware of work being done to accomplish suitable image metrics that encompass motion and change detection, even though these are currently popular imaging modalities.

10.2 LiDAR Parameters

A LiDAR has many parameters that influence its ability to measure object features that allow object detection, recognition, or identification. Resolution in each dimension is one of those interesting parameters. The diffraction limit of the LiDAR resolution is given by the receive aperture size and the wavelength of light, as described in previous chapters. The optical system may not be diffraction limited. It could have aberrations that distort the optical image. The modulation transfer function (MTF) is a measure of the optical resolution limit of a LiDAR. Another factor is sampling. A key sampling parameter is the detector angular subtense (DAS). Often the DAS will be set to match the FWHM angular extent of the optical point spread function (PSF). The PSF can also be oversampled to reduce aliasing of the object. When the PSF is oversampled, some people will perceive the image as being sharper.

For a synthetic-aperture LiDAR (SAL), the angle resolution is given by[6]

$$\vartheta = \frac{\lambda}{2L + D},$$ (10.2)

where D is the diameter of the real aperture, and L is the distance flown. In radar the size of the real aperture can be neglected, but in LiDAR it cannot be neglected. SARs tend to fly kilometers of distance, whereas SALs fly only meters to generate the synthetic aperture.

10.3 Image Parameters: National Imagery Interpretability Rating Scale (NIIRS)

The NIIRS[7] currently exists for visible, infrared, civil, radar, and multispectral imagery. There has been discussion about a potential NIIRS for motion and for 3D imaging, but these scales have not been developed. The NIIRS is mission based. Certain tasks, or missions, are highlighted for each NIIRS LiDAR modality. One of the disconcerting issues, however, is that there are almost no common tasks. Admittedly, each sensor has certain functions it does well. Even in LiDAR, we can say that each LiDAR has certain functions that it does better than other LiDARs. That is one reason that accounts for the differences between functions for each modality. An operational commander has to decide what sensor to use for a given mission, and a service has to decide what sensors to buy to accomplish various missions. Commercial companies have to decide which sensors they need for a certain mission. One potential reason for metrics is to allow users to decide on the relative value of sensor 1, 2, or 3 to accomplish a given function. With different tasks for each sensor modality, this decision becomes more difficult, reducing the value of the metrics to the user. This is a disadvantage to the current NIIRS.

The basis for the NIIRS is the concept that imagery analysts should be able to perform more-demanding interpretation tasks with higher-quality imagery.[8]

NIIRS consists of 10 graduated levels (0 to 9), with several interpretation tasks or criteria forming each level. These criteria indicate the level of information that can be extracted from an image of a given interpretability level. The NIIRS provides a common framework for discussing the interpretability, or information potential, of imagery. The NIIRS, therefore, serves as a standardized indicator of image interpretability for:

- communicating the relative usefulness of the imagery,
- documenting requirements for imagery,
- managing the tasking and collection of imagery,
- assisting in the design and assessment of future imaging systems, and
- measuring the performance of LiDAR systems and imagery exploitation devices.

Currently, however, the NIIRS only really accomplishes the above within a sensor class, since similar tasks are not used in multiple classes. It is the authors' opinion that the value of the NIIRS would be higher if similar tasks could be judged across sensor classes. Another more difficult area to consider is the idea of using more than one sensor modality to accomplish a task. Sensor fusion is a new area of interest, but it is not currently covered by the NIIRS system.

Each unit on the NIIRS results in a doubling of resolution. There are multiple NIIRS, e.g., for visible sensors, radar, and infrared and multispectral imaging. Table 10.2 provides an example of visible NIIRS rating criteria.[9] There is also a civilian NIIRS, as shown in Table 10.3, called the Civil National Imagery Interpretability Rating Scale (October 1995).

Table 10.2 Example visible NIIRS criteria (data from Ref. 9).

Rating	Detect	Identify	Distinguish	Estimated GSD (m)
0				>10
1	Medium-sized port facility		Taxiways from runways	10
2	Large hangers	SA-5 Site based on roads		5
3	String of rail cars	Wing shape on large aircraft		3
4	Large radar antennas	Large fighter by type	Bow shape on medium submarine	1.5
5		Tactical SSM by type	SS-25 TEL and MSV	1
6		Identify the shape of EW antennas	Models of small helicopters	0.5
7	Mount for anti-tank guided missiles	Individual rail ties		0.3
8	Winch cables on deck mounted cranes	Windshield wipers on a vehicle		0.1
9	Individual spikes in railroad ties	Vehicle registration numbers	Cross-slot from straight slot fasteners on aircraft	0.01

SSM is surface-to-surface missile; EW is electronic warfare; TEL is transporter erector launcher; and MSV is modular space vehicle.

Table 10.3 Civil NIIRS criteria.

Rating Level 0	Interpretability of the imagery is precluded by obscuration, degradation, or very poor resolution.
Rating Level 1	Distinguishes between major land-use classes (e.g., urban, agricultural, forest, water, barren). Detects a medium-sized port facility. Distinguishes between runways and taxiways at a large airfield. Identifies large-area drainage patterns by type (e.g., dendritic, trellis, radial).
Rating Level 2	Identifies large (i.e., greater than 160 acre) center-pivot irrigated fields during the growing season. Detects large buildings (e.g., hospitals, factories). Identifies road patterns like clover leafs on major highway systems. Detects ice-breaker tracks. Detects the wake from a large (e.g., greater than 300 ft) ship.
Rating Level 3	Detects large-area (i.e., larger than 160 acre) contour plowing. Detects individual houses in residential neighborhoods. Detects trains or strings of standard rolling stock on railroad tracks (not individual cars). Identifies inland waterways navigable by barges. Distinguishes between natural forest stands and orchards.
Rating Level 4	Identifies farm buildings as barns, silos, or residences. Counts unoccupied railroad tracks along right-of-way or in a railroad yard. Detects basketball court, tennis court, volleyball court in urban areas. Identifies individual tracks, rail pairs, control towers, switching points in rail yards. Detects jeep trails through grassland.
Rating Level 5	Identifies Christmas tree plantations. Identifies individual rail cars by type (e.g., gondola, flat, box) and locomotives by type (e.g., steam, diesel). Detects open bay doors of vehicle storage buildings. Identifies tents (larger than two-person) at established recreational camping areas. Distinguishes between stands of coniferous and deciduous trees during leaf-off condition. Detects large animals (e.g., elephants, rhinoceroses, giraffes) in grasslands.
Rating Level 6	Detects narcotics intercropping based on texture. Distinguishes between row (e.g., corn, soybean) crops and small grain (e.g., wheat, oats) crops. Identifies automobiles as sedans or station wagons. Identifies individual telephone/electric poles in residential neighborhoods. Detects foot trails through barren areas.
Rating Level 7	Identifies individual mature cotton plants in a known cotton field. Identifies individual railroad ties. Detects individual steps on a stairway. Detects stumps and rocks in forest clearings and meadows.
Rating Level 8	Counts individual baby pigs. Identifies a USGS benchmark set in a paved surface. Identifies grill detailing and/or the license plate on a passenger/truck-type vehicle. Identifies individual pine seedlings. Identifies individual water lilies on a pond. Identifies windshield wipers on a vehicle.
Rating Level 9	Identifies individual grain heads on small grain (e.g., wheat, oats, barley). Identifies individual barbs on a barbed wire fence. Detects individual spikes in railroad ties. Identifies individual bunches of pine needles. Identifies an ear tag on large game animals (e.g., deer, elk, moose).

As mentioned, there has been some investigation of a NIIRS for motion-based imagery.[10] While it appears that motion does contribute to higher image interpretability, a NIIRS still needs to be developed for motion imagery.

There was at least one attempt at defining NIIRS for 3D imagery. This was developed under a National Geospatial-Intelligence Agency (NGA) technology fellowship effort. It is not widely accepted, but as of this time is the only attempt the author is aware of to create a 3D NIIRS. As 3D video becomes a possibility, at some point in the future, it may be necessary to develop metrics for sensing that combine 3D imagery and motion.[11]

Figure 10.4 shows digital terrain elevation data (DTED) associated with 3D imaging. DTED at various resolutions is inherently 3D so could relate to a 3D NIIRS. DTED is a standard NGA product that provides medium-resolution quantitative data in a digital format for military system applications that require terrain elevation. Figure 10.4 carries this to finer resolutions in what is called high-resolution terrain elevation (HRTe) data. The figure shows that beyond DTED 2 is HRTe data. Seeing images at various 3D resolutions makes it easier to understand the meaning of a certain DTED level. It is interesting to see correlation with the resolution values for visible NIIRS at certain levels, but no correlation at other levels. For example, DTED 1 has 100-m ground sample distance (GSD), whereas visible NIIRS 1 has >9-m GSD. DTED/HRTe data are not task based like the NIIRS, but instead are based on factors of 3 and 3.3333 in resolution. It would probably be useful if a 3D NIIRS could have some relationship to DTED/HRTe data levels.

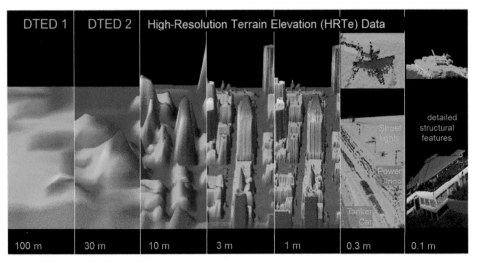

Figure 10.4 Various DTED and HRTe data levels [reprinted from A. Carbonell, "National System for Geospatial Intelligence (NSG) LIDAR Initiatives," presented at the ESRI Conference of the National Geospatial-Intelligence Agency, 16 July 2010; approved for public release 10-173F].

10.4 3D Metrics for LiDAR Images

For monochromatic 2D imagery, the NIIRS is used extensively to characterize image quality. Leachtenauer et al.[12] empirically determined a method to calculate the NIIRS using various image metrics, such as the GSD and the relative edge response (RER). Leachtenauer et al. later used the same empirical methodology applied to infrared images.[13] Thurman and Fienup provide physical explanations for many of the parameters in the general image quality equation (GIQE).[14] Various constants can be used in this equation, representing somewhat different forms of the GIQE. A generic version of the GIQE is given as

$$NIIRS = c_0 + c_1 \log_2(GSD) + c_2 \log_2(RER) + c_3 \frac{G}{SNR} + c_4 J, \quad (10.3)$$

where *NIIRS* is the image quality rating assigned in accordance with the National Image Interpretability Rating Scale, *RER* is the relative edge response, *J* is the mean height overshoot caused by edge sharpening, *G* is the noise gain resulting from edge sharpening, the various *c* values are constants, and *SNR* is the signal-to-noise ratio. No one to date has developed a quality 3D NIIRS or GIQE.

To summarize, sensors are often used for objectives such as detecting, recognizing, or identifying an object. The object to be recognized determines the sensor parameters that will be used to achieve the desired information. The National Imagery Interpretability Rating Scales (NIIRS) was developed. The NIIRS is based on image interpretability for a large variety of tasks as a method to evaluate the relative value of imagery. There is no NIIRS rating yet for 3D LiDAR, but it is anticipated that one will be developed. A NIIRS metric for 3D LiDAR will allow for comparison between various LiDARs. A powerful tool in 3D imaging is the ability to view an object from multiple angles, as shown in Fig. 10.5. A general image quality equation (GIQE) has been developed for visible and near-IR imagery, but has not yet been developed for 3D LiDAR imagery. The GIQE can predict NIIRS ratings.

10.5 General Image Quality Equations

The original GIQE developed for visible imagery is given by

$$NIIRS = 11.81 + 3.32 \times \log_{10} \frac{RER}{GSD} - (1.48 \times H) - \frac{G}{SNR}, \quad (10.4)$$

where *NIIRS* is the image quality rating assigned in accordance with the National Image Interpretability Rating Scale, *RER* is the geometric mean of the normalized RER, *GSD* is the geometric-mean ground sample distance (in inches), *H* is the geometric mean-height overshoot caused by edge

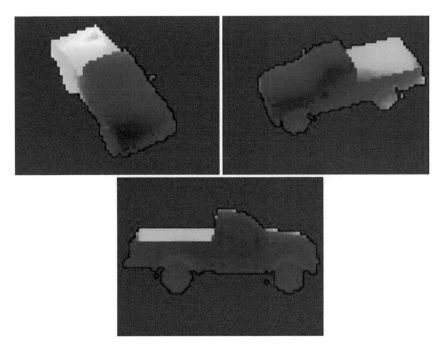

Figure 10.5 3D imaging of a pickup truck viewed from multiple angles (reprinted from Ref. 28).

sharpening, and *G* is the noise gain resulting from edge sharpening. This form of the GIQE was empirically derived by Leachtenauer et al.[12] The GIQE for visible imagery was developed by using a regression modeling approach. Ten imagery analysts provided NIIRS ratings on a large sample of EO imagery. The image quality characteristics of the imagery were used to develop a regression model that predicted NIIRS values as a function of system design (quality) and operating parameters. The GIQE treats three main attributes: scale, expressed as the GSD; sharpness, measured from the system MTF; and the SNR. The terms in the GIQE derive from earlier research relating physical image quality to interpretability. The GSD is a measure of both scale and resolution. RER relates to perceived sharpness or acutance. Because MTF compensation (MTFC) is commonly used with EO system post-processing, it was necessary to account for MTFC effects. The MTFC boosts both spatial frequencies and noise, and, hence, boosts the overshoot and noise gain terms. Noise is modeled in terms of the noise gain derived from the MTFC and the SNR. Noise can reduce contrast, so the SNR term relates to the contrast in the image. The terms in the GIQE account for all of the physical-quality parameters that have been found to affect image interpretability. The GIQE can be applied to any visible LiDAR in the initial GIQE and predicts NIIRS ratings with a standard error of 0.3 NIIRS. A factor to consider is that if you are sensing from a lower angle, one dimension will be spread, as shown in Fig. 10.6.

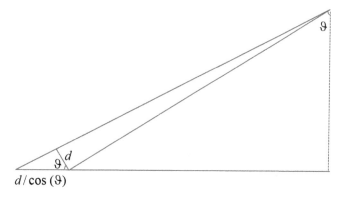

Figure 10.6 Spreading of GSD in the direction away from the LiDAR.

The GSD is the geometric mean of the GSD based on a projection of the pixel pitch distance to the ground. In one dimension, the GSD is spread because of the cosine function due to the LiDAR angle away from nadir. The GSD is given by

$$GSD = \frac{d}{\cos \vartheta}. \tag{10.5}$$

In the other dimension, we do not have this cosine function spreading.

Leachtenauer et al. later attempted to apply the GIQE approach to the infrared.[13] They used the same equation, but changed some of the coefficients. This was not successful in correlating with NIIRS ratings. The best summary relating image quality metrics to LiDAR parameters to date is the paper by Thurman and Fienup,[14] which is an excellent discussion of real LiDAR parameters associated with the GIQE. The authors' use of the eye chart for comparison purposes is novel and useful. This paper provides a more-generalized form of the GIQE:

$$NIIRS = c_0 + c_1 \log_{10}(GSD) + c_2 \log_{10}(RER) + c_3 \frac{G}{SNR} + c_4 H, \tag{10.6}$$

where the various constants can be selected depending on the circumstances; e.g., by selecting different c values, the equation can be the Leachtenauer visible form or the infrared form of the equation.

10.6 Quality Metrics Associated with Automatic Target Detection, Recognition, or Identification

Automatic object recognition could, in theory, have different quality metrics than human recognition. One company uses a mean distance metric between the measured 3D image data and a target model as a metric for recognition quality.[15] Of course, this type of metric would be less precise for a 2D image, but still could be used. The company metric also looks at the effectiveness of segmenting the object of interest from the background (correctly associating a pixel with either target or background). The 3D image adds an additional contrast (in range) to the normal contrasts (in irradiance or color) to allow image segmentation. 2D imagery can only be segmented based on contrast. Range can be a powerful segmentation tool. Another future area of work on image metrics will be image metrics for accomplishing automated tasks. To date, the Johnson criteria and (later) the various NIIRS ratings have all been developed based on a human accomplishing certain tasks. Due to the limited number of analysts and the exploding amount of imagery, at a minimum, machine cueing will become much more prevalent, even if machines do not make final decisions. There will be a need to begin to add machine-based tasks into NIIRS-like image metric ratings.

10.7 Information Theory Related to Image Quality Metrics

The mean information density can be defined as the total information content of the image divided by the total geospatial area of the image, or

$$\overline{K} = \frac{I_{\text{total}}}{A_{\text{total}}}, \tag{10.7}$$

where $\overline{K}$ is the mean information density in bits per square meter, I_{total} is the total information content of the image, and A_{total} is the total physical area covered by the image in square meters.[16] The total information content of that image can be expressed as

$$I_{\text{total}} = N \times H(X), \tag{10.8}$$

where N is the number of pixels in the image, and $H(X)$ is the Shannon entropy[17] of the random variable X, which can be interpreted as the intensity, or grayscale, of an individual pixel for monochromatic images. In this context, each pixel can be regarded as a symbol in a message. The total area can be expressed as

$$A_{\text{total}} = N \times GSD^2, \tag{10.9}$$

where GSD is the mean GSD in meters, and all other terms are as previously defined. In this context, the GSD is the distance between the centers of adjacent pixels when the image is projected to object space. At this time, the application of information theory to image metrics for 2D, 3D, or other sensing modalities is relatively immature, so the discussion here is relatively limited.

10.8 Image Quality Metrics Based on Alternative Basis Sets

The approach used by Johnson described above essentially employs a basis set of mathematical functions to quantify image quality in terms of various detection, classification, and recognition tasks. In his particular case, he used spatial frequencies (complex exponentials) as the mathematical basis set to characterize the image and then derived the spatial frequency content of the image required to perform different tasks. A useful feature of his approach is that the optical system can also be characterized in terms of its spatial frequency response. Hence, the object, system, image, and task all could be characterized using a common basis. There are, however, many other basis sets that can be used to describe an image. Some of these are receiving considerable attention in the literature and are discussed below. To my knowledge, there has been no investigation as to whether or not the spatial frequency basis set is optimal in any sense. Therefore, an investigation into the potential benefit of other basis sets would be a very useful endeavor. A key component in such an investigation would be to do what Johnson did for the spatial frequency basis, which is to characterize the different tasks in terms of level of fidelity in these new basis sets. An additional consideration for 3D imaging is the ability to extract object pixels from background pixels. In 2D imagery, this is done based on the contrast of some sensed parameter, which can include brightness (radiance) or color. In 3D imagery an additional contrast parameter is available: range from LiDAR to object point. Displaying and rotating point clouds can exploit this range contrast to perform object segmentation and imaging.

10.9 Eigenmodes

In another approach,[18] the object, system, and image are described in terms of the eigenmodes associated with optical propagation through the imaging system. Hence, the eigenmodes are defined by the LiDAR. These eigenmodes are then used as the mathematical basis on which to model the object and the image. Image quality is then determined by how many of, and how faithfully, the object eigenmodes are relayed by the imaging system. Such an approach may have benefits in terms of recovery of image content for what some term super-resolution. To date, optical eigenmodes appear to have been used only for 2D imagery. The extension to 3D imagery would require significant modification, as the 3D eigenmodes would include characteristics of the detector and readout circuitry to define the eigenmodes in the range dimension. Description of the imaging problem in terms of optical system eigenmodes inherently includes the characteristics of the optical LiDAR in its formulation. As mentioned above, however, some means would be required to map the eigenmodes of an imaging system onto the required tasks. This may be difficult using optical system eigenmodes as the basis set because they will change from system to system. Hence, this mapping task would be required for each system.

10.10 Compressive Sensing

Compressive sensing is based on the concept that the object and image can be considered sparse in some domain; i.e., when mapped onto a particular basis set, the object and image can be described by a number of basis vectors that is fewer than the number of pixels in a typical image. Compressive sensing to some extent tries to get to the essence of image content by exploiting correlations within the scene being measured. In the example mentioned earlier about a uniform object, there exists strong correlations in irradiance from pixel to pixel; therefore, measurements of all pixels would not be required. Johnson did not consider the fact that objects of interest may have sparseness in the spatial frequency domain; i.e., objects may not have appreciable content at certain spatial frequencies. Again, in the uniform object example above, the higher–spatial-frequency content is small; therefore, expending resources (pixels) to make the higher–spatial-frequency measurements is not worthwhile. In compressive sensing the basis set of choice is often a random basis set, e.g., random pupil plane masks. Image quality using a random basis set may then be assessed as the number of measurements using the random basis that must be made to perform a particular task. It may be possible to produce a mapping of tasks to such a basis in a statistical sense. The object and image may also be mapped to such a basis based on the level of sparseness within the object/image. Papers have been written that document what is called natural scene statistics.[19,20] What is less clear is the ability to characterize the optical system in terms of the basis. Another factor that needs to be considered is how "close" the reconstructed image (from a sparse set of measurements) is to the real object. The required degree of closeness will most likely be related to the particular task.

10.10.1 Knowledge-enhanced compressive sensing

In this approach, *a priori* knowledge, or knowledge that is gained as during the sensing process, is used to move from a random basis set to a sparse basis that is better matched to the object and task. It is possible that the particular task could be included as constraints in the development of the basis as well as the characteristics of the imaging system. Work in this area was initiated under a DARPA program.[21]

10.10.2 Scale-invariant feature transforms

The scale-invariant feature transform (SIFT) technique attempts to extract "features" from an object that are robust to changes in scale and rotation of the object. These features can then be used to track an object, or to classify and recognize objects in clutter. The SIFT technique for image feature generation transforms an image into a collection of local feature vectors. (See "Distinctive image features from scale-invariant keypoints," by D. G. Lowe.[22]) Each of these feature vectors is invariant to any scaling, rotation, or translation of the image. Algorithms have been developed to extract object keypoints that can be used for various exploitation tasks. As a result, the SIFT approach may be able to provide a quantitative description of the various tasks that are typically used in NIIRS assessment. It may be possible to define what separates two tasks such as "Detect large noncombatant ships (e.g., freighters or tankers)" from "Distinguish between large rotary-wing and medium fixed-wing aircraft." It may be possible to extend SIFT to look across sensing modalities as well, although what constitutes a significant feature will probably change from mode to mode. However, identifying those features that persist across modes may aid with LiDAR fusion.

10.11 Machine Learning

Machine learning is receiving considerable attention, and the development of tools that allow automated sorting of images has reached the stage of being implemented in programs such as MATLAB. With machine learning techniques, however, it is sometimes not obvious which characteristics of an image cause the algorithm to sort the images in the manner that it does. While machine learning may provide an automated means of assessing image quality in terms of being able to correctly sort a collection of images, it may not provide a direct link to the typical human tasks that have been used throughout the years for NIIRS assessment. At the same time, machine learning will become more and more important as the ratio of imagery collected to image analysts moves to higher and higher values.

10.12 Processing to Obtain Imagery

Since an objective of this chapter is to consider image metrics for 3D imagery, it might be beneficial to provide the NGA community definitions of various LiDAR processing stages in this section. The definitions below are extracted from Light Detection and Ranging (LIDAR) LiDAR Model Supporting Precise Geopositioning;[23] similar definitions are provided in Chapter 8 on Lidar Processing.

> **Level 0 (L0) – Raw data and metadata:** L0 data consist of the raw data in the form in which they are is stored as collected from the mapping platform. The dataset includes, but is not limited to, data from GPS systems or IMUs, laser measurements (timing, angles), and gimbal(s). Metadata would include items such as the LiDAR type, date, calibration data, coordinate frame, units, and geographic extents of the collection. Other ancillary data would also be included, such as GPS observations from nearby base stations. Typical users of L0 data would include LiDAR builders and data providers, as well as researchers looking into improving the processing of data from L0 to L1.

Level 1 (L1) – Unfiltered 3D point cloud: L1 data consist of a 3D point data (point cloud) representation of the objects measured by the LIDAR mapping system. It is the result of applying algorithms (from LiDAR models, Kalman filters, etc.) in order to project the L0 data into 3D space. All metadata necessary for further processing are also carried forward at this level. Users would include scientists and others working on algorithms for deriving higher-level datasets, such as filtering or registration.

Level 2 (L2) – Noise-filtered 3D point cloud: L2 data differ from L1 in that noisy, spurious data have been removed (filtered) from the dataset, intensity values have been determined for each 3D point (if applicable), and relative registration (among scans, stares, or swaths) has been performed. The impetus behind separating L1 from L2 is due to the nature of Geiger-mode LIDAR (GML) data. Derivation of L1 GML data produces very noisy point clouds, which require specialized processing (coincidence processing) to remove the noise. Coincidence processing algorithms are still in their infancy, so their ongoing development necessitates a natural break in the processing levels. As with L1, all metadata necessary for further processing are carried forward. Typical users include exploitation algorithm developers and scientists developing georegistration techniques.

Level 3 (L3) – Georegistered 3D point cloud: L3 datasets differ from L2 in that the data have been registered to a known geodetic datum. This may be performed by an adjustment using data-identifiable objects of known geodetic coordinates or some other method of control extension for improving the absolute accuracy of the dataset. The primary users of L3 data would be exploitation algorithm developers.

Level 4 (L4) – Derived products: L4 datasets represent LIDAR-derived products to be disseminated to standard users. These products could include digital elevation models (DEMs), viewsheds, or other products created in a standard format and using a standard set of tools. These datasets are derived from L1, L2, or L3 data and are used by the basic user.

Level 5 (L5) – Intel products: L5 datasets are a type of specialized products for users in the intelligence community that may require specialized tools and knowledge to generate. The datasets are derived from L1, L2, or L3 data.

10.13 Range Resolution in EO/IR Imagers

LiDAR data analysis provides a series of unique challenges because of its ability to collect large amounts of precise, spatially dense data. It is often difficult for analysts to efficiently parse and store this data, which is why the recent focus has been on the ability to perform efficient automatic target detection in LiDAR data. Previous work in LiDAR data analysis, specifically region segmentation and object classification, shows an ability to recognize objects such as buildings, vehicles, and power lines with a high level of accuracy.[24,25] The initial results were achieved in point clouds with an average point density of 31.798 ppm² (points per meter squared). Although these results are impressive, there is still room for improvement in the efficiency of the implementation. One way to increase the efficiency is to eliminate the data redundancy or the oversampling of points by reducing the range resolution. In an effort to extend this work, the author proposes an analysis of the success of this algorithm at a variety of different range resolutions, specifically, 0.0625 GSD, 0.25 GSD, 1 GSD, 4 GSD, and 16 GSD.

10.14 Current LiDAR Metric Standards

The U.S. Geological Survey (USGS) National Geospatial Program (NGP) released their Lidar Base Specification in 2014 to define minimum metric requirements for the vertical accuracy, data density, and completeness of LiDAR products for the National, interagency 3D Elevation Program (3DEP). Similarly, the American Society for Photogrammetry and Remote Sensing (ASPRS) released the ASPRS Positional Accuracy Standards for Digital Geospatial Data in 2015 to specify vertical and horizontal accuracy requirements for data providers. These documents describe the current state of the industry in metric performance characterization of LiDAR products. Reference 26 discusses LiDAR data density and completeness, absolute and relative vertical and horizontal accuracy, defining a LiDAR error model, measuring point cloud resolution, and characterizing foliage penetration performance, among other topics.

10.15 Conclusion

While much of what is discussed in this chapter is metrics for lower-dimensionality sensors, we need 3D metric standards, and these are being gradually developed. In the future, we will need metrics for laser vibrometer and possibly other LiDAR modes.

Appendix 10-1 MATLAB code to Fourier transform an image

```
clear all
close all
I=imread('tank1.jpg');
figure (1)
imshow(I)
IG=rgb2gray(I);
```

```
figure (2)
imshow(IG)
IFIG=fftshift(fft2(IG));
FTintensity=abs(IFIG);
figure (3)
imshow(FTintensity, [0 10^5])
```

Problems and Solutions

10-1. What are the latest two versions of the GIQE? Why is the GIQE4 particularly problematic?

10-1 Solution:
The latest two versions of the GIQE are

GIQE4 NIIRS $= 10.251 - a \times \log10(GSD) + b \times \log10(RER) + 0.656 \times H - 0.344 \times G/SNR$, where $a = 3.32$ and $b = 1.559$ if RER ≥ 0.9, or $a = 3.16$ and $b = 2.817$ if $RER < 0.9$;

and

GIQE3 NIIRS $= 11.81 + 3.32 \times \log10(RER/GSD) - 1.48 \times H - G/SNR$.

GIQE4 is particularly problematic because it is dual sloped, with GSD depending on the RER.

10-2. Many imaging systems use a circular aperture with a central obscuration. Take ε as the ratio of the linear obscuration to the outside aperture diameter. What is an equation for the MTF?

10-2 Solution: The MTF is expressed as the following equation:

$$MTF(\rho) = \frac{2}{\pi} \frac{A + B + C}{1 - \varepsilon^2},$$

where:
the frequency of ρ is normalized on the cutoff frequency D/λ,

$$A = \cos^{-1}(\rho) - \rho\sqrt{1 - \rho^2} \text{ for } 0 \leq \rho \leq 1,$$

$$B = \varepsilon^2 \left[\cos^{-1}\left(\frac{\rho}{\varepsilon}\right) - \frac{\rho}{\varepsilon}\sqrt{1 - \left(\frac{\rho}{\varepsilon}\right)^2}\right] \text{ for } 0 \leq \rho \leq \varepsilon,$$

$$B = 0 \text{ for } \varepsilon \leq \rho \leq 1,$$

$C = -\pi\varepsilon^2$ for $0 \leq 2\rho \leq 1 - \varepsilon$,

$$C = -\pi\varepsilon^2 + \left\{ \varepsilon\sin\Theta + \frac{\Theta}{2}(1+\varepsilon^2) - (1-\varepsilon^2)\tan^{-1}\left[\left(\frac{1+\varepsilon}{1-\varepsilon}\right)\tan\left(\frac{\Theta}{2}\right)\right] \right\}$$

for $1 - \varepsilon \leq 2\rho \leq 1 + \varepsilon$,

$C = 0$ for $1 + \varepsilon \leq 2\rho \leq 2$, and

$$\Theta = \cos^{-1}\left(\frac{1 + \varepsilon^2 - 4\rho^2}{2\varepsilon}\right).$$

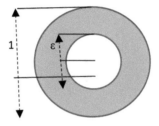

10-3. In the previous problem, plot the MTF for $\varepsilon = 0$, 0.2, 0.5, and 0.8, and show the effect of increasing ε is to increase the higher spatial frequencies at the expense of the low- and mid-spatial frequencies.

10-3 Solution:

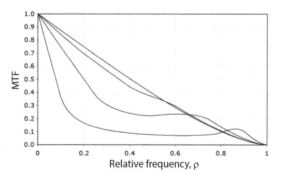

```
rho = 0.001:0.001:1;
eps = [0 0.2 0.5 0.8];
A=acos(rho)-rho.*sqrt(1-rho.^2);
B=zeros(length(rho),length(eps));
C=zeros(length(rho),length(eps));
MTF=zeros(length(rho),length(eps));
for ii = 1:length(eps)
  for jj = 1:length(rho)
    if rho(jj)>=0 && rho(jj)<eps(ii)
```

```
    B(jj,ii)=eps(ii)^2*(acos(rho(jj)/eps(ii))-
rho(jj)/eps(ii)*sqrt(1-(rho(jj)/eps(ii))^2));
  else
    B(jj,ii) = 0;
  end
  if 2*rho(jj)>=0 && 2*rho(jj)<1-eps(ii)
  C(jj,ii) = -pi*eps(ii);
  elseif 2*rho(jj)>=1-eps(ii) && 2*rho(jj)<1+eps(ii)
    theta = acos((1+eps(ii)^2-4*rho(jj)^2)/(2*eps(ii)));
    C(jj,ii) =
    pi*eps(ii)+(eps(ii)*sin(theta)+theta/2*(1+eps(ii)^2)-
(1-eps(ii)^2)*atan((1+eps(ii))/(1-eps(ii))*tan(theta/2)));
  else
    C(jj,ii) = 0;
  end
 end
 MTF(:,ii) = 2*(A'+B(:,ii)+C(:,ii))/(pi*(1-eps(ii)^2));
 figure
 plot(rho,MTF(:,ii))
end
```

10-4. In MATLAB, load 'saturn.png', which is a built-in MATLAB image. You will perform processing with this image in the next few problems. Change the resolution on the image to 10 and 20 pixels, showing the results below. Hint, use 'imresize' to change the resolution of the image.

10-4 Solution:

```
A=imread('saturn.png');
A2=imresize(imresize(A,0.1),10);
A4=imresize(imresize(A,0.05),20);
figure
imagesc(A)
figure
imagesc(A2)
figure
imagesc(A4)
```

10-5. Taking the images you created in the previous problem,

(a) Sharpen the images using 'imsharpen'. Does this help any?
(b) Perform a Gaussian blur first, with a 16 sigma, then sharpen the image. Does this help?

10-5 Solution:

(a) for 10-pixel resolution:

for 20-pixel resolution:

This doesn't help all that much.

(b) for 10-pixel resolution:

for 20-pixel resolution:

This helps somewhat more than just sharpening the image. The reason is that without blurring first, there are only sharp edges to begin with.

10-6. Sharpen the 20-pixel resolution image after performing a Gaussian filter on the image with the following parameters:
$\sigma = 2, 5, 8, 12$
imsharpen with 'Radius' = 16

10-6 Solution:

This image seems almost optimal.

This is still high quality.

These are lower quality.

```
C1=imsharpen(imgaussfilt(A4,2),'Radius',16);
C2=imsharpen(imgaussfilt(A4,5),'Radius',16);
C3=imsharpen(imgaussfilt(A4,8),'Radius',16);
C4=imsharpen(imgaussfilt(A4,12),'Radius',16);

figure
imagesc(C1)
axis 'image'
```

```
figure
imagesc(C2)
axis 'image'
figure
imagesc(C3)
axis 'image'
figure
imagesc(C4)
axis 'image'
```

References

1. J. Johnson, "Analysis of image forming systems," in Image Intensifier Symposium, AD 220160 (Warfare Electrical Engineering Department, U.S. Army Research and Development Laboratories, Ft. Belvoir, Virginia, pp. 244–273 (1958).
2. Wikipedia web page on Johnson's criteria: http://en.wikipedia.org/wiki/Johnson%27s_criteria.
3. Hartford International Group web page on T-72 tank: http://universal-dsg.com/product/t-72/.
4. Wikipedia web page on 1951 USAF resolution test chart: https://en.wikipedia.org/wiki/1951_USAF_resolution_test_chart.
5. P. McManamon, "Review of ladar: a historic, yet emerging, sensor technology with rich phenomenology," *Optical Engineering* **51**(6), 060901 (2012) [doi: 10.1117/1.OE.6.060901].
6. B. D. Duncan and M. P. Dierking, "Holographic aperture ladar," *Applied Optics* **48**(6), 1068–1077 (2009).
7. Federation of American Scientists (FAS) web page on National Image Interpretability Rating Scales: http://www.fas.org/irp/imint/niirs.htm.
8. J. M. Irvine, "National Imagery Interpretability Rating Scales (NIIRS): overview and methodology," *Proc. SPIE* **3128**, 93–103 (1997) [doi: 10.1117/12.279081].
9. From a briefing by Gary Kammerman on July 10, 2012 at the University of Dayton Lidar and Optical Communications Institute (LOCI), Dayton, Ohio.
10. J. M. Irvine, C. Fenimore, D. Cannon, J. Roberts, S. A. Israel, L. Simon, C. Watts, J. D. Miller, M. Brennan, A. I. Aviles, P. F. Tighe, and R. J. Behrens, "Feasibility study for the development of a motion imagery quality metric," *IEEE 33rd Applied Imagery Pattern Recognition Workshop* (AIPR'04), 13–15 Oct., 2004, Washington, D.C., pp. 179–183 (2004).
11. R. Craig, I. Gravseth, R. P. Earhart, J. Bladt, S. Barnhill, L. Ruppert, and C. Centamore, "Processing 3D flash LIDAR point-clouds in real-time for

flight applications," *Proc. SPIE* **6555**, 65550D (2007) [doi: 10.1117/12. 720143].

12. J. C. Leachtenaur, W. Malila, J. Irvine, L. Colburn, and N. Salvaggio, "General image-quality equation: GIQI," *Applied Optics* **36**(32), 8322–8328 (1997).

13. J. C. Leachtenauer, W. Malila, J. Irvine, L. Colburn, and N. Salvaggio, "General image-quality equation for infrared imagery," *Applied Optics* **38**(26), 4826–4828 (2000).

14. S. T. Thurman and J. R. Fienup, "Analysis of the general image quality equation," *Proc. SPIE* **6878**, 68780F (2008) [doi: 10.1117/12.777718].

15. A. M. Burwinkel, S. J. Shelley, and C. M. Ajose, "Extracting intelligence from LIDAR sensing modalities," *Proc. SPIE* **8037**, 80370C (2010) [doi: 10.1017/12.884101].

16. Private communication from Gary Kammerman in August 2016 regarding an unpublished technical paper "Image information density and lidar object interpretability."

17. C. E. Shannon, "The mathematical theory of communication," *The Bell System Technical Journal* **27**, 379–23, July; 623–656, October (1948).

18. K. Piché, J. Leach, A. S. Johnson, J. Z. Salvail, M. I. Kolobov, and R. W. Boyd, "Experimental realization of optical eigenmode super-resolution," *Optics Express* **20**(24), 26424 (2012).

19. A. Mittal, R. Soundararajan, G. S. Muralidhar, A. C. Bovik, and J. Ghosh, "Blind image quality assessment without training on human opinion scores," *Proc. SPIE* **8651**, 86510T (2013) [doi: 10.1117/12. 981761].

20. D. L. Ruderman, "The statistics of natural images," *Network Computation in Neural Systems* **5**(4), 517–548 (1994).

21. DARPA website: https://www.darpa.mil/.

22. D. G. Lowe, "Distinctive image features from scale-invariant keypoints," http://www.cs.berkeley.edu/~malik/cs294/lowe-ijcv04.pdf.

23. Community LiDAR Model Working Group, "Light Detection and Ranging (LIDAR) LiDAR Model Supporting Precise Geopositioning, Version 1.1," NGA.SIG.0004_1.1, August (2010).

24. N. Varney and V. K. Asari, "Target detection in low resolution LiDAR data," presented at the SPIE Conference on Defense + Security: Optical Pattern Recognition XXVI, Baltimore, Maryland, 20–24 April 2015.

25. N. M. Varney and V. K. Asari, "Volumetric features for object region classification in 3D LiDAR point clouds," in 2014 IEEE Applied Imagery Pattern Recognition Workshop (AIPR'14), 14–16 Oct. 2014, Washington D.C. (2014).

26. S. T. Hagstrom and M. Z. Brown, "3D metrics for airborne topographic lidar," *Encyclopedia of Modern Optics*, Second Edition, B. D. Guenther and D. Steel, Eds., Elsevier Ltd., Amsterdam (2018).

27. P. McManamon, G. Kamerman, and M. Huffaker, "A history of laser radar in the United States," *Proc. SPIE*, **7684**, 76840T (2010) [doi: 10.1117/12.862562].

28. P. F. McManamon, *Field Guide to LiDAR*, SPIE Press, Bellingham, Washington (2015) [doi: 10.1117/3.2186106].

Chapter 11
Significant Applications of LiDAR

This chapter provides an overview of auto LiDAR design and brief discussions of the design of a 3D mapping LiDAR, a laser vibrometer, and a wind-sensing LiDAR. Various types of LiDAR are discussed in Chapter 4, which is more about the technology and its applications. This chapter focuses more on design process. Auto LiDAR is currently the dominant application of LiDAR from a monetary point of view. Commercial 3D mapping LiDARs are widespread, and wind LiDARs are growing in use as wind energy becomes more prevalent.

11.1 Auto LiDAR

11.1.1 Introduction

As mentioned, the largest commercial application of LiDAR at this time is auto LiDAR. It is a good practice to start an auto LiDAR design with the field of view (FOV). The azimuth requirements vary. People might have already seen the Velodyne® LiDARs[1] that mount on the roof and rotate 360 deg. These LiDARs represent one end of the spectrum, giving 3D LiDAR coverage in all azimuth directions. On the other end of the spectrum, something like 40 deg may be enough in the forward direction.[2] Some vendors prefer 60 deg or more to provide margin in a forward-looking LiDAR (a bicycle coming in from the side would be a stressing case). It is possible to have a 60-deg, forward-looking, azimuth FOV, but to also have longer range for the LiDAR in the central region of the FOV. Some designers may want more than one LiDAR, possibly a moderate-range LiDAR covering 360 deg together with a longer-range, forward-looking LiDAR. When a car accelerates, the forward-looking LiDAR will need to have a longer range. Some designers may decide that they do not need 3D imagery all around a car; however, obviously, 3D imagery can be useful, so there will be interest in having 360-deg LiDAR coverage with at least a moderate range capability all around a vehicle. There could be an array of LiDARs around the car. With a looking-forward LiDAR, passive cameras could possibly be used to the sides and rear. As mentioned, the Velodyne LiDAR sits on the roof of the car and rotates 360 deg. Four or six imbedded LiDARs could surround the car as an alternative, still providing 360-deg azimuth coverage.

As far as elevation coverage, the LiDAR only needs to see objects on the ground, although, in terms of context, seeing some objects above the ground may be useful, e.g., buildings surrounding the road. These above-ground objects could help with navigation. Hills enlarge the required elevation coverage. The steeper the hills considered in the design the more required elevation coverage. When heading down or up a hill, the LiDAR is tilted down or up, increasing the required elevation FOV. A steep hill might be 7 deg or even more, so the slant of a hill does not require a major increase in elevation FOV for most hill slopes. Velodyne LiDARs, as advertised on their website,[3] show elevation FOVs from 26.8 to 40 deg, depending on the LiDAR. The same data sheet shows the most expensive LiDAR having 64 detectors, with 32- and 16-detector less-expensive versions. Velodyne recently announced a 128–laser-line auto LiDAR, but not all of its specifications are available at this time. The HDL 64 has an elevation FOV (vertical) of +2.0 deg to −24.9 deg (26.9 deg).

Most auto LiDARs do not scan in elevation. Scanning in elevation is certainly possibly and can either increase the elevation FOV or decrease the sample size for each pixel. Sixty-four laser lines could be used multiple times in elevation if elevation scanning is used. For example, two steps up in elevation could double the number of elevation lines available. Elevation scanning, however, adds complexity and cost, so most auto LiDAR developers are going for simplicity instead of the additional capability. Also, the more laser lines a LiDAR has the less incentive there is to scan in elevation. Obviously, scanning in elevation can multiply the number of available elevation points. The HDL-64 LiDAR has 64 detectors in elevation, and the new HDL-128 has 128 laser lines. Of course, more or fewer elevation samples can be used, and it may be unwise to uniformly space the samples in elevation. A foveated vision approach, with more resolution along the region pointing forward, would be wiser. One approach might be to design the elevation coverage to have uniform spacing in distance along the ground from the vehicle, as considered in Problem 11-1. Geometry would imply that this means smaller elevation angular coverage for longer distances, and larger angular coverage closer to the vehicle. Over time, as laser and detector technology improves, it is likely that the number of elevation samples will increase to 128, 256, or even 512.

11.1.2 Resolution

The ability to see what are sometimes referred to as 'corner cases' needs to be emphasized. Every driver has stories from his or her own driving experience that have scared them, e.g., a small child stepping out from between parked cars, a motorcycle zooming between cars, or something falling off a truck right in front of their moving car. Driverless cars needs to handle all of the same corner cases that you handle when you drive. A nominal pedestrian is estimated to be 1.8 m tall by 0.5 m wide and 0.25 m deep. A more-restricted case could be small child, who would be shorter and thinner, so even better resolution would be useful. A toddler wearing a size 2T outfit might be 33–33.5 in. tall, or about 85 cm tall.[4] If an adult is on the side of the road, and you want to see his leg movement to see if he is starting to step into the street, then higher resolution is required. A motorcycle may head toward your car at high speed from head on, requiring a need to see far with high resolution. That said, a LiDAR does not have to be the only sensor used for a driverless car. It is likely that both passive sensors and radars will also be used, although a LiDAR will be the sensor relied on most because of its high-resolution 3D imaging capability. A motorcycle is likely to contain metal, so it would be easily seen by a microwave radar. Microwave radars, however, have an angle resolution issue because of the longer wavelength. In inclement weather, microwave radars will be very useful.

Each company will select its own corner cases to drive the design of its LiDAR. For simplicity, in this book we can pick some round numbers. Assume that we would like to be able to detect a 1-m tall child at 100 m. There will be some dispute as to how many spots on the child are required for detection, but the Johnson criteria can be used. Using the Johnson criteria of points in a given dimension for detection, we need 50-cm resolution in elevation. For simplicity, we have selected a range of 100 m. This would mean an angular resolution of 5 mrad (0.286 deg) in elevation. An elevation FOV that is the same as that of the HDL-64 along with 5-mrad elevation resolution across the whole FOV would imply 94 laser samples in elevation; however, many current auto LiDARs only have 64 or fewer lines. The use of foveated vision may allow one to obtain this required elevation resolution in areas of concern while only using 64 laser lines. In the future, this will not be an issue because LiDARs will go to 128, 256, or even 512 laser lines in elevation. Another option that can be considered is increased sampling by microscanning as a way to use an available number of detectors in elevation. A 2D scanner could also be used for providing more than one pass in azimuth, with each pass at a different elevation during a complete scan.

Microscan is an approach that provides increased sampling by placing the detector angular subtense (DAS) in one scan a fraction of a DAS displaced from the previous scan.[5,6] Some scientists call this super-resolution, although it is not an increase in resolution, but rather an increase in sampling. It can look to the observer like an increase in resolution. For a continuous scanner, oversampling can be done by sampling to finer than the DAS, say at 0.5 or 0.3 DAS, as the scanner moves. For an auto LiDAR, it is anticipated that a continuous scan in the azimuth direction will be used, so this would be simple to implement in azimuth. To implement microscan in elevation, step-up or step-down increments smaller than a full DAS would be required. Implementing elevation microscan requires the ability to have a minor elevation-angle change.

Range imaging can also be used to assist in object recognition. For a person, the smallest dimension is still seeing the person sideways. In that case, the 10-cm resolution we chose for azimuth could also be appropriate for range precision, depending on how the person is oriented toward the LiDAR. There will be a different range to the object in adjacent pixels. One option to consider is to use 3D imaging in good weather, with 5- or 10-cm range resolution, but when weather degrades, to allow poorer range resolution to gather more photons in a given bin.

If the range resolution is larger than the target to be detected, contrast could be negatively influenced. Short pulses using laser diodes can be limited in power because the laser medium does not have internal storage. Laser media that have internal storage will allow Q switching, potentially providing a much higher peak power laser pulse. If we were to increase the laser pulse width by a factor of ten with a peak-power–limited laser, the energy per pulse would increase by a factor or ten, and laser average power would also increase. Therefore, in bad weather, with laser diodes as the laser source, it might be beneficial to increase the laser pulse width if the laser in the LiDAR is capable of that flexibility.

11.1.3 Frame rate

Frame rate is limited by the distance the LiDAR design will allow the car to move forward during a frame period. Table 11.1 shows some possibilities, where the distance traveled is in feet. For a car moving at 45 mph as a design point, and a 10-Hz frame rate, distance traveled per frame is 6.6 feet. Because of higher LiDAR design speeds, companies designing for the Autobahn will require operation at higher frame rates than companies designing for most

Table 11.1   Distance moved per frame.

Frame rate (Hz)	Speed (mph)	Speed (km/hr)	Distance moved per frame (ft)
10	25	40	3.7
10	35	56	5.1
10	45	72	6.6
10	60	96	8.8
10	75	121	11.0
10	90	145	13.2
15	45	72	4.4
15	60	96	5.9
15	75	121	7.3
15	90	145	8.8
15	120	193	11.7

U.S. roads. Companies happy with only having automated driving in the city can plan on going slower than companies that wish to design autonomous cars for the super highway. Companies designing for hilly terrain will want more elevation coverage. There is not perfect auto LiDAR design. It depends on the design goals each company chooses.

Typical human reaction time is 1.5 s, although it used to be considered to be 0.75 s. One of the benefits of autonomous cars should be to decrease the reaction time to under 0.1 s because a machine should be able to react more quickly than a human. An article on braking distance[7] indicates a coefficient of friction of 0.7. Typical stopping distance is given by

$$D_{\text{stopping}} = vT + \frac{v^2}{2ug} \tag{11.1}$$

where v is velocity, T is reaction time, u is the coefficient of friction, and g is the acceleration of gravity, 9.8 m/s^2. Table 11.2 shows that there is a major influence from reducing reaction time, especially at slow speeds; the 10-Hz frame rate is looking very reasonable. It is possible to think of the 10-Hz frame rate as increasing reaction time by 0.1 s. Since human reaction time is 1.5 s, and we are taking machine reaction time as 0.1 s, the total reaction time in both cases would be 1.6 s and 0.2 s using a 10-Hz rate. From this table, it seems that the 10-Hz frame rate should be sufficient for most auto LiDAR applications. A 20-Hz LiDAR repetition rate would only decrease the effective additional reaction time from 0.1 s to 0.05 s. That would mean a human reaction time of effectively 1.65 s and an effective automated reaction time of 0.15 s.

Table 11.2 gives a good idea of the required LiDAR range. Speed will be a major contributor. Of course, we should allow some margin beyond the values in Table 11.2. For one thing, on slippery streets, the coefficient of friction will be much lower.

Table 11.2 Braking distance.

Speed (mph)	Speed (km/hr)	Velocity (m/s)	Reaction time (s)	Braking distance (m)	Braking distance (ft)
25	40	11.2	1.5	26	85
35	56	15.6	1.5	41	135
45	72	20.1	1.5	60	196
60	96	26.8	1.5	93	304
75	121	33.5	1.5	132	433
90	145	40.2	1.5	178	585
25	40	11.2	0.1	10	33
35	56	15.6	0.1	19	64
45	72	20.1	0.1	31	103
60	96	26.8	0.1	55	181
75	121	33.5	0.1	85	279
90	145	40.2	0.1	122	400

11.1.4 Laser options

Two major laser options should be considered for auto LiDAR. Direct diode lasers are highly desirable due to their low cost but cannot be Q switched, so laser peak power will be limited. These lasers often operate at 905 nm, which is not an eye-safe wavelength. The other major alternative is fiber lasers near 1550 nm. Fiber lasers can be Q switched and operate in the eye-safe wavelength regime. Peak power is limited compared to bulk-diode–pumped solid state lasers, but for the ranges associated with auto LiDAR, the limitation will not be restrictive. The peak power limitations of laser diodes will only be a factor for pulsed LiDARs. The challenge with this approach is making a narrow-linewidth diode laser; however, with frequency chirping, it is possible to have very high duty cycle, possibly 100%. This eliminates the need for Q switching. It does impose a low-bandwidth requirement. To make a coherent LiDAR, the linewidth of the laser must be narrow enough that the return signal can beat against the local oscillator. An alternative laser diode approach is to modulate the frequency of a cw laser diode.

Due to the peak power limitations of laser diodes, it is likely that each laser illuminator in a pulsed auto LiDAR will be matched to a DAS of the receiver for the diode lasers at 905 nm. For laser diode use, it is likely that there will be as many lasers as detectors, with each laser diode illuminating the area viewed by a single detector. For fiber lasers, peak power can be high enough to illuminate many DASs with a single laser. Therefore, a fan-shaped laser illumination pattern may be used with a fiber laser. It is possible that only a single fiber laser will be used. Fiber lasers are, however, much more expensive to manufacture. Diode lasers can be made by chips on a wafer. Fiber lasers need to use diode lasers but take more hands-on labor to manufacture.

11.1.5 Eye safety

Most auto LiDARs have a pulsed laser with a high duty cycle. Figure 11.1 shows maximum permissible exposure (MPE) for pulsed lasers. About 1 J/cm^2 is eye safe at 1.55 μm, whereas at 905 nm, the allowable energy in 10 s is about one μJ, a factor of 6 orders of magnitude lower. For a high–duty-cycle laser, this is the exposure over 10 s. This threshold allows about 100 mW/cm^2 average power at 1550 nm, and 6 orders of magnitude lower at 905 nm. There is some controversy over whether the 10-s average is the right way to consider eye safety, but that is the standard. Some scientists promote a shorter time period.

For a scanning LiDAR, if an eye is not placed at the exit aperture, then any given eye will not see the laser for the entire 10 s. This can reduce the average power entering the eye and increase the allowed average power from the laser. For a moving car, a person cannot place their eye close to the transmit aperture without being run over by the car, assuming that the

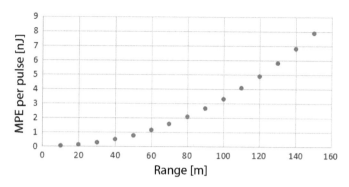

Figure 11.1   Required MPE per pulse versus range for a LMAPD auto LiDAR.

transmit aperture is facing forward. There can be an interlock to determine if the car is moving. There also could be an interlock to determine if someone is close or not, but this seems like overkill. Let us assume that an eye is at least a few meters away from the aperture for any moving vehicle.

For a production unit, there can be issues of multiple lasers from multiple cars illuminating a person, but this will occur after the beam has become larger with distance, so hopefully it will not be a major issue. With scanning, we might want to provide an additional factor of 100 in eye safety just to ensure safety with many cars in the road, all having LiDAR-based guidance. For operation near the 1.5-μm region, eye safety should not be a significant issue. The issue of heavy traffic will have to be faced by auto LiDAR designers.

11.1.6 Unambiguous range

Unambiguous range is given by Eq. (9.1) (repeated here) as

$$R_{\text{unamb}} = \frac{c}{2PRF}.$$

Table 11.3 shows that a laser repetition rate as high as 1,000,000 pulses per second could be used for a range of 150 m without dithering the laser rep rate.

11.1.7 Required laser energy per pulse and repetition rate

To calculate the required energy per pulse, specific LiDAR parameters are assumed. A designer can chose either a 905-nm laser diode approach or a

Table 11.3   Unambiguous range.

Laser repetiton rate (pulse/second)	Unambiguous range (m)
504,000	298
750,000	200
1,000,000	150

1550-nm fiber laser approach. In theory, other options are possible, but these are the most popular. For this discussion, we will select the 905-nm case, and ask the reader to solve for the 1550-nm case in the Problems. We further assume a 2.5-cm receive aperture, 100-m range, and a 5 mrad × 5 mrad DAS. For these calculations, it does not matter if we assume a bistatic or monostatic LiDAR approach. Since we selected laser diode illumination, we will assume that a given laser diode will illuminate 50% more area than the DAS of a single detector. If we had selected a fiber laser, we would have a single laser illuminate the area covered by the array of detectors. In the diode laser case, we match the laser diode illuminators to the detectors. 10% reflectivity will be assumed. A polished car will have lower reflectivity than a person because it will have more specular scattering. This car's reflectivity may vary with angle. For a person, it is possible to use 20% reflectivity, but 10% reflectivity is conservative. Energy per pulse is calculated for 5-km visibility. 23 km visibility could be assumed, which is more of a standard for "good weather," but assuming 5-km visibility is conservative with regard to the LiDAR design. We then increase the energy per pulse by a factor of ten above this value and calculate the LiDAR range for bad weather. For bad weather, we assume 35 m or about 115-ft visibility. This is enough to slow down most drivers on a freeway. In the Problems, the reader is asked to calculate LiDAR range for an even shorter visibility. Having calculated the LiDAR range for bad weather, we use Eq. (11.1) to calculate the maximum speed at which a driverless car can operate with that visibility.

Equation 3.12 for required transmitted energy is repeated here:

$$E_T = E_R \frac{A_{\text{illum}}}{\sigma} \frac{\pi R^2}{A_{\text{rec}}} \frac{1}{\eta_{\text{atm}}^2} \frac{1}{\eta_{\text{sys}}},$$

The atmospheric attenuation has exponential decay. In all cases, exponential decay, or Beer's law is assumed through the obscuring medium. Beer's law [(Eq. 3.13) repeated here] is given as

$$\eta_{\text{atm}} = e^{-\beta R}.$$

Substituting Eq. (3.16) into Eq. (9.1) yields

$$E_T = E_R \frac{A_{\text{illum}}}{\sigma} \frac{\pi R^2}{A_{\text{rec}}} e^{2\beta R} \frac{1}{\eta_{\text{sys}}}. \tag{11.2}$$

Visibility is defined at a wavelength of 550 nm. To adjust to a different wavelength, Eq. (3.14) (repeated here) can be used:[8]

$$\beta(\lambda) \propto \left(\frac{\lambda_0}{\lambda}\right)^q,$$

where Eq. (3.15) defines q as

$$q(\lambda) = 0.1428\lambda - 0.0947.$$

Some over-illumination should be defined. Assume that the illuminated area is 1.5 times the DAS times as many detectors are needed to allow for some illumination inefficiency. This area grows with range. A systems efficiency through the optical train of 40% is assumed. The cross-section is the reflectivity times the area viewed by a given detector, which also grows with range:

$$\sigma = \eta_{\text{reflect}} \cdot DAS_{\text{az}} \cdot DAS_{\text{el}} \cdot R^2. \tag{11.3}$$

The ratio of illuminated area to cross-section is therefore

$$\frac{A_{\text{illum}}}{\sigma} = \frac{1.5N_{\text{det}}}{\eta_{\text{reflect}}}, \tag{11.4}$$

where N_{Det} is the number of detectors. Equation (11.4) assumes 50% over-illumination. The cross-section is an effective area. The LiDAR illuminates a larger area and an area effectively as large as the cross-section that reflects some of the light. For a 32×32 array of detectors we need to illuminate an area at least 1024 times larger than the DAS. Reflectivity of the surface will influence the cross-section, making it smaller with smaller reflectivity. For the diode laser case, only one detector is illuminated per laser. In the Problems for the 1550-nm LiDAR, there will be multiple detectors per illuminated area, and the required transmitted energy is

$$E_T = \frac{Nhc}{\lambda} \tag{11.5}$$

To calculate the required energy per pulse, we use

$$E_T = \frac{6NhcN_{\text{det}}R^2}{\lambda\eta_{\text{refl}}D^2\eta_{\text{sys}}}e^{-2\beta R}. \tag{11.6}$$

Equation (11.6) can be used to calculate the required energy per pulse per laser for our assumptions. Then it is necessary to decide how many photons will be set as a lower sensitivity limit for the calculation. For this calculation, a LMAPD requiring 20 photons per pulse returned into the detector will be assumed. An Excel spreadsheet is used to calculate the energy required versus

Table 11.4 Energy per pulse per laser required.

# of photons received from a 10% reflective object								
Human visibility (m)	$\beta_{0.55}$	$\beta_{1.55}$	Imaging range	One-way transmission	Two-way transmission	Design # of photons	Receiver aperture diameter (mm)	Required transmitter energy (µJ) for 10% reflectivity
5000	0.0006	0.0005	10	99.5%	99.0%	20	25	0.0303
5000	0.0006	0.0005	20	99.0%	97.9%	20	25	0.1225
5000	0.0006	0.0005	30	98.4%	96.9%	20	25	0.2784
5000	0.0006	0.0005	40	97.9%	95.9%	20	25	0.5002
5000	0.0006	0.0005	50	97.4%	94.9%	20	25	0.7899
5000	0.0006	0.0005	60	96.9%	93.9%	20	25	1.1494
5000	0.0006	0.0005	70	96.4%	92.9%	20	25	1.581
5000	0.0006	0.0005	80	95.9%	91.9%	20	25	2.0868
5000	0.0006	0.0005	90	95.4%	91.0%	20	25	2.669
5000	0.0006	0.0005	100	94.9%	90.0%	20	25	3.3299
5000	0.0006	0.0005	110	94.4%	89.1%	20	25	4.0717
5000	0.0006	0.0005	120	93.9%	88.2%	20	25	4.8969
5000	0.0006	0.0005	130	93.4%	87.2%	20	25	5.8077
5000	0.0006	0.0005	140	92.9%	86.3%	20	25	6.8067
5000	0.0006	0.0005	150	92.4%	85.4%	20	25	7.8964

range. Table 11.4 provides the results of the Excel spreadsheet. Figure 1.11 shows the same values plotted.

For a 100-m range, 3.3 nJ per laser per pulse is required. We can then increase the energy per pulse per laser by about a factor of ten by selecting 35 nJ per pulse for this case. We then reverse the process and use 35-m visibility, as shown in Fig. 11.2.

The threshold of 20 photons is reached at a range of between 35 and 40 m. The LiDAR in this case can see a little longer than the visibility. This short visibility will limit the speed for our autonomous car to about 45 mph to make sure the car can stop in time. Because of the response time of the computer, with a 35-m visibility, this is not as limiting as the driver.

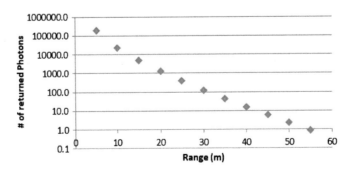

Figure 11.2 Number of returned photons versus range for 35-m visibility.

The assumptions in this calculation are not completely accurate (they are conservative), but we will use them. Some spatial filtering methods allow for the use of more than just the so-called ballistic photons. This calculation only assumes ballistic photons, but it is possible to use some of the scattered illuminator photons as long as we can prevent loss of resolution on return.

Table 11.5 and Fig. 11.3 highlight the challenge in using a 35-m human visibility at 0.55 μm and (assuming this model) imaging under fog-like conditions. The one-way atmospheric transmission is 0.032 out to 35 m, and we increase our laser energy by a factor of 955 to reach the same range, and

Table 11.5 Atmospheric attenuation at range.

Human visibility (m)	$\beta_{0.55}$	$\beta_{1.55}$	Imaging range	One-way transmission	Two-way transmission	Ratio of required increase in LiDAR energy per pulse due to attenuation	Required increase in LiDAR energy per pulse due to attenuation (dB)
35	0.1118	0.0980	10	0.375	0.141	7	8
35	0.1118	0.0980	15	0.230	0.053	19	13
35	0.1118	0.0980	20	0.141	0.020	50	17
35	0.1118	0.0980	25	0.086	0.007	134	21
35	0.1118	0.0980	30	0.053	0.003	358	26
35	0.1118	0.0980	35	0.032	0.00105	955	30
35	0.1118	0.0980	40	0.020	0.00039	2546	34
35	0.1118	0.0980	45	0.012	0.00015	6785	38
35	0.1118	0.0980	50	0.007	0.00006	18084	43
35	0.1118	0.0980	55	0.005	0.00002	48197	47
35	0.1118	0.0980	60	0.003	0.00001	128455	51
35	0.1118	0.0980	65	0.00171	2.921E-06	342359	55
35	0.1118	0.0980	70	0.00105	1.096E-06	912454	60
35	0.1118	0.0980	75	0.00064	4.112E-07	2431869	64

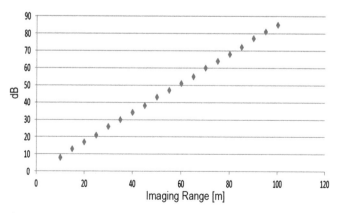

Figure 11.3 Atmospheric attenuation versus range (link budget margin required) for 35-m visibility.

only increase it by a factor of 358 to reach 30 m. If we decide we would like to double the human visibility range, we would need to increase the laser power by a factor of almost 1,000,000.

All of values in Table 11.5 can of course be partially mitigated by a larger receive aperture. A larger–receive-aperture area is inversely proportional to the laser energy per pulse. Increasing the aperture diameter by a factor of 2.5 decreases the required laser power by a factor of 6.25, and increasing the aperture diameter by a factor of 5 decreases the laser power by a factor of 25.

11.1.8 Obscurants considered for auto LiDAR

For poor and moderate visibility, scattering can be a major contributor to loss and to backscatter. We saw earlier that a LiDAR has exponential loss through the atmosphere to and from the target. Lower visibility means increased loss and increased exponential loss. In poor visibility, the following obscurants are the major possible contributors to loss and backscatter:

Rain or truck spray. Rain has large water particles, but not as many as in a cloud or fog. Rain droplets are large even with respect to millimeter-wave wavelengths. Rain can have a size distribution from 500 μm to 5000 μm in diameter. Truck spray may be a mix of particle sizes, with some sizes smaller than rain droplets. It is also, of course, very dynamic.

Fog. Fog is composed of a large number of spherical water particles. The scattering of light in fog is primarily Mie scattering. This scattering is highly dependent on the relationship of particle size to wavelength. Fog is a cloud on the ground, so there is no real difference between fog and a cloud, except in location. Particle sizes on the order of 1–20 μm are assumed. Of course, there are many different particle size distributions, so each fog will be different.

Snow. As is generally known, snowflakes come in many shapes and sizes. In general, they are much larger than fog droplets. We can sometimes see individual snowflakes. A snowflake may be similar in size to a rain drop, but it is likely that it will extend more in two dimensions and less in the third dimension compared to the rain drop. Snowflakes are very large compared to optical wavelengths.

11.1.9 Keeping the auto LiDAR aperture clear

When driving in inclement weather, one condition that can cause issues is whether or not the windshield is clear. It is the same in the case of a LiDAR. Windshields use wipers; some approach will be required to keep the LiDAR transmit and receive apertures clear. Designing an auto LiDAR will not be complete without addressing this issue. This requirement is not a basic LiDAR design issue but is nonetheless very important.

11.2 3D Mapping LiDAR

11.2.1 Introduction to 3D mapping LiDAR

Multiple companies sell 3D scanning LiDARs for mapping. Two of them are Teledyne Optech and RIEGL. Harris is providing data collection services and selling the data collected with a Geiger-mode (3D flash) LiDAR. The objective of these LiDARs is to 3D map the ground. The Teledyne Optech website describes five LiDARs that they sell for 3D mapping. The most recent LiDAR is multispectral, including laser lines at 532 nm, 1064 nm, and 1550 nm. RIEGL also has a full line of 3D laser scanners. Both RIEGL and Teledyne Optech make use of a limited number of laser detectors at a time and scan to new angular locations. The Harris LiDAR has a larger-area FPA so could be called a flash LiDAR. They use a Geiger-mode APD (GMAPD) purchased from Princeton Lightwave. This is a commercial 32×128 detector array GMAPD camera. At the time of this writing, no one is selling or leasing a 3D mapper using a linear-mode APD flash camera, but that could occur in the future as the technology evolves. Until they were purchased by Argo AI, Princeton Lightwave had a commercial line of GMAPD area-based cameras for sale, making an area-based camera using GMAPDs an attractive option. It is expected that another company (Ball Aerospace) will soon be selling the (previous) Princeton Lightwave commercial LiDAR camera product line.

3D mapping LiDAR systems, such as by Teledyne Optech and RIEGL, have been used for mapping the ground (terrain) and other natural and manmade features since the mid-1990s on various platforms, such as helicopters, unmanned aerial systems (UAS), and lightweight fixed-wing aircraft at altitudes ranging from 1,000 ft to 5,000 ft above ground level (AGL) for most engineering and survey mapping applications. NASA has flown one of the RIEGL LiDARs as high as 22,000 ft above sea level.[9] Since it was flown in Colorado, it was not as high above ground level. Starting around the year 2010, the U.S. Department of Defense (DOD) used high-altitude 3D LiDAR mapping systems in Afghanistan that were based on MIT/LL GMAPD technology.[10,11] As a technology spin off in about 2013, Harris started selling data from a single-photon–sensitive LiDAR that allows medium altitude collections (5,000 to 30,000 ft AGL) on pressurized turbo jet aircraft. A typical flight altitude might be between 12,000 to 27,000 ft AGL, depending on the application.

3D mapping applications include wide-area terrain mapping, distribution utility mapping, and vegetation management; transportation planning and asset management; flood and fire risk modeling; and water, landslide, and forest inventory management.

Aerial LiDAR systems are generally grouped into two broad categories: (1) scanning LiDAR, including full-waveform systems and (2) flash LiDAR, including single-photon LiDAR (SPL) and GMAPD LiDAR. The author expects that at some future time, competing, flash, area-based, 3D mapping LiDARs will emerge using 2D arrays of LMAPDs; however, to date, no attractive LMAPD flash camera is available.

Conventional LiDAR systems often use one or two detector elements. The spot is scanned in a line pattern, typically via a bidirectional scan pattern (sinusoidal or sawtooth), or a spinning polygonal mirror. At the end of each swath, the aircraft turns and flies in the opposite direction, with the new swath overlapping the previous swath at a nominal sidelap of 20% or higher to eliminate the gap between swaths. 3D mapping LiDAR performance is governed by detection sensitivity. As a general rule, we can use 450–500 required photons for detection using a p–i–n diode camera. Chapter 6 explains that LMAPDs require 20–40 photons per detection, although Fig. 6.16 shows the probability of detection versus the number of photons for a specific InGaAs LMAPD camera, and a required 5-cm range precision. For most mapping cases, we do not need range precision to be this high, so fewer photons would be required. HgCdTe LMAPDs can be much more sensitive, approaching single-photon sensitivity. GMAPD LiDARs often operate using 0.4 to 0.8 photons per detection, but they require multiple pulses when performing coincidence detection. Therefore, for a real detection, it might take say 15 photons total to archive sufficient coincidence detection. More information on this process is provided in Chapter 6 on receivers.

Scanning, linear-mode LiDAR technology is mature and has been successfully used for numerous mapping applications for more than 30 years. These systems are based on helicopter and unmanned aerial vehicles usually operating from an altitude of 1000–5000 ft AGL. Because of their data capture rates, they are often used for corridor mapping, such as utility transmission line mapping, pipeline surveys, and site inspections. Other conventional LiDAR systems operate at mid-altitude levels (i.e., altitudes of a few thousand feet), typically from a slow-moving aircraft for engineering grade surveys, small-area surveys, and mapping transportation networks. The area coverage rate of these systems is generally on the order of 100 km^2 per hour or less. During recent years, linear-mode LiDAR systems have been commercially produced, allowing improved collection rates that are supported by the technology advancements in higher–pulse-energy and higher–pulse-repetition-frequency laser sources. The scanning, linear-mode LiDAR has two variants: (1) a discrete return capture with offline waveform analysis and (2) an online full-waveform processing. Figure 11.4 shows the number of returned photons from a RIEGL 3D mapper with 23-km visibility and for a 3-km visibility. Of course, usually a company would make 3D maps during good weather. We can see from Fig. 11.4 that with 23-km visibility, the RIEGL 3D mapper has enough photons to operate at 3-km range—about 10,000-ft altitude or farther. The spreadsheet used to develop this figure is shown in Table 11.6.

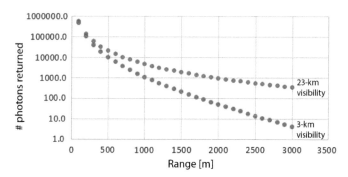

Figure 11.4 Photons returned versus range for two visibilities, using parameters that are constant with one of the RIEGL 3D mappers.

The commercial Harris GMAPD flash, single-photon LiDAR system operates at higher altitudes ranging from 5000 to 30,000 ft AGL, providing collection rates approaching 1000 km^2 per hour, which is roughly a factor of 10 higher compared to a conventional 3D mapping LiDAR.

The Geiger-mode and single-photon LiDAR systems share a common characteristic: they both provide sensitivity to detect a single photon as a returned signal from various objects. In addition, these systems are flash LiDARs, using multi-element detector arrays to collect data simultaneously from many ground points. In effect, this permits operation of these two systems from higher altitudes with higher point densities; therefore, they cover a wider instantaneous ground swath than what is feasible with a scanning, linear-mode system. New technology developments and the commercial availability of large-format-array–based cameras has led to the commercialization of these new LiDAR systems. The GMAPD-3D-mapping–based LiDAR uses a solid state 128×32 detector FPA camera that operates at high frame rates, and has low timing jitter and high time resolution but requires the detection of multiple coincident pulses from the same location to declare a detection. This type of detection is discussed in Chapter 6.

Current GMAPD LiDARs for commercial mapping operate with a 1064-nm near-IR spectral band.[12] By contrast, the currently available SPL system uses a 10×10 detector photocathode array (operationally akin to a photomultiplier tube) that operates in the 532-nm visible spectral band.[13] Due to its green laser wavelength, the SPL has the ability to penetrate shallow, clear water and thus can support bathymetric as well as topographic mapping missions. The SPL is also more commonly flown during the day because getting permits to fly is easier with the green laser; therefore, it is commonly co-collected with passive visible imagery. In comparison, an operational Geiger-mode LiDAR offers a higher ($\sim 40\times$) detection rate and higher instantaneous area coverage than the SPL system on the market today.[14] The GMAPD-based LiDAR can support day or night collection, but is most frequently used for night-time mapping collections in order to take advantage of more commonly

Table 11.6 Spreadsheet to calculate number of returned photons.

Laser wavelength (μm): 1.06
Pulse repetition rate (kHz): 400
Target reflectance: 10.0%
Speed of light (m/s): 299,792,458
Planck's constant (Js): 6.626E-34

Calculation of the number of received photons from a diffusely reflecting target of 10% reflectance (Lambertian scatterer)

Visibility (m)	$\beta_{1.06}$	One-way transmission	Two-way transmission	Visibility (m)	$\beta_{1.06}$	One-way transmission	Two-way transmission	Range to target (m)	Energy per pulse (μJ)	Receiver aperture diameter (mm)	Pulse energy received from a diffusely reflecting target of 10% reflectance @ 23-km vis.	Pulse energy received from a diffusely reflecting target of 10% reflectance @ 3 km vis.	# of photons received from a diffusely reflecting target of 10% reflectance @ 23 km vis.	# of photons received from a diffusely reflecting target of 10% reflectance @ 3 km vis.
23000	0.00011	98.9%	97.8%	3000	0.00084	91.9%	84.5%	100	25.00	42	1.07851E-13	9.31381E-14	575516.6	497005.7
23000	0.00011	97.8%	95.7%	3000	0.00084	84.5%	71.4%	200	25.00	42	2.6376E-14	1.96706E-14	140748.4	104966.4
23000	0.00011	96.8%	93.6%	3000	0.00084	77.6%	60.3%	300	25.00	42	1.14676E-14	7.38555E-15	61193.7	39410.9
23000	0.00011	95.7%	91.6%	3000	0.00084	71.4%	50.9%	400	25.00	42	6.31016E-15	3.50958E-15	33672.4	18727.9
23000	0.00011	94.6%	89.6%	3000	0.00084	65.6%	43.0%	500	25.00	42	3.95063E-15	1.89751E-15	21081.4	10125.5
23000	0.00011	93.6%	87.6%	3000	0.00084	60.3%	36.3%	600	25.00	42	2.68379E-15	1.11319E-15	14321.3	5940.2
23000	0.00011	92.6%	85.7%	3000	0.00084	55.4%	30.7%	700	25.00	42	1.92886E-15	6.90916E-16	10292.8	3686.9
23000	0.00011	91.6%	83.9%	3000	0.00084	50.9%	25.9%	800	25.00	42	1.44465E-15	4.4688E-16	7709.0	2384.6
23000	0.00011	90.6%	82.0%	3000	0.00084	46.8%	21.9%	900	25.00	42	1.11661E-15	2.98287E-16	5958.5	1591.7
23000	0.00011	89.6%	80.3%	3000	0.00084	43.0%	18.5%	1000	25.00	42	8.84777E-16	2.04112E-16	4721.4	1089.2
23000	0.00011	88.6%	78.5%	3000	0.00084	39.5%	15.6%	1100	25.00	42	7.15309E-16	1.42506E-16	3817.0	760.4
23000	0.00011	87.6%	76.8%	3000	0.00084	36.3%	13.2%	1200	25.00	42	5.8798E-16	1.01159E-16	3137.6	539.8
23000	0.00011	86.7%	75.1%	3000	0.00084	33.4%	11.2%	1300	25.00	42	4.90099E-16	7.28163E-17	2615.3	388.6
23000	0.00011	85.7%	73.5%	3000	0.00084	30.7%	9.4%	1400	25.00	42	4.1339E-16	5.30406E-17	2205.9	283.0
23000	0.00011	84.8%	71.9%	3000	0.00084	28.2%	8.0%	1500	25.00	42	3.52273E-16	3.90329E-17	1879.8	208.3
23000	0.00011	83.9%	70.3%	3000	0.00084	25.9%	6.7%	1600	25.00	42	3.02877E-16	2.89816E-17	1616.2	154.7
23000	0.00011	82.9%	68.8%	3000	0.00084	23.8%	5.7%	1700	25.00	42	2.62455E-16	2.16877E-17	1400.5	115.7
23000	0.00011	82.0%	67.3%	3000	0.00084	21.9%	4.8%	1800	25.00	42	2.29009E-16	1.63424E-17	1222.0	87.2

(*continued*)

Table 11.6 *(Continued)*

Visibility (m)	$\beta_{1.06}$	One-way transmission	Two-way transmission	Visibility (m)	$\beta_{1.06}$	One-way transmission	Two-way transmission	Range to target (m)	Energy per pulse (μJ)	Receiver aperture diameter (mm)	Pulse energy received from a diffusely reflecting target of 10% reflectance @ 23-km vis.	Pulse energy received from a diffusely reflecting target of 10% reflectance @ 3 km vis.	# of photons received from a diffusely reflecting target of 10% reflectance @ 23 km vis.	# of photons received from a diffusely reflecting target of 10% reflectance @ 3 km vis.
23000	0.00011	81.1%	65.8%	3000	0.00084	20.1%	4.1%	1900	25.00	42	2.01065E-16	1.23909E-17	1072.9	66.1
23000	0.00011	80.3%	64.4%	3000	0.00084	18.5%	3.4%	2000	25.00	42	1.77513E-16	9.4471E-18	947.2	50.4
23000	0.00011	79.4%	63.0%	3000	0.00084	17.0%	2.9%	2100	25.00	42	1.57506E-16	7.23884E-18	840.5	38.6
23000	0.00011	78.5%	61.6%	3000	0.00084	15.6%	2.4%	2200	25.00	42	1.4039E-16	5.572E-18	749.1	29.7
23000	0.00011	77.6%	60.3%	3000	0.00084	14.4%	2.1%	2300	25.00	42	1.25652E-16	4.30675E-18	670.5	23.0
23000	0.00011	76.8%	59.0%	3000	0.00084	13.2%	1.7%	2400	25.00	42	1.12888E-16	3.34143E-18	602.4	17.8
23000	0.00011	76.0%	57.7%	3000	0.00084	12.1%	1.5%	2500	25.00	42	1.01774E-16	2.6015E-18	543.1	13.9
23000	0.00011	75.1%	56.4%	3000	0.00084	11.2%	1.2%	2600	25.00	42	9.20482E-17	2.03192E-18	491.2	10.8
23000	0.00011	74.3%	55.2%	3000	0.00084	10.3%	1.1%	2700	25.00	42	8.34988E-17	1.59175E-18	445.6	8.5
23000	0.00011	73.5%	54.0%	3000	0.00084	9.4%	0.9%	2800	25.00	42	7.59516E-17	1.25036E-18	405.3	6.7
23000	0.00011	72.7%	52.8%	3000	0.00084	8.7%	0.8%	2900	25.00	42	6.92632E-17	9.84699E-19	369.6	5.3
23000	0.00011	71.9%	51.7%	3000	0.00084	8.0%	0.6%	3000	25.00	42	6.33143E-17	7.77331E-19	337.9	4.1

advantageous weather patterns, and to avoid the backscatter solar radiation that is present during daylight hours. Backscatter from the sun can be an issue for GMAPD LiDARs, but as long as the individual DAS is kept small enough, blocking loss is not a significant limitation.

The Harris Geiger-mode LiDAR supports mapping wide areas at higher point density by incorporating a 128×32 pixel GMAPD array to capture single-photon (on average) returns from near-ground targets instantaneously at ~205 million samples per second (one signal return per pulse for each detector element at 50,000 laser pulse repetition rate).[15] This GML system collects time of flight, or range measurements, multiple times over every ground location. A Palmer scanner sweeps the LiDAR FPA ground footprint in a doughnut hole pattern (see Fig. 11.5). The Palmer scanner angle is fixed at ±15 deg, and the scan rate is adjusted to a preset value for the data collections. The forward motion of the aircraft, flying straight and level at a pre-determined flight altitude, produces a swath on the ground as the aircraft moves forward. This 4- to 5-km swath width (at ~8 km flight altitude) scans every ground location. At the end of each swath, the aircraft makes a turn and flies in the opposite direction. The overlapping swaths capture the target signature at four look angles (forward arc, aft arc, left side, and right side). Collections are made either in wide-area mapping (WAM) mode, or in single- or multiple–line-of-collection (LOC) mode. WAM collection mode (with four looks) is typically used for terrain mapping, and electrical utility distribution mapping in areas with dense utility poles and wires, as expected in a transmission line network. Although a typical sidelap at nominally 20% supports wide-area Quality Level 2 (QL2) mapping programs, the benefit of

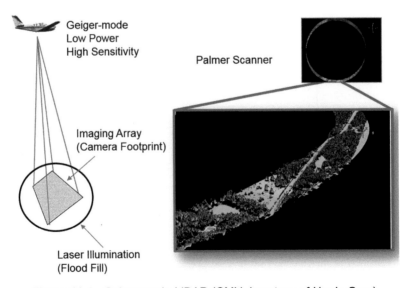

Figure 11.5   Geiger-mode LiDAR (GML) (courtesy of Harris Corp).

collecting with approximately a 50% sidelap between swaths results in every ground location being illuminated (measured) multiple times from four directions. This reduces voids and occlusions (shadows) in dense urban areas and enables greater foliage penetration for optimal ground measurements, even from higher altitudes. The LOC mode is typically used for corridor mapping, as expected in an electrical transmission line network or a transportation highway system. LOC acquisitions can be a single line (with two looks) or a set of lines with some sidelap, depending on the mapping requirement.

11.2.2 3D Mapping LiDAR design

A good starting point for designing a 3D mapping LiDAR is to define the altitude at which the LiDAR will fly, the required data rate, and the required angle/angle/range resolution for the mapping. Higher-altitude requirements, or a larger-area rate, may drive the LiDAR designer to a flash LiDAR mapping approach, but low cloud decks in an area may require a lower-altitude design. Current, flash, 3D-mapping LiDAR can use the GMAPD 3D LiDAR because of available commercial camera technology, but a linear-mode, area-based APD could also be used if the right camera becomes available. Eye-safety considerations will need to be considered in designing a 3D mapping LiDAR. Commercial 3D mapping LiDARs are not usually at eye-safer wavelengths of 1.5 μm or higher, although there are some 3D mapping LiDARs at eye-safer wavelengths. The 3D mapping LiDAR has to be eye safe on the ground. The designer should also consider the possibility that a ground observer has a pair of binoculars and decides to watch the mapping aircraft. In general, the diffraction limit will not be a significant concern, as can be seen in Table 11.7.

11.3 Laser Vibrometers

Laser vibrometers can be used to characterize the motion of machines or bridges, or of the ground. Single-point laser vibrometers are commercially available. One of the sources is a company called Polytec GmbH.[16] OMS

Table 11.7 Minimum apertures sizes when limited by diffraction.

Altitude (km)	Distance 30 deg off nadir (km)	Swath width (km)	Diameter of diffraction-limited aperture (cm)	2 × diffraction-limited aperture diameter (cm)	Max. unambiguous range PRF (Hz)
1	1.15	1.00	0.41	0.82	12,990
2	2.31	2.00	0.82	1.64	64,952
4	4.62	4.00	1.64	3.28	32,476
8	9.24	8.00	3.28	6.55	16,238

Corp. is another company in this market.[17] Area-based vibrometers have a much wider application space because you can see an entire vibration mode simultaneously, but they suffer from the need for larger, coherent-capable FPAs. Laser vibrometers are discussed in Chapter 4 on Types of LiDAR. The discussion from Chapter 4 is repeated here.

The Doppler shift changes the frequency of the return laser signal based on Eq. (4.1), repeated here:

$$\Delta f = \frac{2V}{\lambda},$$

where V is velocity, and λ is wavelength. The Doppler shift is used to measure velocity. It can provide a very accurate measure of velocity and as a result is often used to measure the small back-and-forth velocity caused by vibration. Table 4.2 provides the Doppler frequencies resulting from a wavelength of 1.55 µm. A 10-µm/s velocity yields a frequency shift of 12.9 Hz, which can be measured in a fraction of a second. As an object vibrates, it creates velocities toward and away from the sensors. For vibration, we see a frequency-modulated return signal, with the frequency of the modulation as the vibration frequency and the magnitude of the frequency shifts equal to the varying Doppler shifts. Equation (4.2) (repeated here) shows the variation in position that results from a vibration:

$$x = A \sin(2\pi f t),$$

where A is the amplitude of the vibration, and f is its frequency. A Doppler vibrometer measures velocity to and from the LiDAR. The velocity is the derivative of Eq. (4.2) for the surface position of a vibrating object, which is given by Eq. (4.3) and is repeated here:

$$V = \frac{dx}{dt} = 2\pi f A \cos(2\pi f t).$$

Equation (4.3), along with Eq. (4.1), can be used to generate a table showing the frequencies resulting from illuminating a vibrating object with light at a wavelength of 1.55 µm. In Table 4.3 the required (minimum) sample

Table 4.2 Doppler frequency resulting from a certain velocity when using 1550-nm illumination.

Velocity (µm/s)	Frequency (Hz)
1	1.29
10	12.9
100	129.03
1000	1290.32

Table 4.3 Doppler frequencies resulting from surface vibration.

Vibration frequency (Hz)	Vibration amplitude (μm)	Max. velocity (μm/s)	Max. Doppler shift frequency (Hz)	Required sample time (ms)
10	1	63	81	24.7
50	1	314	405	4.9
100	1	628	810	2.5
200	1	1257	1622	1.2
10	0.1	6	8	250.0
50	0.1	31	40	50.0
100	0.1	63	81	24.7
200	0.1	126	163	12.3

time is twice one over the maximum Doppler shift frequency. Even for a 1-μm magnitude of vibration at 10 Hz, there will be a maximum frequency for 81 Hz, which should be measurable.

For high Doppler sensitivity, detection is usually performed using temporal heterodyning, where the return signal is combined on the detector with a local oscillator to create a frequency-downshifted signal, called a beat frequency. Laser vibrometers can be used to characterize many types of vibration. The type of an engine in a vehicle can be identified; it is easy to tell the difference between a turbine engine and a piston engine. It should be easy to determine how many cylinders a piston engine has. Almost any type of vibration can be identified. The vibrational image of an internal combustion engine can be used to identify combustion pressure pulses and inertial acceleration of the pistons and drive trains. It can also help to identify mechanical imbalances and misfires. Power through a transformer, or liquid through a pipe, can be characterized. If the return Doppler shift is sent to headsets, we will hear noise. A diesel engine will sound like a diesel engine, and the same for other common vibrations.

Flash illumination for vibration detection would be powerful because it allows simultaneous vibration measurement over an area, but requires a 2D array of photodetectors operating at very high frame rates. Such imaging cameras exist at low frame rates, in the range of a few hundred hertz. Area vibrometers are just now being developed. Point vibrometers such as Polytec's laser vibrometer are currently commercially available.

The use of area laser vibrometers will enable a significant number of new application areas. For example, if you look at a bridge while traffic is going over it, you will see spatial vibration modes.[18] As the bridge decays, it will begin to crack. Those cracks will change the vibration modes. Also, area laser vibrometers will be ideal for mapping vibration sources across any machinery or underground seismic sources.[19] These applications can be attempted with point laser vibrometers, but you cannot see a flash image of the vibration modes like you can with an area-based laser vibrometer.

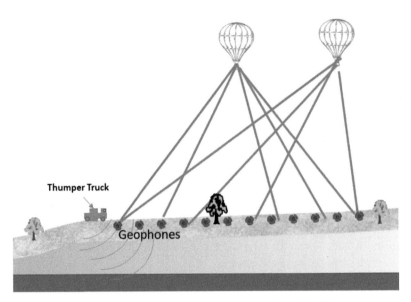

Figure 11.6 Current approach to explore for oil.

Figure 11.6 shows how engineers currently explore for oil, but with two balloons added where LiDARs could be housed. A thumper truck sends acoustic shock waves into the ground. Geophones are carefully placed over a wide area, e.g., an array of 30×30 geophones on 50-m spacing. The geophones pick up the vibrations, and then it is possible to map the underground structure. A flash laser vibrometer could serve the same function, but without the labor of placing all of the geophones and then moving them to a new location. The two balloons housing LiDARs are shown above the area so that 3D vectors of velocity can be calculated.

11.3.1 Designing a laser vibrometer

One immediate question when designing a laser vibrometer is: What velocity sensitivity is desired? A second question is addresses the required range resolution. A third consideration is whether the designer needs to use a flash laser vibrometer, or if the signal can be built up over time. For the seismic case depicted in Fig. 11.7, the engineer would like to see the pattern of vibrations at an instant, so it would be useful to have an array camera. A Princeton Lightwave, Inc. 32×128 pixel camera could be used. This is a GMAPD camera so would need to use the unique method of coherent LiDAR design required for use with GMAPDs, as is discussed in Chapter 4. Alternatively, a designer could obtain a HgCdTe LMAPD array from DRS, or an InGaAs 128×128 pixel p–i–n diode array, or possibly a LMAPD, from Voxtel. At

Figure 11.7 Illustration of a microburst. The air moves in a downward motion until it hits ground level. It then spreads outward in all directions [reprinted from Wikipedia web page on microburst: https://en.wikipedia.org/wiki/Microburst; image created by NASA (public domain)].

this time, the deepest depth in range recorded by any of the Voxtel arrays is in the VX 819, which has 24 range capture events stored. This is marginal for coherent LiDAR.

Another design constraint to consider is that the Doppler shift is only recorded to and from the LiDAR. Therefore, to record 3D vibration signals, more than a single LiDAR will be required.

11.4 Wind Sensing

Wind sensing is being used at airports and around wind turbines. It has even been used in sailboat races to find the best wind. A really good book describing wind LiDAR in great detail is *Laser Remote Sensing*, edited by Takashi Fuji and Tetsuo Fukuchi,[20] to which Dr. Sammy Henderson contributed a >250-page chapter on wind sensing. LiDAR wind sensing primarily uses coherent Doppler imaging to measure wind speed. To obtain a 3D wind profile, it is desirable to do a conic scan, possibly below an aircraft, since Doppler shift is only in the direction toward and away from the LiDAR. (Doppler shift is only along the vector between the LiDAR and the target.) A conic scan allows calculation of a 3D wind field. Because the reader can obtain an extensive discussion of wind sensing in the referenced book,[20] this discussion will be limited in scope. Wind LiDARs are commercially available from multiple companies. Some promising applications of LiDAR for wind sensing are:

- Wind shear detection and warning for
  - microbursts,
  - gust fronts, and
  - terrain-induced shears;
- Aircraft wake vortex detection and tracking;
- Clear air turbulence detection and warning;
- ballistic winds correction;
- precision air drop;
- global winds from space; and
- wind farm optimization—this could be a large application.

Figure 11.7 shows a microburst wind pattern. A microburst can be very dangerous on approach for an aircraft. The aircraft is flying at a certain wind speed, say 10–20 knots above stall speed, but it is doing that with a 20-knot head wind from the microburst. When it gets below the microburst, it loses the 20-knot head wind and has a wind pushing it down. Then, once it is beyond the microburst, it has lost a total of 40 knots of speed compared to the air because now it has a 20-knot tail wind. This means that the aircraft is likely to stall and crash.

Another interesting wind-sensing application is for seeing wind vortices, as shown in Fig. 11.8. Wing-tip vortices are a major reason for aircraft separation when taking off or landing. A wing-tip vortex can create control issues for following aircraft.

Figure 11.8 Photograph showing aircraft wing-tip vortices (photo credit: Ryoh Ishihara, University of Illinois College of Engineering).

Problems and Solutions

11-1. Assume that a LiDAR is 2 m above the ground, that it has 64 laser lines, and that we are on a flat surface. Let the elevation coverage be up to +2 deg. Assume a LiDAR range from 6 m to 120 m, and that we want an equal range on the ground. Assume 6 laser lines above the elevation angle for 120-m range, and the remaining laser lines spread over the ground.

(a) What is the separation in ground distance between the laser lines?
(b) What is the angular spacing for each of the laser lines?

11-1. Solution:

(a) Range coverage is $120 - 6 = 114$ m. Number of laser lines is $64 - 6 = 58$. If the separation is equal, then it is $114/58 = 1.97$-m separation.

(b) The separation in angle varies with distance. It is larger close to the LiDAR, and smaller far away from the LiDAR. The Excel table below shows the results.

Laser Line Separation		
Laser line location	**Angle (rad)**	**Angle difference (rad)**
6	0.34625355	
7.97	0.256495054	0.089758496
9.93	0.204156404	0.05233865
11.90	0.169717873	0.034438531
13.86	0.14528813	0.024429743
15.83	0.127038529	0.018249601
17.79	0.11287889	0.01415964
19.76	0.101568763	0.011310127
21.72	0.092324477	0.009244285
23.69	0.084626192	0.007698285
25.66	0.078115297	0.006510896
27.62	0.072536305	0.005578992
29.59	0.067702224	0.004834081
31.55	0.063473013	0.004229211
33.52	0.059741704	0.003731309
35.48	0.056425171	0.003316533
37.45	0.053457834	0.002967337
39.41	0.050787256	0.002670578
41.38	0.048371006	0.00241625
43.34	0.046174381	0.002196625
45.31	0.044168719	0.002005661
47.28	0.042330143	0.001838577
49.24	0.040638596	0.001691547
51.21	0.039077111	0.001561484
53.17	0.037631237	0.001445874
55.14	0.036288587	0.00134265
57.10	0.035038483	0.001250104
59.07	0.033871672	0.001166811

(*continued*)

Laser Line Separation		
Laser line location	Angle (rad)	Angle difference (rad)
61.03	0.032780095	0.001091577
63.00	0.031756701	0.001023394
64.97	0.030795292	0.000961409
66.93	0.029890401	0.000904891
68.90	0.029037186	0.000853215
70.86	0.028231341	0.000805845
72.83	0.027469027	0.000762314
74.79	0.026746809	0.000722218
76.76	0.026061603	0.000685205
78.72	0.025410636	0.000650968
80.69	0.024791402	0.000619234
82.66	0.024201636	0.000589766
84.62	0.023639284	0.000562353
86.59	0.023102476	0.000536808
88.55	0.022589511	0.000512965
90.52	0.022098834	0.000490677
92.48	0.021629024	0.00046981
94.45	0.021178777	0.000450247
96.41	0.020746896	0.000431881
98.38	0.020332279	0.000414617
100.34	0.019933911	0.000398368
102.31	0.019550856	0.000383055
104.28	0.019182246	0.000368609
106.24	0.018827281	0.000354966
108.21	0.018485215	0.000342066
110.17	0.018155359	0.000329856
112.14	0.01783707	0.000318289
114.10	0.017529749	0.00030732
116.07	0.01723284	0.000296909
118.03	0.016945823	0.000287018
120.00	0.01666821	0.000277613

11-2. Calculate the power required for a LiDAR with 0.15-m range resolution, 100-m range, 5% object reflectivity, and a 2.5-cm–diameter monostatic transmit/receive aperture.

11-2 Solution:

$$\frac{P_R}{P_T} = \frac{\sigma}{A_{\text{illum}}} \frac{A_{\text{rec}}}{\pi R^2} \eta_{\text{atm}}^2 \eta_{\text{sys}}.$$

We aren't considering atmospheric attenuation at such close range, so we ignore that factor. Keeping in mind a single detector ($N_{\text{det}} = 1$),

$$A_{\text{illum}} = \pi \left(\frac{\lambda R}{2D}\right)^2 = 3.02 \times 10^{-5}\,\text{m}^2.$$

Assuming an ideal LiDAR,

$$P_R = \frac{\sigma}{A_{\text{illum}}} \frac{A_{\text{rec}}}{\pi R^2} = \frac{0.15}{3.02 \times 10^{-5}} \frac{0.025^2}{100^2} = 3.104 \times 10^{-4} P_T.$$

11-3. For the case of Problem 11-2:

    (a) Calculate the eye-safe range for a Gaussian illuminator laser with 10% clipping, a 10-mm–diameter aperture, and a wavelength of 1550 nm. For simplicity, assume that the laser is illuminating a single spot 10% of the time, so the amount of energy in 10 s is the average power.

    (b) Calculate the eye-safe distance for a super-Gaussian illuminator beam with $b = 4$ instead of 2, and 10% clipping.

11-3 Solution:

    (a)

$$E_T = MPE \times A_{\text{rec}}[1 - \exp(-2(1.1)^2)] = 0.001789 \, \text{J},$$

$$N = 20, \ \lambda = 1550 \, \text{nm},$$

$$E_T = \frac{Nhc}{\lambda} \frac{A_{\text{illum}}}{\sigma} \frac{\pi R^2}{A_{\text{rec}}},$$

$$R = \sqrt{\frac{E_T \lambda \sigma A_{\text{rec}}}{\pi Nhc A_{\text{illum}}}} = 12.09 \, \text{km}.$$

    (b) For the super-Gaussian ($b = 4$),

$$E_T = MPE \times A_{\text{rec}}[1 - \exp(-2(1.1)^4)] = 0.001789 \, \text{J},$$

$$\lambda = 1550 \, \text{nm}.$$

This is the amplitude of the Gaussian.

At the aperture, the energy per exposure time per area is

$$MPE = \frac{E_T[1 - \exp\left(-2(\frac{1.1r}{w_0})^2\right)]}{A_{\text{illum}}}.$$

Solving for w,

$$1 - \exp\left[-2\left(\frac{1.1}{w}\right)^2\right] = \frac{MPE A_{\text{illum}}}{E_T},$$

$$-2\left(\frac{1.1r}{w_0}\right)^2 = \ln\left[1 - \frac{MPE A_{\text{illum}}}{E_T}\right],$$

$$r = w_0 \frac{1}{1.1} \sqrt{-\frac{1}{2} \ln\left[1 - \frac{MPEA_{\text{illum}}}{E_T}\right]}.$$

For $b = 2$, $1.0542 w_0 = w(z)$.

For $b = 4$, $1.220 w_0 = w(z)$.

Assume the beam waist to be 10 mm:

$$w_0 = 10\,\text{mm}$$

$$1.0055 = \sqrt{1 + \left(\frac{z\lambda}{\pi w_0^2}\right)^2} \rightarrow z = \pi w_0^2 \frac{\sqrt{1.0055^2 - 1}}{\lambda}.$$

$$z = 4.72\,\text{m}.$$

$$1.0758 = \sqrt{1 + \left(\frac{z\lambda}{\pi w_0^2}\right)^2} \rightarrow z = \pi w_0^2 \frac{\sqrt{1.0758^2 - 1}}{\lambda}.$$

$$z = 9.55\,\text{m}.$$

Since more energy per pulse is contained in the flat-top Gaussian case, the range at which it is eye safe is further out, roughly doubled.

11-4. Assume a LiDAR range of 15 m, and calculate the maximum safe speed for a driverless car, allowing a 20% range margin; in other words, the range calculated for 15-m visibility should be reduced by 20% to estimate the maximum safe driving speed for a driverless car. Assume a 0.7 coefficient of friction.

11-4 Solution:

$$D_{\text{stopping}} = 15 \times 0.8 = 12\,\text{m}.$$

We will need to be able to stop within 12 m of the car's location when the detection occurred. Considering a reaction time of 1.5 s, and a 0.7 coefficient of friction,

$$vT + \frac{v^2}{2ug} = D_{\text{stopping}},$$

$$\frac{v^2}{2ug} + vT - D_{\text{stopping}} = 0,$$

$$v = ug\left(-T \pm \sqrt{T^2 - \frac{4D_{\text{stopping}}}{2ug}}\right) = 6.15 \, \text{m/s},$$

$$v = 13.76 \, \text{mph}.$$

11-5. Assume a 5-cm–diameter receive aperture, an illumination pattern of 3 mrad × 384 mrad, an operating wavelength of 1550 nm, 10% reflectivity, and 200-m range. We will use a detector array that is 1 × 128, with each individual detector 3 mrad × 3 mrad.

(a) Calculate the energy per pulse for 5-km visibility. Assume that we need 20 photons per detector per laser pulse for detection, and that we over-illuminate the required area by 10%. Plot the required energy versus range for 50 to 300 m.

(b) Calculate the safe driving speed for 50 m visibility, assuming a link budget additional margin of 10 dB based on the original 5-km visibility calculation.

11-5 Solution:

(a) $\sigma = \eta_r DAS_{\text{oc}} DAS_{\text{ei}} R^2 = 0.1 \times 0.003 \times 0.384 \times 200^2 = 4.6 \, \text{m}^2,$

$$A_{\text{illum}} = \frac{1.1\sigma}{\eta_r} = 50.7 \, \text{m}^2,$$

$$q = 0.1428\lambda - 0.0947 = 0.12664,$$

$$\beta = \left(\frac{\lambda_0}{\lambda}\right)^q = 0.87703,$$

$$E_T = \frac{4NhcA_{\text{illum}}R^2}{\lambda\eta_r D^2\sigma} e^{-2\beta R}.$$

# of photons received from a 10% reflective object								
Human visibility (m)	$\beta_{0.55}$	$\beta_{1.55}$	Imaging range	One-way Transmission	Two-way transmission	Design # of photons	Receiver aperture diameter (mm)	Required transmitter energy (µJ) for 10% reflectivity
5000	0.0006	0.0005	15	99.2%	98.4%	20	50	0.0022
5000	0.0006	0.0005	30	98.4%	96.9%	20	50	0.0089
5000	0.0006	0.0005	45	97.7%	95.4%	20	50	0.0204
5000	0.0006	0.0005	60	96.9%	93.9%	20	50	0.0368

(*continued*)

# of photons received from a 10% reflective object								
Human visibility (m)	$\beta_{0.55}$	$\beta_{1.55}$	Imaging range	One-way Transmission	Two-way transmission	Design # of photons	Receiver aperture diameter (mm)	Required transmitter energy (μJ) for 10% reflectivity
5000	0.0006	0.0005	75	96.1%	92.4%	20	50	0.0584
5000	0.0006	0.0005	90	95.4%	91.0%	20	50	0.0854
5000	0.0006	0.0005	105	94.6%	89.6%	20	50	0.1181
5000	0.0006	0.0005	120	93.9%	88.2%	20	50	0.1567
5000	0.0006	0.0005	135	93.2%	86.8%	20	50	0.2015
5000	0.0006	0.0005	150	92.4%	85.4%	20	50	0.2527
5000	0.0006	0.0005	165	91.7%	84.1%	20	50	0.3106
5000	0.0006	0.0005	180	91.0%	82.8%	20	50	0.3755
5000	0.0006	0.0005	195	90.3%	81.5%	20	50	0.4477
5000	0.0006	0.0005	200	90.0%	81.0%	20	50	0.4735
5000	0.0006	0.0005	210	89.6%	80.2%	20	50	0.5275
5000	0.0006	0.0005	225	88.8%	78.9%	20	50	0.6152
5000	0.0006	0.0005	240	88.2%	77.7%	20	50	0.711
5000	0.0006	0.0005	255	87.5%	76.5%	20	50	0.8155
5000	0.0006	0.0005	270	86.8%	75.3%	20	50	0.9287
5000	0.0006	0.0005	285	86.1%	74.1%	20	50	1.0512
5000	0.0006	0.0005	300	85.4%	73.0%	20	50	1.1833

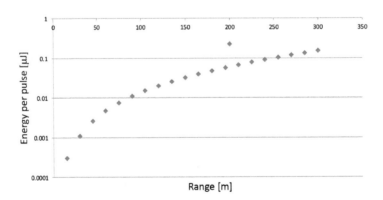

(b) At 200 m, we need an energy per pulse of 0.4735 μJ. Ten times that would be 4.735 μJ. We have a range of 60 m, using 20 photons for detection.

Human visibility (m)	$\beta_{0.55}$	$\beta_{1.55}$	Imaging range	One-way transmission	Two-way transmission	Energy per pulse (mJ)	Receiver aperture diameter (mm)	# of photons received from a 10% reflective hard object
50	0.0599	0.0525	5	76.9%	59.1%	4.7	50	1021161.4
50	0.0599	0.0525	10	59.1%	35.0%	4.7	50	150947.2
50	0.0599	0.0525	15	45.5%	20.7%	4.7	50	39667.4
50	0.0599	0.0525	20	35.0%	12.2%	4.7	50	13193.1
50	0.0599	0.0525	25	26.9%	7.2%	4.7	50	4992.5
50	0.0599	0.0525	30	20.7%	4.3%	4.7	50	2050.0
50	0.0599	0.0525	35	15.9%	2.5%	4.7	50	890.5
50	0.0599	0.0525	40	12.2%	1.5%	4.7	50	403.1
50	0.0599	0.0525	45	9.4%	0.9%	4.7	50	188.3
50	0.0599	0.0525	50	7.2%	0.5%	4.7	50	90.2
50	0.0599	0.0525	55	5.6%	0.3%	4.7	50	44.1
50	0.0599	0.0525	60	4.3%	0.2%	4.7	50	21.9
50	0.0599	0.0525	65	3.3%	0.1%	4.7	50	11.0

(Table heading: # of photons received from a 10% reflective object)

11-6. Assume an airplane flying at 75 m/s and a wind velocity of 5 m/s below the aircraft. We have a LiDAR with a 30-deg conical scan pattern below the aircraft. The LiDAR is at 1550-nm wavelength. Plot the LiDAR frequency versus time for a full rotation of the conical scan. Also plot the measured velocity versus time for a full scan cycle.

11-6 Solution:

$$v = (v - v_{\text{wind}}) \cos \theta \cos \varphi = 60.62 \, \text{m/s} \cos \theta,$$

$$\Delta f = \frac{2v}{\lambda} = 78.22 \, \text{MHz} \cos \theta = 78.22 \, \text{MHz} \cos[2\pi t_{\text{norm}}],$$

$$\theta = \omega t = t_{\text{norm}},$$

where $0 \leq t_{\text{norm}} \leq 1$,

```
f0 = 78.22e6;
t = (0.001:0.001:1);
f = f0*cos(2*pi*t);
v = 60.62*cos(2*pi*t);
figure
plot(t,f)
figure
plot(t,v)
```

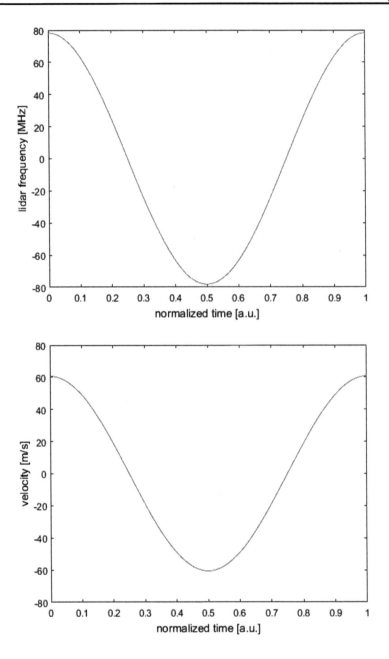

11-7. For a LMAPD, plot the number of returned photons for 7-km visibility versus range considering a 2-in. receiver aperture and 1550-nm fiber laser.

11-7 Solution:

```
h = 6.63e-34;
c = 3e8;
r = 1.0;
V = 7;
l = 1.55e-6;
D = 5.08e-3;
DAS = 5e-3;
R = linespace(5,100,1000);
l0 = 0.55;
q = 0.1428*l*1e6-0.0947;
b = (l0/(l*1e6))^q;
na = 1-exp(-b*V);
s = max(r*DAS^2*R.^2);
N = 20;
Nd = 1./(D^2*r./(1.5*s));
E = max(6*N*h*c*Nd.*R.^2*na^2/(l*r*D^2));
N = E*l*r*D^2./(6*h*c*Nd.*R.^2*na^2);
semilogy(R,N)
ylabel('Photon Count')
xlabel('Range [m]')
grid on
```

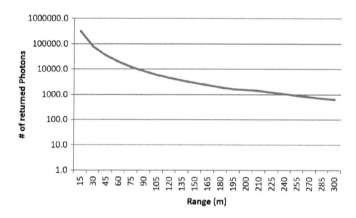

11-8. For 7-km visibility and 200-m range, calculate the required energy per pulse considering a 2-in. receive aperture and 10% reflectivity. The suggested laser is 1550-nm fiber laser, and you may assume that 30 photons per pulse will return to our LMAPD.

11-8 Solution:

						Design	Receiver aperture diameter (mm)	Required transmitter energy (nJ) for 10% reflectivity

# of photons received from a 10% reflective object								
Human visibility (m)	$\beta_{0.55}$	$\beta_{1.55}$	**Imaging range**	**One-way transmission**	**Two-way transmission**	**Design # of photons**	**Receiver aperture diameter (mm)**	**Required transmitter energy (nJ) for 10% reflectivity**
7000	0.0004	0.0004	15	99.4%	98.9%	30	50	0.025579
7000	0.0004	0.0004	30	98.9%	97.8%	30	50	0.103476
7000	0.0004	0.0004	45	98.3%	96.7%	30	50	0.235457
7000	0.0004	0.0004	60	97.8%	95.6%	30	50	0.423330
7000	0.0004	0.0004	75	97.2%	94.5%	30	50	0.668944
7000	0.0004	0.0004	90	96.7%	93.5%	30	50	0.974187
7000	0.0004	0.0004	105	96.1%	92.4%	30	50	1.340992
7000	0.0004	0.0004	120	95.6%	91.4%	30	50	1.771333
7000	0.0004	0.0004	135	95.1%	90.4%	30	50	2.267230
7000	0.0004	0.0004	150	94.5%	89.4%	30	50	2.830744
7000	0.0004	0.0004	165	94.0%	88.4%	30	50	3.463987
7000	0.0004	0.0004	180	93.5%	87.4%	30	50	4.169112
7000	0.0004	0.0004	195	92.9%	86.4%	30	50	4.948322
7000	0.0004	0.0004	200	92.8%	86.1%	30	50	5.224910
7000	0.0004	0.0004	210	92.4%	85.4%	30	50	5.803868
7000	0.0004	0.0004	225	91.9%	84.5%	30	50	6.738049
7000	0.0004	0.0004	240	91.4%	83.5%	30	50	7.753215
7000	0.0004	0.0004	255	90.9%	82.6%	30	50	8.851765
7000	0.0004	0.0004	270	90.4%	81.7%	30	50	10.036153
7000	0.0004	0.0004	285	89.9%	80.7%	30	50	11.308881
7000	0.0004	0.0004	300	89.4%	79.8%	30	50	12.672510

11-9. For the system defined in Problem 11-8, plot the energy per pulse in mJ versus range up to 300 m.

11-9 Solution:

```
h = 6.63e-34;
c = 3e8;
r = 0.1;
V = 7;
l = 1.55e-6;
D = 5.08e-3;
DAS = 5e-3;
R0 = 200;
R = linspace(5,300,1000);
l0 = 0.55;
q = 0.1428*l*1e6-0.0947;
b = (l0/(l*1e6))^q;
na = 1-exp(-b*V);
```

```
s = r*DAS^2*R0.^2;
N = 30;
Nd = 1./(D^2*r./(1.5*s));
E = 6*N*h*c*Nd.*R.^2*na^2/(1*r*D^2);
N = E*1*r*D^2./(6*h*c*Nd.*R.^2*na^2);
plot(R,E*1e3)
ylabel('Required Energy per Pulse [mJ]')
xlabel('Range [m]')
grid on
```

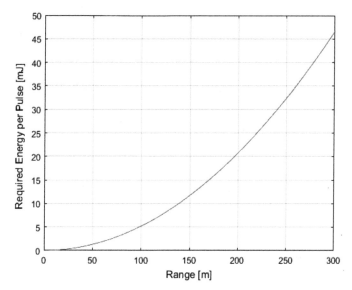

11-10. Consider 5 km of visibility, a 1550-nm fiber laser, and a 10-mm receive aperture. If we assume 2-μJ energy per pulse of a GMAPD, plot the number of photons received from a 10% reflective hard object versus range of the LiDAR.

11-10 Solution:

```
h = 6.63e-34;
c = 3e8;
r = 0.1;
V = 5;
l = 1.55e-6;
D = 10e-3;
DAS = 5e-3;
R0 = 200;
R = linrspace(5,300,1000);
l0 = 0.55;
q = 0.1428*l*1e6-0.0947;
```

```
b = (10/(1*1e6))^q;
na = 1-exp(-b*V);
s = r*DAS^2*R0.^2;
N = 30;
Nd = 1./(D^2*r./(1.5*s));
E = 6*N*h*c*Nd.*R.^2*na^2/(1*r*D^2);
E = 2e-6;
N = E*1*r*D^2./(6*h*c*Nd.*R.^2*na^2);
semilogy(R,N)
ylabel('Number of Photons Received')
xlabel('Range [m]')
grid on
```

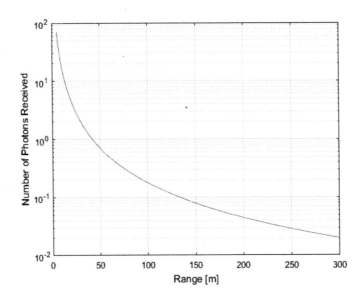

11-11. Calculate the required energy per pulse for a 2-μm laser diode, 2-in. receiver aperture, 100-m range, and a 5 mrad × 5 mrad DAS. Supposed that the system is designed for 20 photons and 10% reflectivity of the target. Plot the required pulse energy versus range.

11-11 Solution:
For 100-m range, the required energy per pulse is 1.688 mJ. The plot is as follows:

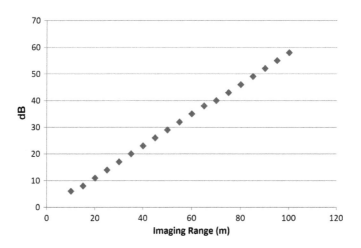

$$E = 8.84 \times 10^{-10}\,\mathrm{J}$$

```
N = 20;
h = 6.63e-34;
c = 3e8;
r = 0.1;
V = 23;
l = 2.0e-6;
D = 150e-3;
DAS = 5e-3;
R = 100;
l0 = 0.55;
q = 0.1428*l*1e6-0.0947;
b = (l0/(l*1e6))^q;
na = 1-exp(-b*V);
s = r*DAS^2*R.^2;
Nd = 1./(D^2*r./(1.5*s));
E = 6*N*h*c*Nd.*R.^2*na^2/(l*r*D^2);
R2 = linspace(5,100,1000);
s2 = r*DAS^2*R2.^2;
Nd2 = 1./(D^2*r./(1.5*s2));
E2 = 6*N*h*c*Nd2.*R2.^2*na^2/(l*r*D^2);
plot(R2,E2)
ylabel('Energy per Pulse [J]')
xlabel('Range [m]')
```

11-12. Plot the atmospheric attenuation versus imaging range for a 2-μm diode laser considering 50-m human visibility.

11-12 Solution:

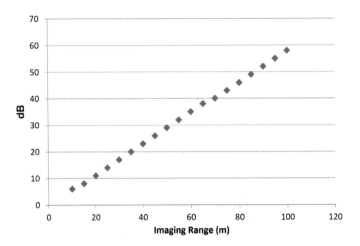

References

1. Velodyne® LiDAR web page: http://velodyneLiDAR.com/products.html.
2. E. Rosén, J.-E. Källhammer, D. Eriksson, M. Nentwich, R. Fredriksson, and K. Smith, "Pedestrian injury mitigation by autonomous braking," *Accident Analysis & Prevention* **42**(6), 1949–1957 (2010).
3. Velodyne LiDAR Product Guide: http://velodyneLiDAR.com/docs/datasheet/LiDAR%20Comparison%20chart_Rev-A_Web.pdf.
4. Overstock.com Children's Clothing Buying Guide: https://www.overstock.com/guides/childrens-clothing-buying-guide.
5. E. A. Watson, R. A. Muse, and F. P. Blommel, "Aliasing and blurring in microscanned imagery," *Proc. SPIE* **1689**, 242–250 (1992) [doi: 10.1117/12.137955].
6. R. C. Hardie, K. J. Barnard, J. G. Bognar, E. E. Armstrong, and E. A. Watson, "High-resolution image reconstruction from a sequence of rotated and translated frames and its application to an infrared imaging system," *Optical Engineering* **37**(1), 247–260 (1998) [doi.org/10.1117/1.601623].
7. Wikipedia page on Braking distance: https://en.wikipedia.org/wiki/Braking_distance.
8. I. I. Kim, B. McArthur, and E. Korevaar, "Comparison of laser beam propagation at 785 nm and 1550 nm in fog and haze for optical wireless communications," *Proc. SPIE* **4214**, 26–37 (2001) [doi: 10.1117/12.417512].
9. YouTube vide on RIEGL LMS-Q1560 supporting NASA Mission: https://www.youtube.com/watch?v=X1BU5VqrrmA.
10. SPIE Press Release: "Better, faster solutions needed, says DARPA head Regina Dugan in talk at SPIE Defense, Security, and Sensing,"

27 April 2011: https://spie.org/about-spie/press-room/press-releases/dugan-4-27-2011?SSO=1.

11. DARPA web page on high-altitude LIDAR: https://www.darpa.mil/about-us/timeline/LiDAR.

12. M. Itzler, M. Entwistle, X. Jiang, M. Owens, K. Slomkowski, and S. Rangwala, "Geiger-mode APD single-photon cameras for 3D laser radar imaging," 2014 IEEE Aerospace Conference, 1–4 March, Big Sky, Montana (2014).

13. J. J. Degnan, "Scanning, multibeam, single photon lidars for rapid large scale, high resolution, topographic and bathymetric mapping," *Remote Sensing* **2016**(8), 958 (2016).

14. T. Bahr and P. Smith, "Airborne Geiger-mode LiDAR for large-area, high-resolution wide-area mapping," *GI_Forum* **1**, 85–93 (2016).

15. U.S. Geological Survey LiDAR Base Specification version 1.3 (2018): https://www.usgs.gov/news/lidar-base-specification-version-13-released

16. Polytec GmbH web page on basic principles of vibrometry: https://www.polytec.com/us/vibrometry/technology/.

17. OMS Corp. corporate products: http://www.omscorporation.com/products/.

18. D. Killinger, P. Mamidipudi, J. Potter, J. Daly, E. Thomas, and S. Chen, "Laser Doppler vibration lidar sensing of structural defects in bridges," presented at the 10th Biennial Coherent Laser Radar Technology and Applications Conference, June 28–July 2, 1999, Mount Hood, Oregon (1999).

19. D. Killinger, "System and method for multi-beam laser vibrometry triangulation mapping of underground acoustic sources," U.S. Patent 7,190,635 B1 (2007).

20. T. Fujii and T. Fukuchi, Eds., *Laser Remote Sensing*, CRC Press, Boca Raton, Florida (2005).

Index

Dr. **Paul F. McManamon** started Exciting Technology LLC after he retired from being Chief Scientist for the Air Force Research Lab (AFRL) Sensors Directorate. He is also Technical Director of the Lidar and Optical Communications Institute (LOCI) at the University of Dayton. He chaired the 2014 U.S. National Academy of Sciences Study "Laser Radar: Progress and Opportunities in Active Electro-Optical Sensing." He was the main LiDAR expert witness for Uber in the lawsuit Uber versus Google/Waymo. He cochaired the 2012 U.S. NAS study "Optics and Photonics, Essential Technologies for Our Nation," which recommended a National Photonics Initiative (NPI). Dr. McManamon was also vice chair of the 2010 NAS study "Seeing Photons: Progress and Limits of Visible and Infrared Sensor Arrays."

Dr. McManamon is a Fellow of SPIE, IEEE, OSA, AFRL, the Directed Energy Professional Society (DEPS), the Military Sensing Symposia (MSS), and the American Institute of Aeronautics and Astronautics (AIAA). Dr. McManamon received the IEEE W.R.G. Baker Award in 1998 for the best paper in *any* refereed IEEE journal or publication (>20,000 papers). He was president of SPIE in 2006. He served on the SPIE Board of Directors for seven years and on the SPIE Executive Committee from 2003 through 2007. Dr. McManamon worked with Dr. Fenner Milton and Dr. Gerry Trunk to found the MSS, combining IRIS and the Tri-Service Radar Symposia. He worked as a civilian employee of the Air Force at WPAFB from May 1968 through May 2008. His last position for the Air Force was chief scientist for the AFRL Sensors Directorate, where he was responsible for the technical aspects of all AFRL sensing technologies, including RF and EO sensing, automatic object recognition, infrared countermeasure (IRCM), electronic warfare, and device technologies. Prior to that, he also was senior scientist for EO/IR Sensors, and acting chief scientist for the Avionics Directorate for >2.5 years. In 2006 he received the Presidential Rank Award of Meritorious Executive. He was the co-recipient of the SPIE President's Award in 2013.